SECOND ORDER LINEAR DIFFERENTIAL EQUATIONS IN BANACH SPACES

NORTH-HOLLAND MATHEMATICS STUDIES 108
Notas de Matemática (99)

Editor: Leopoldo Nachbin

*Centro Brasileiro de Pesquisas Físicas,
Rio de Janeiro
and University of Rochester*

SECOND ORDER LINEAR DIFFERENTIAL EQUATIONS IN BANACH SPACES

H. O. FATTORINI
University of California at Los Angeles
U.S.A.

1985

NORTH-HOLLAND – AMSTERDAM • NEW YORK • OXFORD

ISBN: 0 444 87698 7

Publishers:

ELSEVIER SCIENCE PUBLISHERS B.V.
P.O. BOX 1991
1000 BZ AMSTERDAM
THE NETHERLANDS

Sole distributors for the U.S.A. and Canada:

ELSEVIER SCIENCE PUBLISHING COMPANY, INC.
52 VANDERBILT AVENUE
NEW YORK, N.Y. 10017
U.S.A.

Library of Congress Cataloging in Publication Data

Fattorini, H. O. (Hector O.), 1938-
Second order linear differential equations in Banach spaces.

(North-Holland mathematics studies ; 108)
(Notas de matemática ; 99)
Bibliography: p.
1. Differential equations, Partial. 2. Differential equations, Linear. 3. Banach spaces. I. Title. II. Series. III. Series: Notas de matemática (Amsterdam, Netherlands) ; 99.
QA1.N86 no. 108 510 s [515.3'54] 84-28658
[QA377]
ISBN 0-444-87698-7

PRINTED IN THE NETHERLANDS

PREFACE

An initial value or initial-boundary value problem

$$u_t = Au, \ u = u_0 \quad \text{for} \quad t = 0 \tag{1}$$

where A is a partial differential operator in the space variables $x_1, \ldots x_m$ can be recast in the form of an ordinary differential initial value problem

$$u'(t) = Au(t), \ u(0) = u_0, \tag{2}$$

where A is thought of as an operator in a function space E and the boundary conditions, if any, are included in the definition of the space E or of the domain of A. If E is suitably chosen, solutions of (2) will exist for sufficiently many initial data u_0 and will depend continuously on u_0 in the norm of E. This way of looking at (1) was initiated by Hille and Yosida in the forties and resulted in the creation and development of semigroup theory, now an integral part of most advanced treatments of parabolic and hyperbolic partial differential equations.

A second order initial value or initial-boundary value problem

$$u_{tt} = Au, \ u = u_0, \ u = u_1 \quad \text{for} \quad t = 0 \tag{3}$$

can be reduced in the same way to an ordinary differential initial value problem

$$u''(t) = Au(t), \ u(0) = u_0, \ u'(0) = u_1 \tag{4}$$

where A is defined as in (2). This, however, can often be avoided reducing (3) to a first order system following the "take the derivative as a new function" rule one learns in elementary theory of partial differential equations. That this trick always works, at least if one measures the derivative in a new norm, is in fact one of the results in Chapter III of this book. Moreover, the choice of this norm is usually natural and has physical meaning. However, reduction to first order is of no particular help in a problem as elementary as the growth of solutions of $u''(t) = (A + cI)u(t)$

in terms of the growth of the solutions of $u''(t) = Au(t)$. In other problems, such as singular perturbation, direct consideration of second order equations leads to simpler and more inclusive theories. Finally, the formalism associated with (4) has proven useful in other fields, such as the control theory of hyperbolic equations. These and other reasons give motivation to the development of a theory of second order differential equations in Banach spaces.

This work presents a few facts on that theory and some applications. No claim of completeness is made, either in the text or in the references; many important results have been left out and many important papers are not mentioned. Chapter I expounds semigroup theory; Chapter II presents cosine function theory, which stands in relation to the second order equation (4) as semigroup theory stands in relation to the first order equation (2). Chapter III deals with the reduction of (4) to a first order system mentioned above and other related topics. The next four chapters are on applications; in Chapter IV we treat the initial-boundary value problem (3) with A a second order uniformly elliptic partial differential operator in a domain of m-dimensional Euclidean space, with either the Dirichlet boundary condition or a variational boundary condition. Chapter V treats the second order equation (4) in Hilbert spaces, where many special results are available; there are applications to equations with almost periodic and periodic solutions. Chapters VI and VII are on singular perturbation problems, with applications to diverse physical situations. Finally, in Chapter VIII we touch upon the theory of the "complete" second order equation

$$u''(t) + Bu'(t) + Au(t) = 0 \tag{5}$$

without going too far into it; mostly, we search for the correct definition of correctly posed initial value problem for (5). Some shortcuts through the book are possible, and we do not bother to indicate them explicitly; for instance, Chapter III is only briefly needed in Chapters IV and V and not used at all in Chapters VI and VII.

Some effort has been made to make this book as self-contained as possible; nothing is needed except the elementary theory of Banach and Hilbert spaces and some acquaintance with parabolic partial differential equations. The functional calculus for self adjoint operators is only used in Chapters IV and V and in exercises in other chapters.

The exercises throughout the book cover parts of the theory not in the text or related facts of interest; references are included for the less immediate.

I am glad to acknowledge my thanks to many colleagues who read parts of the book and suggested improvements and to the Instituto Argentino de Matemática, Consejo Nacional de Investigaciones Científicas y Técnicas, Argentina, for its hospitality during March 1983 and August 1984, at which time the actual writing was concluded.

Finally, and most important of all, the undertaking of this project would have been impossible without the understanding support of the National Science Foundation, which support extended during the entire time it took to complete it.

As always, my wife Natalia was encouraging, patient and understanding and to her go my very special thanks.

Buenos Aires,
August 1984

CONTENTS

LIST OF SYMBOLS

CHAPTER I

THE CAUCHY PROBLEM FOR FIRST ORDER EQUATIONS. SEMIGROUP THEORY

§I.1 The Cauchy problem for first order equations.

We denote by E a general Banach space and by A a linear operator with domain $D(A)$ in E and range in E. Unless otherwise stated E will be complex and $D(A)$ will be dense in E. We shall use in the sequel the symbols $u(\hat{t})$, $f(\hat{t})$, ... to indicate functions $t \to u(t)$, $t \to f(t), \ldots$; their individual values at t are $u(t)$, $f(t)$, etc. A function $u(\hat{t})$ with values in E is continuously differentiable in $t \geq 0$ if and only if the limit as $h \to 0$ of the quotient of increments $h^{-1}(u(t+h) - u(t))$ exists and is a continuous function of t in the norm of E; of course, the limit is one-sided at $t = 0$. Similar definitions are used in intervals other than $[0, \infty)$. A (strong or genuine) solution of the abstract differential equation

$$u'(t) = Au(t) \tag{1.1}$$

in $[0, \infty)$ is a continuously differentiable E-valued function $u(t)$ such that $u(t) \in D(A)$ and (1.1) is satisfied for $t \geq 0$. Solutions are correspondingly defined in other intervals.

The Cauchy or initial value problem for (1.1) in $t \geq 0$ is that of finding solutions satisfying the initial condition

$$u(0) = u_0 . \tag{1.2}$$

The Cauchy problem for (1.1) is well posed (or properly posed) in $t \geq 0$ if and only if the following two assumptions hold:

(a) (Existence). There exists a dense subspace D of E such that for any $u_0 \in D$ there exists a solution $u(\hat{t})$ of (1.1) in $t \geq 0$ satisfying (1.2).

(b) (Continuous dependence). *There exists a nonnegative, finite function* $C(\hat{t})$ *defined in* $t \geq 0$ *such that*

$$\|u(t)\| \leq C(t)\|u(0)\| \tag{1.3}$$

for any solution of (1.1).

Note that (1.3) is assumed to hold for *any* solution of (1.1), whether or not $u(0) \in D$. Also, since the equation must hold in particular for $t = 0$ we must have

$$D \subseteq D(A) . \tag{1.4}$$

Finally, the Cauchy problem for (1.1) is *uniformly well posed* (or *uniformly properly posed*) in $t \geq 0$ if (a) and (b) hold and the function $C(\hat{t})$ is nondecreasing in $t \geq 0$ (obviously, it is enough to assume that $C(\hat{t})$ is bounded on compacts of $t \geq 0$ since we may replace it by $\tilde{C}(t) = \sup \{C(s);\ 0 \leq s \leq t\}$.

We note that condition (b) in the well posed case is equivalent to pointwise continuous dependence of the solutions on their initial values, since no relation between different values of $C(\hat{t})$ is postulated. In the uniformly well posed case, the dependence is continuous on bounded subintervals of $t \geq 0$ due to the fact that $C(t) \leq C(T)$ in $0 \leq t \leq T < \infty$.

Let $u \in D$, $t \geq 0$. Define

$$S(t)u = u(t) , \tag{1.5}$$

where $u(\hat{t})$ is the solution of (1.1) with $u(0) = u$. Obviously, $S(t)$ is a bounded operator with $\|S(t)\| \leq C(t)$. Since $D(S(t)) = D$ is dense in E, $S(t)$ can be extended to a bounded operator in E (which we denote by the same symbol) without increase of norm. We call the operator function $S(\hat{t})$ the *propagator* (or *solution operator* or *fundamental solution* or *evolution operator*) of (1.1). If $u(\hat{t})$ is an arbitrary solution of (1.1) then

$$u(t) = S(t)u(0) \qquad (t \geq 0) . \tag{1.6}$$

This is nothing but the definition of $S(t)$ when $u(0) \in D$. To prove

(1.6) in the general case, let $\{u_n\}$ be a sequence of elements in D such that $u_n \to u(0)$, $u_n(\hat{t})$ the solution of (1.1) with $u_n(0) = u_n$. Then $S(t)u(0) = \lim S(t)u_n = \lim u_n(t) = u(t)$ by (b).

THEOREM 1.1. Let the Cauchy problem for (1.1) be well posed in $t \geq 0$. Then (i)

$$S(0) = I, \quad S(s + t) = S(s)S(t) \qquad (s, t \geq 0). \tag{1.7}$$

(ii) There exists a constant ω such that, for every $\delta > 0$ we have

$$\|S(t)\| \leq C_\delta e^{\omega t} \qquad (t \geq \delta > 0) \tag{1.8}$$

with C_δ depending on δ (if the Cauchy problem for (1.1) is uniformly well posed there exists C_0 such that

$$\|S(t)\| \leq C_0 e^{\omega t} \qquad (t \geq 0)\ ). \tag{1.9}$$

(iii) $S(\hat{t})$ is strongly continuous in $t > 0$ (strongly continuous in $t \geq 0$ if the Cauchy problem for $t \geq 0$ is uniformly well posed).

Proof. The first equality (1.7) is obvious. To prove the second pick $u \in D$ and let $u(\hat{t})$ be the solution of (1.1) with $u(0) = u$. Fix $t \geq 0$ and consider the function $u(\hat{s}) = S(\hat{s} + t)u$. Since $u(\hat{s})$ is a solution of (1.1) with $u(0) = S(t)u$ we deduce that $u(s) = S(s)S(t)u$ from (1.6). Noting that all operators involved are bounded and D is dense in E, the second equality (1.7) follows.

We prove next (iii). Let $u_0 \in E$, $\{u_n\}$ a sequence in D with $u_n \to u_0$. Since

$$S(t)u_n \to S(t)u_0 \tag{1.10}$$

pointwise in $t \geq 0$, $S(\hat{t})u_0$ is strongly measurable there. Accordingly, $S(\hat{t})u_0$ is almost separably valued, that is, there exists a null subset d_0 of $[0, \infty)$ such that $X_0 = \{S(t)u_0;\ t \notin d_0\}$ is separable (for this and other results on measurable vector valued functions see for instance HILLE-PHILLIPS [1957:1, Ch. III]. It follows that the closed subspace E_0 generated by $\{u_0\} \cup X_0$ is separable and there exists a sequence $\{t_n;\ n \geq 1\}$ contained in the complement of d_0 such that the set $Y_0 = \{u_0\} \cup \{S(t_n)u_0;\ n \geq 1\}$ is fundamental in E_0 (finite linear combinations of elements of Y_0 are dense in E_0). Let now $t \notin d_0$ such that $t + t_n \notin d_0$ $(n = 1,2,\ldots)$; then $S(t)u_0$ and $S(t)S(t_n)u_0 = S(t + t_n)u_0$ belong to E_0. Define $e_0 = d_0 \cup (d_0 - t_1) \cup (d_0 - t_2) \cup \ldots$; e_0 is a null set and it follows from the preceding arguments that

$$S(t)E_0 \subseteq E_0 \qquad (t \notin e_0) . \tag{1.11}$$

We show next that if $0 < \alpha < \beta < \infty$ there exists $C = C_{\alpha,\beta}$ such that

$$\|S(t)\| \le C \qquad (\alpha \le t \le \beta) . \tag{1.12}$$

If this were not the case we could find a sequence $\{u_n\} \subseteq E$, $\|u_n\| = 1$ and a numerical sequence $\{t_n\}$ in $[\alpha, \beta]$ such that $\|S(t_n)u_n\| \ge n$ $(n = 1,2,\ldots)$. Applying the argument leading to (1.11) we can construct for each n a null set e_n and a separable subspace E_n with $S(t)E_n \subseteq E_n$ for $t \neq e_n$; hence if E_∞ is the closed subspace generated by the union of the E_n, E_∞ is separable and $S(t)E_\infty \subseteq E_\infty$ for t in the complement of the null set $e_\infty = e_1 \cup e_2 \cup \ldots$. Given $t \ge 0$ denote by $m(t) = \|S(t)\|_\infty$ the norm of the restriction of $S(t)$ to E_∞; since E_∞ is separable, $m(\hat{t})$ is the supremum of the sequence $\|S(\hat{t})v_n\|$ with $\{v_n\}$ a dense sequence in the unit sphere of E_∞ and is itself measurable. Moreover, $m(t_n) > n$ and if $t \neq e_\infty$ we have $m(s + t) = \sup\{\|S(s)S(t)u\|;\ u \in E_\infty,\ \|u\| \le 1\} \le \sup\{\|S(s)v\|;\ v \in E_\infty,\ \|v\| \le m(t)\} \le m(s)m(t)$. Accordingly, a contradiction results from:

LEMMA 1.2. _Let_ $m(\hat{t})$ _be a nonnegative finite measurable function defined in_ $t > 0$ _and such that_

$$m(s + t) \le m(s)m(t) \qquad (s > 0,\ t \notin e), \tag{1.13}$$

where e _is a null set in_ $(0, \infty)$. _Then_ $m(\hat{t})$ _is bounded in every interval_ $[\alpha, \beta]$, $0 < \alpha < \beta < \infty$.

Proof. Let $a > 0$. If $s + t = a$, $t \notin e$, (1.13) implies $m(s)m(t) \ge m(a)$, thus either $m(s) \ge \sqrt{m(a)}$ or $m(t) \ge \sqrt{m(a)}$. Hence if d is the set of all t in $[0, a]$ with $m(t) \ge \sqrt{m(a)}$ we have $e \cup d \cup (a - d) \supseteq [0, a]$ so that $|d| + |a - d| \ge a$, where $|\cdot|$ indicates Lebesgue measure. But $|d| = |a - d|$, hence $|d| \ge a/2$.

Assume now that $m(\hat{t})$ is unbounded in $[\alpha, \beta]$, so that there exists a sequence $\{a_n\}$ there with $m(a_n) \to \infty$. Applying the argument above we deduce the existence of a measurable set d_n in $[0, \beta]$ with $|d_n| \geq a_n/2 \geq \alpha/2$ and $m(t) \geq \sqrt{m(a_n)}$ in d_n, which contradicts the fact that $m(t)$ is everywhere finite. This completes the proof of Lemma 1.2.

End of proof of Theorem 1.1. Let $u \in E$, $t_0 > 0$, $0 < r < t_0$, $|h| \leq r$, $0 < \alpha < \beta < t_0 - r$. We obtain from (1.7) the equality $(S(t_0 + h) - S(t_0))u = S(t)(S(t_0 + h - t) - S(t_0 - t))u$, valid at least in $0 \leq t \leq \beta$. In view of (1.12) the function on the right hand side is bounded in $\alpha \leq t \leq \beta$; it is also easily seen to be strongly measurable. Integrating we obtain

$$(\beta - \alpha)\|S(t_0 + h)u - S(t_0)u\| \leq C \int_\alpha^\beta \|S(t_0 + h - t)u - S(t_0 - t)u\|dt, \quad (1.14)$$

where the right hand side tends to zero in view of the mean continuity of Bochner integrable functions (HILLE-PHILLIPS [1957:1, p. 86]). We have then completed the proof of (iii) in the general case. If the Cauchy problem for (1.1) is uniformly well posed the convergence in (1.10) is uniform on finite subintervals of $[0, \infty)$ thus $C(\hat{t})u$ is continuous in $t \geq 0$.

Finally, we show (ii). Observe first that, by virtue of (iii) for each $u \in E$ and $\delta > 0$ the function $S(\hat{t})u$ is continuous, hence bounded in $\delta \leq t \leq 1$; it follows from the uniform boundedness principle (DUNFORD-SCHWARTZ [1958:1, p. 52]) that $\|S(t)\| \leq C_\delta$ in $\delta \leq t \leq 1$. Let $t \geq 1/2$ be arbitrary, n the largest integer with $n/2 \leq t$. Using (1.7) we have $S(t) = S(n/2)S(t - n/2) = S(1/2)^n S(t - n/2)$ so that if $C = C_{1/2}$, $\|S(t)\| \leq C \exp(n \log C) \leq C \exp(\omega t)$ with $\omega = 2 \log C$, which is (1.8) in $t \geq 1/2$; obviously, the inequality can be extended to $t \geq \delta$ (possibly with a different constant) or to $t \geq 0$ when the Cauchy problem is uniformly well posed. This completes the demonstration of Theorem 1.1. ⊠

We comment briefly on several obvious consequences. In the uniform case, the function $C(\hat{t})$ in (1.3) can be replaced by $C_0 \exp(\omega\hat{t})$ so that if $\{u_n(\hat{t})\}$, $u(\hat{t})$ are solutions of (1.1) and $u_n(0) \to u(0)$ then $\exp(-\omega t)u_n(t) \to \exp(-\omega t)u(t)$ uniformly in $t \geq 0$. When the Cauchy problem for (1.1) is simply well posed, $\exp(-\omega t)u(t) \to \exp(-\omega t)u(t)$ uniformly in $t \geq \delta$ for each $\delta > 0$. This is considerably more than we bargained for in the definition.

§I.2 The Cauchy problem in $(-\infty, \infty)$.

We declare the Cauchy problem for (1.1) well posed in $-\infty < t < \infty$ if (a) and (b) in the previous section hold with the following modifications: the solutions in (a) are solutions in $(-\infty, \infty)$ rather than in $t \geq 0$, and (1.3) holds in $(-\infty, \infty)$ with a finite function $C(t)$ defined for all t. If $C(\hat{t})$ and $C(-\hat{t})$ are both nondecreasing in $t \geq 0$ (or just bounded on compacts of $(-\infty, \infty)$) the Cauchy problem for (1.1) is uniformly well posed in $(-\infty, \infty)$. Plainly, the propagator $S(t)$ can now be defined for all t. It is remarkable that the notions of well posed and uniformly well posed problem coalesce in the present situation, as the next result shows.

THEOREM 2.1. *Let the Cauchy problem for* (1.1) *be well posed in* $(-\infty, \infty)$. *Then* (i)

$$S(0) = I, \qquad S(s + t) = S(s)S(t) \qquad (-\infty < s, t < \infty). \tag{2.1}$$

(ii) *There exist constants* C_0, ω *such that*

$$\|S(t)\| \leq C_0 e^{\omega|t|} \qquad (-\infty < t < \infty) \tag{2.2}$$

(thus the Cauchy problem for (1.1) *is uniformly well posed in* $(-\infty, \infty)$). (iii) $S(\hat{t})$ *is strongly continuous for all* t.

The proof of (2.1) is the same as that for the case $t \geq 0$. To show (iii) we note that the assumptions imply that the Cauchy problems for both (1.1) and

$$u'(t) = -Au(t) \tag{2.3}$$

are well posed in $t \geq 0$, since the correspondence $u(\hat{t}) \to u(-\hat{t})$ maps solutions of (1.1) into solutions of (2.3) and vice versa. Let $S_-(t)$ be the propagator of (2.3) in $t \geq 0$. We verify easily that

$$S_-(t) = S(-t) \qquad (t \geq 0).$$

Applying Theorem 1.1(iii) to $S(\hat{t})$ and to $S(-\hat{t}) = S_-(\hat{t})$ in $t > 0$ we deduce that $S(\hat{t})$ is strongly continuous in $t < 0$ and $t > 0$. However, $S(h)u = S(t + h)S(-t)u$, thus strong continuity at $t = 0$ follows. In particular, the Cauchy problems for (1.1) and (2.3) are uniformly well posed in $t \geq 0$ which yields (2.2) by intervention of Theorem 1.1(ii). Obviously, this inequality implies that the Cauchy problem is well posed in $(-\infty, \infty)$, thus completing the proof of Theorem 2.1.

§I.3 The Hille-Yosida theorem.

Assume that the Cauchy problem for (1.1) is well posed in $t \geq 0$ and that A is closed. Let ω be the constant in (1.9), λ a complex number with $\operatorname{Re}\lambda > \omega$. Define an operator in E by

$$R(\lambda)u = \int_0^\infty e^{-\lambda t} S(t)u\, dt \quad (u \in E). \tag{3.1}$$

Since the norm of the integrand is bounded by $C_0\|u\| \exp(\omega - \operatorname{Re}\lambda)t$, $R(\lambda)$ is a bounded operator in E. Assume now that $u \in D$. Then $S(\hat{t})u$ is a solution of (1.1), so that $S'(t)u = AS(t)u$.(1) Using a well known result on introduction of closed operators under the integral sign (HILLE-PHILLIPS [1957:1, p. 83]) we obtain

$$A\int_0^T e^{-\lambda t} S(t)u\, dt = \int_0^T e^{-\lambda t} AS(t)u\, dt = \int_0^T e^{-\lambda t} S'(t)u\, dt$$
$$= e^{-\lambda T} S(T)u - u + \lambda \int_0^T e^{-\lambda t} S(t)u\, dt$$

for $T > 0$. Letting $T \to \infty$ and using closedness of A we deduce that $R(\lambda)u \in D(A)$ and $AR(\lambda)u = \lambda R(\lambda)u - u$. Now, if u is an arbitrary element of E, choose a sequence $\{u_n\}$ in D such that $u_n \to u$. Then $R(\lambda)u_n \to R(\lambda)u$, $AR(\lambda)u_n = \lambda R(\lambda)u_n - u_n \to \lambda R(\lambda)u - u$ so that $R(\lambda)E \subseteq D(A)$ and

$$(\lambda I - A)R(\lambda) = I. \tag{3.2}$$

In particular, (3.2) shows that $(\lambda I - A): D(A) \to E$ is onto. It is as well one-to-one. To see this assume there exists $u \in D(A)$ with $Au = \lambda u$. Then $u(t) = (\exp(\lambda t))u$ is a solution of (1.1) with $\|u(t)\| = (\exp(\operatorname{Re}\lambda)t)\|u\|$ which contradicts (1.9) unless $u = 0$. We have then proved that $\lambda \in \rho(A)$, the resolvent set of A and that $R(\lambda;A) = (\lambda I - A)^{-1} = R(\lambda)$, so that $\rho(A)$ contains the half plane $\operatorname{Re}\lambda > \omega$.

The fact that $\rho(A)$ is nonempty allows for the complete identification of the subspace D_0 of "admissible initial data" consisting of all $u_0 \in E$ such that $S(\hat{t})u_0$ is a solution of (1.1). This is done as follows. If $\lambda \in \rho(A)$, $R(\lambda;A)S(t)u$ is a solution of (1.1), hence it follows from (1.6) that

$$R(\lambda;A)S(t)u = S(t)R(\lambda;A)u \quad (t \geq 0) . \tag{3.3}$$

This equality extends to all $u \in E$ by denseness of D and implies that

$$S(t)D(A) \subseteq D(A) \quad (t \geq 0) . \tag{3.4}$$

Write (3.3) for $u = (\lambda I - A)v$, $v \in D(A)$ and then apply $(\lambda I - A)$ to both sides. The result is

$$AS(t)u = S(t)Au \quad (t \geq 0) , \tag{3.5}$$

valid for $u \in D(A)$. Making use of (3.5) we obtain

$$S(t)u - u = \int_0^t S'(s)u\, ds = \int_0^t S(s)Au\, ds$$

for $u \in D$. Apply $R(\lambda;A)$ to both sides and use (3.3): the result is

$$R(\lambda;A)(S(t)u - u) = \int_0^t S(s)AR(\lambda;A)u\, ds .$$

Since both sides are bounded operators of u the equality can be extended to all $u \in E$; using (3.3) again and applying $(\lambda I - A)$ to both sides, we obtain

$$S(t)u - u = \int_0^t AS(s)u\, ds \tag{3.6}$$

for all $u \in D(A)$. Obviously, this implies that $S(\hat{t})u$ is a solution of (1.1) for any $u \in A$, hence

$$D_0 = D(A) . \tag{3.7}$$

We can obtain estimates for $R(\lambda;A)$ and its powers using the well known formula $R(\lambda;A)^{(n)} = (-1)^n n! R(\lambda;A)^{n+1}$ and differentiating (3.1) under the integral sign (this can be easily justified on the basis of the dominated convergence theorem). The formula

$$R(\lambda;A)^n u = \frac{1}{(n-1)!} \int_0^\infty t^{n-1} e^{-\lambda t} S(t)u\, dt \tag{3.8}$$

results for $u \in E$ and $\operatorname{Re} \lambda > \omega$. Using (1.9) we obtain

$$\|R(\lambda;A)^n\| \leq C_0(\operatorname{Re} \lambda - \omega)^{-n} \quad (\operatorname{Re} \lambda > \omega, \ n = 0,1,\dots) , \tag{3.9}$$

where C_0 and ω are the constants in (1.9). We show next that inequalities (3.9) are as well sufficient for uniform well posedness of the

Cauchy problem for (1.1).

THEOREM 3.1 (Hille-Yosida). Let A be closed. The Cauchy problem for (1.1) is uniformly well posed in $t \geq 0$ with propagator $S(t)$ satisfying (1.9) if and only if $\sigma(A)$ is contained in the half-space $\operatorname{Re} \lambda \leq \omega$ and $R(\lambda;A)$ satisfies inequalities (3.9).

We have already proved one half. The proof of the other half below is perhaps not the shortest available but can be adapted with minor changes to equations other than (1.1) (see Chapter II). We begin by constructing solutions of (1.1) given sufficiently "smooth" initial conditions. Let $u \in D(A^3)$, $\omega' > \omega, 0$. Define[2]

$$\tilde{u}(t;u) = u + tAu + \frac{t^2}{2} A^2 u + \frac{1}{2\pi i} \int_{\omega'-i\infty}^{\omega'+i\infty} \frac{e^{\lambda t}}{\lambda^3} R(\lambda;A)A^3 u \, d\lambda \qquad (t \geq 0) . \tag{3.10}$$

A deformation of contour shows that the integral in (3.10) vanishes for $t \leq 0$, thus

$$\tilde{u}(0,u) = u . \tag{3.11}$$

Obviously, $\|\tilde{u}(t;u)\| = 0(\exp(\omega' t))$ as $t \to \infty$. An easily justifiable differentiation under the integral sign shows that $\tilde{u}(\hat{t};u)$ has a continuous derivative in $t \geq 0$; on the other hand A can be introduced under the integral sign with convergence of the resulting integral, so that $\tilde{u}(t;u) \in D(A)$ and

$$\begin{aligned} \tilde{u}'(t;u) - A\tilde{u}(t;u) &= -\frac{t^2}{2} A^3 u \\ &\quad + \frac{1}{2\pi i} \int_{\omega'-i\infty}^{\omega'+i\infty} \frac{e^{\lambda t}}{\lambda^3} (\lambda I - A)R(\lambda;A)A^3 u \, d\lambda \\ &= 0 \qquad (t \geq 0) . \end{aligned} \tag{3.12}$$

Accordingly, $\tilde{u}(t;u)$ is a solution of (1.1). An argument similar to the one leading to the proof that $R(\lambda)$ in (3.1) coincides with $R(\lambda;A)$ (or a direct integration of (3.10)) shows that

$$\int_0^\infty e^{-\lambda t} \tilde{u}(t;u) \, dt = R(\lambda;A)u . \tag{3.13}$$

This equality will be used to obtain bounds on $u(\cdot)$ by use of the Post inversion formula below.

LEMMA 3.2. Let $f(\hat{t})$ be a E-valued continuous function defined in $t \geq 0$ such that $f(t) = O(\exp(\alpha t))$ as $t \to \infty$ for some α and let $(\mathcal{L}f)(\hat{\lambda})$ be the Laplace transform of t,

$$(\mathcal{L}f)(\lambda) = \int_0^\infty e^{-\lambda t} f(t)\, dt .$$

Then

$$f(t) = \lim \frac{(-1)^n}{n!} \left(\frac{n}{t}\right)^{n+1} (\mathcal{L}f)^{(n)}\left(\frac{n}{t}\right) \tag{3.14}$$

uniformly on compacts of $t > 0$.

The proof in the scalar case can be found in WIDDER [1946:1, Chapter VII]. The extension to vector valued functions is immediate.

End of proof of Theorem 3.1. We apply the inversion formula (3.14) to the function $\tilde{u}$ in (3.10); in view of (3.13) we obtain

$$\begin{aligned} \tilde{u}(t;u) &= \lim \frac{(-1)^n}{n!} \left(\frac{n}{t}\right)^{n+1} R\left(\frac{n}{t};A\right)^{(n)} u \\ &= \lim \left(\frac{n}{t}\right)^{n+1} R\left(\frac{n}{t};A\right)^{n+1} u , \end{aligned} \tag{3.15}$$

so that, using inequalities (3.9) we obtain

$$\begin{aligned} \|\tilde{u}(t;u)\| &\leq C_0 \|u\| \lim \left(1 - \frac{\omega t}{n}\right)^{-(n+1)} \\ &= C_0 \|u\| e^{\omega t} \qquad (t \geq 0) . \end{aligned} \tag{3.16}$$

Define

$$\tilde{S}(t)u = \tilde{u}(t;u) \qquad (t \geq 0) .$$

for $u \in D(A^3)$, $u(t)$ the solution in (3.10). Arguing as in the treatment of $S(\hat{t})$ we can extend each $\tilde{S}(t)$ to a bounded everywhere defined operator such that $\tilde{S}(\hat{t})$ is strongly continuous in $t \geq 0$ and

$$\|\tilde{S}(t)\| \leq C_0 e^{\omega t} \qquad (t \geq 0) . \tag{3.17}$$

Clearly, the proof of Theorem 3.1 will be complete if we show that any solution $u(\hat{t})$ of (1.1) admits the representation

$$u(t) = \tilde{S}(t)u(0) . \tag{3.18}$$

This is done as follows. It results from the definition of $\tilde{S}(t)$ that $R(\lambda;A)\tilde{S}(t)u = \tilde{S}(t)R(\lambda;A)u$ for all $u \in D(A^3)$ thus for all $u \in E$. Accordingly,

$$\tilde{S}(t)Au = A\tilde{S}(t)u \qquad (u \in D(A),\ \ t \geq 0) . \tag{3.19}$$

On the other hand, $\mathfrak{m}(\hat{t}) = S(\hat{t})R(\lambda;A)^3$ is a continuously differentiable operator valued function that satisfies $\mathfrak{m}'(t) = A\,\mathfrak{m}(t)$. Accordingly, if $u(\hat{t})$ is an arbitrary solution of (1.1) we have

$$\frac{d}{ds}\,\mathfrak{m}(t - s)u(s) = -A\,\mathfrak{m}(t - s)u(s)$$

$$+ \mathfrak{m}(t - s)Au(s) = 0 \qquad (0 \leq s \leq t) \tag{3.20}$$

after (3.19), hence $R(\lambda;A)^3u(t) = \mathfrak{m}(0)u(t) = \mathfrak{m}(t)u(0) = R(\lambda;A)^3\tilde{S}(t)u(0)$. Applying $(\lambda I - A)^3$ to both sides (3.18) results.

Corresponding to the case $(-\infty, \infty)$ we have

THEOREM 3.3. Let A be closed. The Cauchy problem for (1.1) is uniformly well posed in $(-\infty, \infty)$ with propagator $S(\hat{t})$ satisfying (2.2) if and only if $\sigma(A)$ is contained in the strip $|\mathrm{Re}\,\lambda| \leq \omega$ and $R(\lambda;A)$ satisfies

$$\|R(\lambda;A)^n\| \leq C_0(|\mathrm{Re}\,\lambda| - \omega)^{-n} \qquad (|\mathrm{Re}\,\lambda| > \omega,\ \ n = 0,1,\ldots) . \tag{3.21}$$

Proof. We have already observed that if the Cauchy problem for (1.1) is well posed in $(-\infty, \infty)$ then the Cauchy problem for (2.3) (of course also for (1.1)) is well posed in $t \geq 0$. Since $R(\lambda;-A) = -R(-\lambda;A)$, inequalities (3.21) follow from Theorem 3.1. Conversely, we deduce from the same theorem that if (3.21) holds then the Cauchy problem for (1.1) and (3.2) is well posed in $t \geq 0$ with S and S_- satisfying (1.9) in $t \geq 0$. This is easily seen to imply that the Cauchy problem for (1.1) is well posed in $(-\infty, \infty)$, thus completing the proof of Theorem 2.1.

For the sake of brevity in future statements, we introduce the following notation. A closed, densely defined operator A is said to belong to $\mathfrak{C}_+^1(C_0,\omega)$ if the Cauchy problem for (1.1) is uniformly well posed in $t \geq 0$ and the solution operator $S(\hat{t})$ satisfies (1.9). We also write $\mathfrak{C}_+^1(\omega) = \cup \{\mathfrak{C}_+^1(C_0,\omega);\ C_0 \geq 1\}$ (note that $C_0 < 1$ is impossible) and $\mathfrak{C}_+^1 = \cup \{\mathfrak{C}_+^1(\omega);\ -\infty < \omega < \infty\}$. The notations for the case $-\infty < t < \infty$ are $\mathfrak{C}^1(C_0,\omega)$ (C_0 the constant in (2.2)), $\mathfrak{C}^1(\omega)$, $\mathfrak{C}^1$; we shall not employ any special symbol for operators that make the Cauchy problem only well posed.

The following result is an immediate consequence of the Hille-Yosida theorem.

THEOREM 3.4. <u>Let</u> A <u>be an operator such that</u> $\rho(A)$ <u>contains the half plane</u> $\operatorname{Re}\lambda \geq \lambda_0$, $S(\hat{t})$ <u>an operator valued function strongly continuous in</u> $t \geq 0$ <u>and such that</u>

$$\|S(t)\| \leq C_0 e^{\omega t} \qquad (t \geq 0)\ . \tag{3.22}$$

<u>Assume that, for each</u> $u \in E$

$$\int_0^\infty e^{-\lambda t} S(t)u\,dt = R(\lambda;A)u \qquad (\operatorname{Re}\lambda > \lambda_0)\ . \tag{3.23}$$

<u>Then</u> $A \in \mathfrak{C}_+^1(C_0,\omega)$ <u>and</u> $S(t)$ <u>is the solution operator of</u> (1.1).

<u>Proof</u>. We obtain the inequalities (3.9) from (3.23) in the same way as from (3.1); applying Theorem 3.1 it results that $A \in \mathfrak{C}_+^1(C_0,\omega)$. Let $\bar{S}(t)$ be the propagator of (1.1). Then (3.23) holds for both S and $\bar{S}$ whence $S = \bar{S}$ follows from uniqueness of Laplace transforms.

REMARK 3.5. Inequalities (3.9) follow from their real counterparts

$$\|R(\lambda;A)^n\| \leq C_0(\lambda - \omega)^{-n} \qquad (\lambda > \omega,\ n = 0,1,\dots)\ . \tag{3.24}$$

The proof is an elementary exercise in Taylor series. This allows us to weaken, say Theorem 3.4 by postulating (3.23) only for $\lambda > \omega$, although the advantage does not seem to be very significant.

REMARK 3.6. Replacing powers of the resolvent by derivatives, inequalities (3.9) can be written thus:

$$\|R(\lambda;A)^{(n)}\| \leq C_0 n!(\operatorname{Re}\lambda - \omega)^{-(n+1)} \qquad (\operatorname{Re}\lambda > \omega,\ n = 0,1,\dots)\ . \tag{3.25}$$

Although this form is less practical, similarity with generation theorems for other equations becomes more apparent (see Section II.2).

§I.4 Semigroup theory.

Given two Banach spaces E and F we denote by $(E;F)$ the space of all linear bounded operators from E into F endowed with its usual (uniform operator) topology. We usually abbreviate $(E;E)$ to (E).

A E-valued function $S(\hat{t})$ defined in $t \geq 0$ is called a semigroup if (1.7) holds, that is if

$$S(0) = I, \qquad S(s+t) = S(s)S(t) \qquad (s, t \geq 0) . \tag{4.1}$$

Equations (4.1) are often called the exponential (functional) equations. We have seen in the preceding sections that the solution operator of a well posed Cauchy problem is a strongly continuous semigroup. As the following result shows, the converse is as well true.

THEOREM 4.1. *Let $S(\cdot)$ be a semigroup strongly continuous in $t \geq 0$. Then there exists a unique closed, densely defined operator $A \in \mathfrak{C}_+^1$ such that $S(t)$ is the evolution operator of*

$$u'(t) = Au(t) . \tag{4.2}$$

Proof. We define the infinitesimal generator A of $S(t)$ by the formula

$$Au = \lim_{h\to 0+} \frac{1}{h}(S(h) - I)u . \tag{4.3}$$

The domain of A consists of all $u \in E$ such that (4.3) exists. For u arbitrary and $a > 0$ define

$$u^a = \frac{1}{a}\int_0^a S(s)u\,ds . \tag{4.4}$$

The second equation (4.1) implies that

$$\frac{1}{h}(S(h) - I)u^a = \frac{1}{a}\left\{\frac{1}{h}\int_a^{a+h} S(s)u\,ds - \frac{1}{h}\int_0^h S(s)u\,ds\right\},$$

hence $u^a \in D(A)$ (with

$$Au^a = \frac{1}{a}(S(a) - I)u) . \tag{4.5}$$

But $u^a \to u$ as $a \to 0$, thus the set of all elements of the form u^a (a fortiori, $D(A)$) is dense in E.

We prove next that A is closed. If $u \in E$ we have

$$\frac{1}{h}(S(h) - I)S(t)u = S(t)\frac{1}{h}(S(h) - I)u$$

for $h > 0$, $t \geq 0$ thus if $u \in D(A)$ then $S(t)u \in D(A)$ as well and

$$AS(t)u = S(t)Au . \tag{4.6}$$

Hence, for any $u \in E$,

$$(Au)^a = Au^a = \frac{1}{a}(S(a) - I)u$$

after (4.5). Let $\{u_n\}$ be a sequence in $D(A)$ such that $u_n \to u$, $Au_n \to v$. Then

$$\frac{1}{h}(S(h) - I)u = \lim_{n\to\infty} \frac{1}{h}(S(h) - I)u_n = \lim_{n\to\infty} (Au_n)^h = v^h .$$

Taking limits as $h \to 0+$ we see that $u \in D(A)$ and that $Au = v$, which shows the closedness of A.

Let now $u \in D(A)$. Integrating (4.6) in $0 \leq s \leq t$ we obtain

$$\int_0^t S(s)Au\,ds = \int_0^t AS(s)u\,ds = A\int_0^t S(s)u\,ds = tAu^t = S(t)u - u$$

using (4.6) in the last equality. This shows that $S(\hat{t})u$ is continuously differentiable in $t \geq 0$ with

$$S'(t)u = S(t)Au = AS(t)u ,$$

so that $S(\hat{t})u$ is a solution of (4.2) in $t \geq 0$. This proves (a) in the definition of uniformly well posed Cauchy problem. The continuous dependence property (b) is checked as follows. Since $S(\hat{t})u$ is continuous for all u, $\|S(t)u\|$ must be bounded on bounded subsets of $t \geq 0$ thus by the uniform boundedness theorem $\|S(t)\|$ must be as well bounded on compacts of $t \geq 0$. Using this we can show that if $u(\hat{t})$ is an arbitrary solution of (4.2) then $v(s) = S(t - s)u(s)$ is continuously differentiable in $0 \leq s \leq t$ and $v'(s) = S(t - s)Au(s) - AS(t - s)u(s) = 0$, so that

$$u(t) = S(t)u(0) .$$

This completes the proof.

The case $(-\infty, \infty)$ is handled in a similar way. A (E)-valued function $S(\hat{t})$ defined for all t and satisfying (4.1) for all s, t is called a group. The propagator of (4.2) is a strongly continuous group when $A \in \mathfrak{C}^1$; conversely, we have

THEOREM 4.2. Let $S(\cdot)$ be a group strongly continuous in $-\infty < t < \infty$. Then there exists a unique closed, densely defined operator $A \in \mathfrak{C}^1$ such that $S(\hat{t})$ is the evolution operator of (4.2).

We omit the proof, which imitates that of Theorem 4.1.

In the future, "strongly continuous semigroup" means a semigroup strongly continuous in $t \geq 0$, while "strongly continuous group" indicates continuity in $-\infty < t < \infty$. We shall make no use of semigroups only continuous in $t > 0$, which are naturally associated with Cauchy problems only well posed in $t \geq 0$.

§I.5 The inhomogeneous equation.

Let $f(\hat{t})$ be a continuous E-valued function defined in $t \geq 0$. Solutions (strong or genuine) of the inhomogeneous equation

$$u'(t) = Au(t) + f(t) \tag{5.1}$$

are defined in the same way as solutions of (1.1). Since the difference of two solutions of (5.1) is a solution of (1.1), if $A \in \mathfrak{C}_+^1$ there is at most one solution of (5.1) satisfying the initial condition

$$u(0) = u_0 \ . \tag{5.2}$$

Let $u(\hat{t})$ be an arbitrary solution of (5.1) satisfying (5.2) and let $S(\hat{t})$ be the propagator of (1.1). Given $t > 0$ fixed we show easily that the function $S(t - \hat{s})u(\hat{s})$ is continuously differentiable in $0 \leq s \leq t$ with derivative $-S'(t - \hat{s})u(\hat{s}) + S(t - \hat{s})u'(\hat{s}) = S(t - \hat{s})(u'(\hat{s}) - Au(\hat{s})) = S(t - \hat{s})f(\hat{s})$. Integrating we obtain

$$u(t) = S(t)u_0 + \int_0^t S(t - s)f(s)\, ds \ . \tag{5.3}$$

However, the converse is not in general true: to make (5.3) a genuine solution of (5.1) more than continuity of f is needed. We prove below a classical result.

LEMMA 5.1. *Assume that one of the following two conditions is satisfied*:

(a) $f(t) \in D(A)$ *and* $f(\hat{t})$, $Af(\hat{t})$ *are continuous in* $0 \leq t \leq T$, *or*

(b) $f(\hat{t})$ *is continuously differentiable in* $0 \leq t \leq T$.

Suppose in addition that $u_0 \in D(A)$. *Then* (5.3) *is a genuine solution of* (5.1) *in* $0 \leq t \leq T$.

Proof. Since $S(\hat{t})u_0$ is a solution of (1.1) when $u_0 \in D(A)$ we may disregard the first term on the right hand side of (5.3). To prove our claim it is enough to show that $u(t) \in D(A)$, $Au(\hat{t})$ is continuous and

$$u(t) = \int_0^t (Au(s) + f(s))\, ds \qquad (0 \leq t \leq T) . \tag{5.4}$$

Assume (a) holds. Then

$$\begin{aligned} \int_0^t Au(s)ds &= \int_0^t \left(\int_0^s S(s - r)Af(r)dr \right) ds \\ &= \int_0^t \left(\int_r^t AS(s - r)f(r)ds \right) dr \\ &= \int_0^t (S(t - r)f(r) - f(r))\, dr , \end{aligned}$$

which is nothing if not (5.4).

On the other hand, suppose that (b) is verified. Then, integrating by parts,

$$\begin{aligned} u(t) &= -\int_0^t \left(\frac{d}{ds} \int_0^{t-s} S(r)f(s)dr \right) ds \\ &= \int_0^t S(s)f(0)ds + \int_0^t \left(\int_0^{t-s} S(r)f'(s)dr \right) ds . \end{aligned}$$

Hence

$$\begin{aligned} Au(t) &= S(t)f(0) - f(0) + \int_0^t (S(t - s) - I)f'(s)\, ds \\ &= S(t)f(0) - f(t) + \int_0^t S(t - s)f'(s)\, ds . \end{aligned}$$

Integrating and reversing the order of integration we obtain (5.4). Details are omitted.

Obviously, formula (5.3) makes sense with much weaker conditions on f. Let $1 \leq p < \infty$. The space $L^p(0,T;E)$ consists of all (equivalence classes of) strongly measurable functions $f(\hat{t})$ defined in $0 \leq t \leq T$ such that

$$\|f\|_p = \left(\int_0^T \|f(t)\|^p \, dt \right)^{1/p} < \infty ,$$

endowed with the norm $\|\cdot\|_p$; $L^\infty(0,T;E)$ consists of all (equivalence classes of) essentially bounded strongly measurable functions endowed with the norm

$$\|f\|_\infty = \operatorname*{ess.\,sup}_{0 \leq t \leq T} \|f(t)\| .$$

All the spaces $L^p(0,T;E)$ are Banach spaces (HILLE-PHILLIPS [1957:1, p. 89]).

If f is, say, a function in $L^1(0,T;E)$ and u_0 is an arbitrary element of E, (5.3) is not in general a solution of (5.1) ($u(\hat{t})$ may not be differentiable anywhere or $u(t)$ may fail to belong to $D(A)$ for any t). We shall declare $u(t)$ to be a <u>generalized solution</u> of (5.1). The following result concerns continuity properties of $u(\hat{t})$.

LEMMA 5.2. <u>Assume</u> $u_0 \in E$, $f \in L^1(0,T;E)$. <u>Then</u> $u(\hat{t})$ <u>is continuous in</u> $0 \leq t \leq T$. <u>Moreover, we have</u>

$$\|u(t)\| \leq C_0 e^{\omega t}\|u_0\| + C_0 e^{\omega T}\|f\|_1 \qquad (0 \leq t \leq T) , \tag{5.5}$$

C_0, ω <u>the constants in</u> (1.9).

<u>Proof</u>. (5.5) is rather obvious. To prove the rest of Lemma 5.2 we note that

$$\|u(t') - u(t)\| \leq \int_t^{t'} \|S(t' - s)\| \, \|f(s)\| \, ds + \int_0^t \|(S(t' - s) - S(t - s))f(s)\| \, ds$$

for $0 < t < t'$. The first integral can be made small when $t' - t \to 0$ by continuity of the indefinite integral of summable functions; the second yields to strong continuity of $S(\hat{t})$ and the uniform boundedness theorem.

§I.6 Miscellaneous comments.

Theorem 3.1 was discovered independently by HILLE [1948:1] and YOSIDA [1948:1] in the particular case where $C_0 = 1$. The proof for the general case was discovered, also independently, by FELLER [1953:1], MIYADERA [1952:1] and PHILLIPS [1953:1]. In all these papers, Theorem 3.1 is formulated in the language of strongly continuous semigroups and their generators rather than that of abstract differential equations and their propagators. The proof of Theorem 3.1 presented here follows HILLE [1952:1]. Earlier results on semigroups of operators were obtained by NATHAN, FUKAMIYA and GELFAND (for more information on the history of the subject see HILLE-PHILLIPS [1957:1] or the author [1983:1, p. 91]. Theorem 1.1 is due independently to MIYADERA [1951:1] and PHILLIPS [1951:1], who improved on less general statements of DUNFORD and HILLE.

In its present form, the notion of uniformly well posed (abstract) Cauchy problem is due to LAX (see LAX-RICHTMYER [1956:1]. The notion of abstract Cauchy problem was introduced by HILLE [1952:2] (see also HILLE [1953:1], [1953:2], [1954:1], [1954:2], [1957:1], PHILLIPS [1954:1]) although his definition of a well posed Cauchy problem is somewhat different from that employed here. For a thorough discussion of the abstract Cauchy problem and the classical Cauchy problem of solving a partial differential equation with initial data prescribed on a noncharacteristic surface see the author [1983:1, p. 54]. In particular, the present formulation of the Cauchy problem originates in the concept of properly posed Cauchy problem formulated by HADAMARD [1923:1].

EXERCISE 1. Let A be a bounded operator in a Banach space E. Show that $A \in \mathfrak{C}^1$, that is, that the Cauchy problem for

$$u'(t) = Au(t) \tag{6.1}$$

is uniformly well posed in $-\infty < t < \infty$. The group $S(\hat{t})$ generated by A (or, equivalently, the propagator of (6.1)) is given by

$$S(t) = e^{tA} = \sum_{n=0}^{\infty} \frac{t^n}{n!} A^n, \tag{6.2}$$

the series (6.2) converging uniformly on compact subsets of $-\infty < t < \infty$. Prove that $S(\hat{t})$ can also be expressed in the form

$$S(t) = \frac{1}{2\pi i} \int_{\Gamma} e^{\lambda t} R(\lambda;A)\, d\lambda , \tag{6.3}$$

where Γ is a simple closed curve (or the union of a finite number of these) oriented counterclockwise and enclosing the spectrum $\sigma(A)$ in its (their) interior. Show that

$$\|S(t)\| \le Ce^{\omega_1 t} \quad (t \ge 0) , \tag{6.4}$$

$$\|S(t)\| \le Ce^{\omega_2 |t|} \quad (t \le 0) , \tag{6.5}$$

if $\omega_1 > \omega_+$, $\omega_2 > -\omega_-$, where

$$\omega_+ = \sup \{\mathrm{Re}\,\lambda;\ \ \lambda \in \sigma(A)\} , \tag{6.6}$$

$$\omega_- = \inf \{\mathrm{Re}\,\lambda;\ \ \lambda \in \sigma(A)\} . \tag{6.7}$$

Produce an example where (6.4) (resp. (6.5)) does not hold with $\omega_1 = \omega_+$(resp. $\omega_2 = -\omega_-$). Show that $S(\hat{t})$ can be extended to a function $S(\hat{\zeta})$ holomorphic (as a (E)-valued function) for all ζ; in particular $S(t)$ is continuous in the norm of (E).

EXERCISE 2. Let $S(\hat{t})$ be a semigroup continuous in the topology of (E) (continuity at $t = 0$ suffices). Show that the infinitesimal generator A of $S(\hat{t})$ is a bounded operator (Hint: use formula (3.8) for $n = 1$ and continuity of $S(\hat{t})$ at $t = 0$ to show that $\|\lambda R(\lambda;A) - I\| < 1$ for λ sufficiently large so that $\lambda R(\lambda;A)$ has a bounded inverse). Show that $S(\hat{t})$ admits the representations (6.2) and (6.3).

EXERCISE 3. Let $E = C_0(-\infty,\infty)$ be the Banach space of all (complex-valued) continuous functions $u(x)$ defined in $-\infty < x < \infty$ and such that $u(x) \to 0$ as $|x| \to \infty$, endowed with the norm

$$\|u\| = \sup_{-\infty < x < \infty} |u(x)| . \tag{6.8}$$

Define

$$S(t)u(x) = u(x + t) \qquad (-\infty < t < \infty) . \tag{6.9}$$

Show that $S(\hat{t})$ is a strongly continuous group in E. Identify its infinitesimal generator.

EXERCISE 4. Prove the result of Exercise 6.3 (and identify the infinitesimal generator) in the space $L^p(-\infty,\infty)$, $1 \le p < \infty$. Show that (6.9) is not strongly continuous (or even strongly measurable) in $L^\infty(-\infty,\infty)$.

EXERCISE 5. Define

$$S(t)u(x) = u(x+t) \qquad (t \ge 0) \tag{6.10}$$

in the space $C_0[0,\infty)$ of all continuous functions defined in $x \ge 0$ tending to zero at infinity, endowed with the supremum norm. Show that $S(\hat{t})$ is a strongly continuous semigroup and identify its infinitesimal generator. Do the same in the spaces $L^p(0,\infty)$, $1 \le p < \infty$.

EXERCISE 6. Define

$$S(t)u(x) = \begin{cases} u(x-t) & (t \le x < \infty) \\ 0 & (0 \le x < t) \end{cases} \qquad (t \ge 0). \tag{6.11}$$

Show that $S(\hat{t})$ is a strongly continuous semigroup in the spaces $L^p(0,\infty)$ $(1 \le p < \infty)$ and in the space $C_{0,0}[0,\infty)$ of all continuous functions defined in $x \ge 0$, tending to zero at infinity and such that $u(0) = 0$ (but not in $C_0[0,\infty)$). Identify the infinitesimal generator in each case.

EXERCISE 7. Let A be a normal operator in a Hilbert space H. Show that $A \in \mathfrak{C}^1_+$ if and only if ω_+ (defined by (6.6) is finite. Prove the formula

$$S(t)u = \int_{\sigma(A)} e^{\lambda t} P(d\lambda)u \qquad (u \in H,\ t \ge 0) \tag{6.12}$$

for the semigroup $S(\hat{t})$ generated by A, where $P(d\lambda)$ is the resolution of the identity associated with A. Using the functional calculus for normal operators show that

$$\|S(t)\| = e^{\omega_+ t} \qquad (t \ge 0)\ . \tag{6.13}$$

Show that if A is self adjoint then each $S(t)$ is self adjoint.

EXERCISE 8. Let A be a normal operator in a Hilbert space E. Show that $A \in \mathfrak{C}^1$ if and only if ω_+ and ω_- (defined by (6.7)) are finite. Prove that the formula (6.12) holds in $-\infty < t < \infty$ and that (6.13) and

$$\|S(t)\| \le Ce^{-\omega_- t} = e^{|\omega_-| |t|} \qquad (t \le 0) \tag{6.14}$$

hold. Show that if iA is a self adjoint operator then each $S(t)$ is unitary.

EXERCISE 9. Let $S(\hat{t})$ be a strongly continuous semigroup in a Hilbert space H, A its infinitesimal generator. Assume that each $S(t)$ is normal. Show that A is normal (Hint: using the first formula (3.8) show that $R(\lambda;A)$ is normal).

EXERCISE 10. (HILLE [1938:1], SZ.-NAGY [1938:1]). Under the hypoteses of Exercise 9, assume that each $S(t)$ is self adjoint. Then A is self adjoint (Hint: show that $R(\lambda;A)$ is self adjoint for λ real).

EXERCISE 11. (STONE [1932:1]) Under the hypoteses of Exercise 9, assume that each $S(t)$ is unitary. Then iA is self adjoint.

EXERCISE 12. Let $S(\hat{t})$ be a strongly continuous semigroup in a Banach space E. Define

$$\omega = \overline{\lim_{t\to\infty}} \frac{1}{t} \log \|S(t)\| . \tag{6.15}$$

(a) Show that the limit exists (the value $\omega = -\infty$ is allowed) and $\omega < \infty$. (b) If $\omega' > \omega$ then

$$\|S(t)\| \le Ce^{\omega' t} \qquad (t \ge 0) , \tag{6.16}$$

although (6.16) may not hold for $\omega' = \omega$ itself. (c) (6.16) cannot hold for $\omega' < \omega$. (d) Using Exercise 6.1 show that if the infinitesimal generator of A is a bounded operator then

$$\omega = \omega_+, \tag{6.17}$$

ω_+ defined by (6.6). Show that (6.17) holds as well if E is a Hilbert space and A is a normal operator.

EXERCISE 13. Show, by means of an example, that there exists a semigroup $S(\hat{t})$ in separable Hilbert space H with infinitesimal generator A such that $\omega > \omega_+$. More generally, we can construct $S(\hat{t})$ in such a way that

$$\omega > \omega_+ + h ,$$

where $h > 0$ is arbitrarily preassigned. See ZABCZYK [1975:1]; an earlier (although less elementary) example can be found in HILLE-PHILLIPS [1957:1, p. 665] where $\sigma(A)$ is in fact empty (!) so that $\omega_+ = -\infty$.

EXERCISE 14. Prove Lemma 3.2 (Hint: use the fact that

$$\frac{(-1)^n}{n!}\left(\frac{n}{t}\right)^{n+1}(\mathcal{S}f)^{(n)}\left(\frac{n}{t}\right) = \int_0^\infty \gamma_n(t,s)f(s)\, ds$$

with

$$\gamma_n(t,s) = \frac{1}{n!}\left(\frac{n}{t}\right)^{n+1} s^n e^{-ns/t}) . \tag{6.18}$$

EXERCISE 15. Prove that the "real" inequalities (3.24) imply their complex counterparts (3.21). (Hint: if $\mathrm{Re}\,\lambda > \omega$ express $R(\lambda;A)$ by means of its Taylor series

$$R(\lambda;A) = \sum_{n=0}^{\infty} (\mu - \lambda)^n R(\mu;A)^{n+1} \tag{6.19}$$

with μ real. Then let $\mu \to \infty$).

EXERCISE 16. Let $S(\hat{t})$ be a strongly continuous semigroup, A its infinitesimal generator, ω the number defined by (6.15), $\omega' > \omega, 0$. Show that if $u \in D(A)$ then

$$S(t)u = \lim_{r\to\infty} \frac{1}{2\pi i}\int_{\omega'-ir}^{\omega'+ir} e^{\lambda t} R(\lambda;A)u\, d\lambda , \tag{6.20}$$

the limit being uniform on compact subsets of $t > 0$. For $t = 0$ the limit is $\frac{1}{2}u$ (compare with (6.3)).

EXERCISE 17. Let A be a (not necessarily densely defined) linear operator in a Banach space E such that $R(\lambda;A)$ exists for $\lambda > \omega$ and

$$\|R(\lambda;A)\| \le C/\lambda \quad (\lambda > \omega) . \tag{6.21}$$

Show that if $u \in D(A)$ then

$$\lim_{\lambda\to+\infty} \lambda R(\lambda;A)u = u . \tag{6.22}$$

FOOTNOTES TO CHAPTER I

(1) We denote here by $S'(\hat{t})u$ the derivative of the function $S(\hat{t})u$ not the derivative of $S(\hat{t})$ (which may fail to exist in the norm of (E)) applied to u. The same observation applies to other operator valued functions. The correct notation $(S(\hat{t})u)'$ becomes cumbersome later.

(2) Roughly speaking, $u(t)$ is the inverse Laplace transform of $R(\hat{\lambda};A)u$. However, the corresponding integral may not be convergent thus we use the well known formula $R(\lambda;A)u = \lambda^{-1}u + \lambda^{-2}Au + \ldots + \lambda^{-m}A^{m-1}u + \lambda^{-m}R(\lambda;A)A^{m}u$ for $m = 3$, which allows differentiation with respect to t.

CHAPTER II

THE CAUCHY PROBLEM FOR SECOND ORDER EQUATIONS
COSINE FUNCTION THEORY

§II.1 The Cauchy problem for second order equations.

A theory for the equation

$$u''(t) = Au(t) \tag{1.1}$$

that parallels closely that of (I.1.1) can be developed without undue difficulty. We prove the fundamental results in this chapter. A *solution* of (1.1) in $t \geq 0$ is a twice continuously differentiable E-valued function $u(\hat{t})$ such that $u(t) \in D(A)$ and (1.1) holds there; solutions in different intervals are defined accordingly. The *Cauchy problem* is now that of finding solutions of (1.1) that satisfy the initial conditions

$$u(0) = u_0,\ u'(0) = u_1 \ . \tag{1.2}$$

The Cauchy problem is *well posed* or *properly posed* in $t \geq 0$ if and only if

(a) *There exists a dense subspace* D *of* E *such that for any* $u_0, u_1 \in D$ *there exists a solution* $u(\hat{t})$ *of* (1.1) *in* $t \geq 0$ *satisfying* (1.2).

(b) *There exists a nonnegative, finite function* $C(\hat{t})$ *defined in* $t \geq 0$ *such that*

$$\|u(t)\| \leq C(t)(\|u(0)\| + \|u'(0)\|) \quad (t \geq 0). \tag{1.3}$$

The definitions of well posed problem in $(-\infty, \infty)$ and uniformly well posed problem in $[0, \infty)$ and $(-\infty, \infty)$ are obvious analogues of the first order case and we omit the details.

We define now two *propagators* or *solution operators* $\mathcal{C}(\hat{t})$ and $\mathcal{S}(\hat{t})$

as follows:

$$\mathcal{C}(t)u = u(t) \tag{1.4}$$

for $u \in D$, where $u(\hat{t})$ is the solution of (1.1) with $u(0) = u$, $u'(0) = 0$; $\mathcal{S}(\hat{t})$ is defined in the same way in relation to the solution $v(\hat{t})$ satisfying $v(0) = 0$, $v'(0) = u$. Both $\mathcal{C}(t)$ and $\mathcal{S}(t)$ are extended by continuity to bounded, everywhere defined operators E satisfying $\|\mathcal{C}(t)\| \leq C(t)$, $\|\mathcal{S}(t)\| \leq C(t)$. It follows from the definitions that

$$\mathcal{S}(t)u = \int_0^t \mathcal{C}(s)u \, ds. \tag{1.5}$$

If $u(\hat{t})$ is an arbitrary solution of (1.1) we have

$$u(t) = \mathcal{C}(t)u(0) + \mathcal{S}(t)u'(0) . \tag{1.6}$$

The proof is the same as in the first order case. Applying this equality to $u(\hat{s}) = \mathcal{C}(\hat{s} + t)u$ and $v(\hat{s}) = \mathcal{S}(\hat{s} + t)u$ (and using (1.5) in the second instance) we instantly obtain

$$\mathcal{C}(s + t)u = \mathcal{C}(s)\mathcal{C}(t)u + \mathcal{S}(s)\mathcal{C}'(t)u, \tag{1.7}$$

$$\mathcal{S}(s + t)u = \mathcal{C}(s)\mathcal{S}(t)u + \mathcal{S}(s)\mathcal{C}(t)u , \tag{1.8}$$

for $u \in D$ (obviously, (1.8) may be extended to all $u \in E$ by continuity).

The next result shows (among other things) that the two notions of properly posed problem coalesce in the second order case.

THEOREM 1.1. *Let the Cauchy problem for* (1.1) *be well posed in* $t \geq 0$. *Then* (i) *The Cauchy problem for* (1.1) *is uniformly well posed in* $(-\infty, \infty)$, *so that* $\mathcal{C}(\hat{t})$ *is strongly continuous in* $(-\infty, \infty)$. (ii) $\mathcal{C}(\hat{t})$ *satisfies the cosine functional equations*

$$\mathcal{C}(0) = I, \ \mathcal{C}(s + t) + \mathcal{C}(s - t) = 2\mathcal{C}(s)\mathcal{C}(t) \quad (-\infty < s, \ t < \infty) . \tag{1.9}$$

(iii) *There exist constants* C_0, ω *such that*

$$\|\mathcal{C}(t)\| \leq C_0 e^{\omega|t|} \quad (-\infty < t < \infty) . \tag{1.10}$$

Proof: Assume the Cauchy problem for (1.1) is well posed in $t \geq 0$. The assignation $u(\hat{t}) \to u(-\hat{t})$ maps solutions into solutions, thus it is obvious that (1.3) can be extended to $-\infty < t < \infty$. On the other hand, a solution in $t \geq 0$ with $u'(0) = 0$ (resp. with $u(0) = 0$) can be extended to a solution in $-\infty < t < \infty$ by $u(t) = u(-t)$, $t > 0$ (resp. by $u(t) = -u(-t)$, $t < 0$) thus (a) is verified in $-\infty < t < \infty$ if it is verified in $t \geq 0$. It follows from the above considerations that

$$\mathcal{C}(-t) = \mathcal{C}(t),\ \mathcal{S}(-t) = -\mathcal{S}(-t)\ . \tag{1.11}$$

Write identity (1.7) (for $u \in D$) for t and $-t$ and add the two resulting equalities using (1.11): the result is (1.9) applied to $u \in D$, and the equality can be extended to all $u \in E$ by continuity.

The proof of (iii) imitates closely that of Theorem I.1.1 (iii). Given $u_0 \in E$ there exists a null subset d_0 of $[0,\infty)$ such that the subspace E_0 generated by $\{u_0\} \cup \{\mathcal{C}(t)u_0;\ t \notin d_0\}$ is separable and a sequence $\{t_n\} \notin d_0$ such that the set $\{u_0\} \cup \{\mathcal{C}(t_n)u_0\}$ is fundamental in E_0; accordingly, if $e_0 = (d_0 - t_1) \cup (d_0 + t_1) \cup (d_0 - t_2) \cup \ldots$, e_0 is a null set and it follows from the second equation (1.9) that

$$\mathcal{C}(t)E_0 \subseteq E_0 \quad (t \notin e_0)\ . \tag{1.12}$$

We show next that $\|\mathcal{C}(t)\|$ is bounded on intervals $[-\beta,\beta]$, $\beta < \infty$. Assuming the opposite we obtain a bounded sequence $\{t_n\}$ and a sequence $\{u_n\} \subseteq E$, $\|u_n\| = 1$ with $\|\mathcal{C}(t_n)u_n\| \geq n$. Operating exactly as in the proof of Theorem I.1.1 we can then construct a separable subspace E_∞ of E containing all the u_n and a null set e_∞ such that $\mathcal{C}(t)E_\infty \subseteq E_\infty$ for $t \notin e_\infty$; since $\mathcal{C}(-t) = \mathcal{C}(t)$ this relation can be extended to negative t not in $-e_\infty$. As in the semigroup case in Chapter I, the function $m(t)$, the norm of the restriction of $\mathcal{C}(t)$ to E_∞ is defined and measurable for all t; moreover, $m(t_n) > n$ and if $t \neq e_\infty$, $m(s + t) \leq \sup\{2\|\mathcal{C}(s)\mathcal{C}(t)u\|;\ u \in E_\infty,\ \|u\| \leq 1\} + \sup\{\|\mathcal{C}(s - t)u\|;\ u \in E_\infty,\ \|u\| \leq 1\} \leq 2m(s)m(t) + m(s - t)$. The contradiction is this time obtained with the aid of the following result:

LEMMA 1.2. <u>Let</u> $m(\hat{t})$ <u>be a nonnegative finite measurable function defined in</u> $-\infty < t < \infty$ <u>such that</u> $m(-t) = m(t)$ <u>and</u>

$$m(s + t) \leq 2m(s)m(t) + m(s - t)\ (-\infty < s < \infty,\ t \notin e), \tag{1.13}$$

where e is a null set in $(-\infty,\infty)$. Then $m(\hat{t})$ is bounded on intervals $[-\beta,\beta]$, $\beta < \infty$.

Proof: Let $a > 0$, $s + t = a$, $t \notin e$. Since $2m(s)m(t) + m(s-t) = 2m(a-t)m(t) + m(a-2t) \geq m(a)$ one of the following three inequalities must occur: a) $m(a-2t) \geq m(a)/2$ (b) $m(a-t) \geq \sqrt{m(a)}/2$ (c) $m(t) \geq \sqrt{m(a)}/2$. Accordingly, if $m(a) \geq 1$ and d is the set of all $t \in [-a,a]$ with $m(t) \geq \sqrt{m(a)}/2$ we have $e \cup d \cup (a-d) \cup (a-2d) \supseteq [-a,a]$, thus $4|d| = |d| + |a-d| + |a-2d| \geq 2a$. Arguing as in Lemma I.1.2 we deduce that $m(\hat{t})$ is bounded on every set of the form $[-\beta,-\alpha] \cup [\alpha,\beta]$ with $0 < \alpha < \beta < \infty$. To show boundedness near the origin we note that, since $m(\hat{t})$ is even we deduce from (1.13) the inequality $m(s-t) \leq 2m(s)m(t) + m(s+t)$ for $t \notin -e$; taking $t > 0$ fixed and s near t the proof of Lemma 1.2 is complete.

End of proof of Theorem 1.1. Let $u \in E$, t_0 and h arbitrary. A few manipulations with the cosine functional equation (1.9) show that $C(t_0+h) - C(t_0) = 2C(t)(C(t_0+h-t) - C(t_0-t)) - (C(t_0+h-2t) - C(t_0-2t))$. Take $\alpha < \beta$ and integrate in $\alpha \leq t \leq \beta$; we obtain

$$(\beta-\alpha)\|C(t_0+h)u - C(t_0)u\|$$
$$\leq C\int_{t_0-\beta}^{t_0-\alpha} \|C(t+h)u - C(t)u\|\,dt$$
$$+\frac{1}{2}\int_{t_0-2\beta}^{t_0-2\alpha} \|C(t+h)u - C(t)u\|\,dt\,,$$

thus again continuity of $C(\hat{t})u$ for all t follows from its mean continuity.

To show (ii) we apply once more the uniform boundedness principle and conclude that $\|C(t)\|$ is bounded in $0 \leq t \leq 1$. Choose $_0,\omega$ such that $\|C(t)\| \leq C_0e^{\omega t}$ in $0 \leq t \leq 1$ and $2\|C(1)\|e^{-\omega} + e^{-2\omega} \leq 1$. Assume that (1.10) holds for $0 \leq t \leq n$. Then

$$\|C(t+1)\| \leq 2\|C(1)\|\,\|C(t)\| + \|C(t-1)\| \leq C_0e^{\omega(t+1)}\,,$$

thus it holds for $0 \leq t \leq n+1$. By induction, (1.10) is valid for all $t \geq 0$; since $C(-t) = C(t)$, it is valid for all t. This ends the proof.

The comments after Theorem I.1.1 have an obvious counterpart here: we omit the details. Note also that (1.10) yields an exponential bound on $\mathcal{S}$ through (1.5):

$$\|\mathcal{S}(t)\| \le C_0(e^{\omega|t|} - 1)/\omega \qquad (-\infty < t < \infty) \tag{1.14}$$

if $\omega \neq 0$; for $\omega = 0$ we obtain

$$\|\mathcal{S}(t)\| \le C_0|t| \qquad (-\infty < t < \infty) . \tag{1.15}$$

§II.2 The generation theorem.

The following analogue of Theorem I.3.1 holds.

THEOREM 2.1. Let A be closed. The Cauchy problem for (1.1) is uniformly well posed in $(-\infty,\infty)$ with propagator $C(\hat{t})$ satisfying (1.10) if and only if $R(\lambda^2;A)$ exists in the half space $\operatorname{Re}\lambda > \omega$ and

$$\|(\lambda R(\lambda^2;A))^{(n)}\| \le C_0 n!(\operatorname{Re}\lambda - \omega)^{-(n+1)} \qquad (\operatorname{Re}\lambda > \omega,\ n = 0,1,\ldots)^{(1)}. \tag{2.1}$$

Proof: The first task is to show that $\rho(A)$ is nonempty (in the first order case, this was a byproduct of the proof and did not require a separate argument). Choose a twice continuously differentiable function $\varphi(t)$ such that $\varphi(t) = 1$ in $0 \le t \le 1$, $\varphi(t) = 0$ in $t \ge 2$ and define

$$Q(\lambda)u = \int_0^\infty e^{-\lambda t}\varphi(t)C(t)u\,dt . \tag{2.2}$$

If $u \in D$ then $C(\hat{t})u$ is twice continuously differentiable and $C''(t)u = AC(t)u$, thus we obtain integrating by parts twice that

$$\begin{aligned} AQ(\lambda)u &= \int_0^\infty e^{-\lambda t}\varphi(t)C''(t)u\,dt = -\lambda u + \lambda^2 Q(\lambda)u \\ &+ \int_1^2 (e^{-\lambda t}\varphi''(t) - 2\lambda e^{-\lambda t}\varphi'(t))C(t)u\,dt = \\ &= -\lambda u + \lambda^2 Q(\lambda)u + M(\lambda)u . \end{aligned} \tag{2.3}$$

Since A is closed and $Q(\lambda)$ and $M(\lambda)$ are bounded we can easily show that $Q(\lambda)E \subseteq D(A)$ and extend (2.3) to all $u \in E$; we rewrite it in the form

$$\lambda^{-1}(\lambda^2 I - A)Q(\lambda) = I - \lambda^{-1}M(\lambda) . \tag{2.4}$$

A simple estimation yields

$$\|\lambda^{-1}M(\lambda)\| \leq Ce^{-\lambda} \tag{2.5}$$

for $\lambda \geq \delta > 0$. Take now λ real so large that the right hand side of (2.5) is less than 1. Then $(I - \lambda^{-1}M(\lambda))^{-1}$ exists and multiplying (2.4) on the right by it we obtain

$$\lambda^{-1}(\lambda^2 I - A)P(\lambda) = I , \tag{2.6}$$

where $P(\lambda) = Q(\lambda)(I - \lambda^{-1}M(\lambda))^{-1}$, so that $\lambda^2 I - A$ is onto. We can show that $P(\lambda)$ is 1 - 1 if we assume in addition that $\lambda > \omega$; then if $(\lambda^2 I - A)u = 0$ the function $u(t) = \cosh(\lambda t)$ solves (1.1) but does not obey (1.3) with $C(t) = C_0 \exp(\omega t)$, a contradiction unless $u = 0$. We have then shown that $R(\lambda^2;A)$ exists and equals $\lambda^{-1}P(\lambda)$. Arguing now exactly as in Theorem I.1.1 we show that $R(\lambda;A)C(t) = C(t)R(\lambda;A)$ which again implies that $C(t)D(A) \subseteq D(A)$ and

$$AC(t)u = C(t)Au . \tag{2.7}$$

Making use of (2.7) we obtain

$$C(t)u - u = \int_0^t (t-s)C''(s)u\,ds = \int_0^t (t-s)C(s)Au\,ds$$

for $u \in D$; the equality is extended to $u \in D(A)$ in the same way as (I.3.6) and implies that $C(\hat{t})u$ is a solution of (1.1) for any $u \in D(A)$; in other words,

$$D_0 = D(A) , \tag{2.8}$$

where $D_0 \supseteq D$ is the space of all "admissible initial values" consisting of every u such that $C(\hat{t})u$ is a solution of (1.1). (The identification of the space $D_1 \supseteq D$ of all "admissible initial derivatives" u such that $S(\hat{t})u$ is a solution is considerably more complex and less satisfactory and will be postponed until the next chapter).

Let now $\mathrm{Re}\,\lambda > \omega$. Define

$$R(\lambda)u = \int_0^\infty e^{-\lambda t}C(t)u\,dt$$

for $u \in E$. If $u \in D(A)$ and $T > 0$ we have

$$A\int_0^T e^{-\lambda t}C(t)u\,dt = \int_0^T e^{-\lambda t}C''(t)u\,dt$$

$$= e^{-\lambda T}C'(T)u + \lambda e^{-\lambda T}C(T)u - \lambda u \tag{2.9}$$

$$+ \lambda^2\int_0^T e^{-\lambda t}C(t)u\,dt\ .$$

Since $C''(t)u = C(t)Au$,

$$C'(T)u = \int_0^T C(t)Au\,dt\ ,$$

and we can take limits in (2.9) as $T \to \infty$ to conclude that $R(\lambda)u \in D(A)$ with $AR(\lambda)u = \lambda^2 R(\lambda)u - \lambda u$; again, this equality can be extended to all $u \in E$ so that $R(\lambda)E \subseteq D(A)$ and

$$\lambda^{-1}(\lambda^2 I - A)R(\lambda) = I\ . \tag{2.10}$$

showing that $(\lambda^2 I - A)$ is onto. We have already proved before that $\lambda^2 I - A$ is one-to-one, so that $\lambda \in \rho(A)$ if $\mathrm{Re}\,\lambda > \omega$ and $R(\lambda) = \lambda R(\lambda^2;A)$. Differentiating under the integral sign we deduce that

$$(\lambda R(\lambda^2;A))^{(n)}u = (-1)^n\int_0^\infty t^n e^{-\lambda t}C(t)u\,dt \tag{2.11}$$

(see the justification of (I.3.8)) and all the inequalities (2.1) follow using (1.10) in $t \geq 0$.

We prove now the other half of Theorem 2.1. Solutions of (1.1) are defined by the formula[2]

$$\tilde{u}(t;u) = u + \frac{t^2}{2!}Au + \frac{t^4}{4!}A^2u + \frac{1}{2\pi i}\int_{\omega'-i\infty}^{\omega'+i\infty}\frac{e^{\lambda t}}{\lambda^5}R(\lambda^2;A)A^3u\,d\lambda \tag{2.12}$$

$$(t \geq 0)$$

for $u \in D(A^3)$ and $\omega' > \omega, 0$. The checking that $\tilde{u}(\hat{t};u)$ is actually a solution of (1.1) is done just as in Theorem I.3.1; also, we prove in the same way that $\|\tilde{u}(t;u)\| = 0(\exp(\omega' t))$ as $t \to \infty$,

$$\tilde{u}(0,u) = u, \quad \tilde{u}'(0,u) = 0 \tag{2.13}$$

and

$$\int_0^\infty e^{-\lambda t}\tilde{u}(t;u)\,dt = \lambda R(\lambda^2;A)u$$

for $\mathrm{Re}\,\lambda > \omega'$. It follows from this equality and from Lemma I.3.2 that

$$\tilde{u}(t;u) = \lim_{n\to\infty} \frac{(-1)^n}{n!}\left(\frac{n}{t}\right)^{n+1}\left(\frac{n}{t}R\left(\left(\frac{n}{t}\right)^2;A\right)\right)^{(n)} u , \tag{2.14}$$

hence, using inequalities (2.1),

$$\|\tilde{u}(t;u)\| \le C_0\|u\|\lim_{n\to\infty}\left(1-\frac{\omega t}{n}\right)^{-(n+1)} = C_0\|u\|e^{\omega t} \quad (t \ge 0) . \tag{2.15}$$

Condition (a) in the definition of uniformly well posed problem for (1.1) is verified as follows: if $u_0, u_1 \in D(A^3)$, a solution satisfying (1.2) is given by

$$u(t) = \tilde{u}(t;u_0) + \int_0^t \tilde{u}(s;u_1)\, ds .$$

To check the continuous dependence statement (b) we define

$$\tilde{C}(t)u = \tilde{u}(t;u) \tag{2.16}$$

for $u \in D(A^3)$ and extend it by continuity to all of E, obtaining a (E) - valued strongly continuous function with

$$\|\tilde{C}(t)\| \le C_0 e^{\omega t} . \tag{2.17}$$

A second operator valued function $\tilde{S}(t)$ is defined by

$$\tilde{S}(t)u = \int_0^t \tilde{C}(s)u\, ds .$$

We shall show that if $u(\hat{t})$ is an arbitrary solution of (1.1) in $t \ge 0$ we must have

$$u(t) = \tilde{C}(t)u(0) + \tilde{S}(t)u'(0). \tag{2.18}$$

This is done as in the first order case and we only sketch the details. The first step is to show that $\tilde{C}(t)$ and $\tilde{S}(t)$ commute with $R(\lambda;A)$ and consequently with A; the second is to note that the operator valued functions $\mathfrak{m}(t) = \tilde{C}(t)R(\lambda;A)^3$ and $\mathfrak{n}(t) = \tilde{S}(t)R(\lambda;A)^3$ are (twice) continuously differentiable and satisfy $\mathfrak{n}'(t) = \mathfrak{m}(t)$, and the equality $\mathfrak{m}'(t) = \tilde{S}''(t)R(\lambda;A)^3 = A\tilde{S}(t)R(\lambda;A)^3 = A\mathfrak{n}(t)$. Accordingly

$$\begin{aligned} &\frac{d}{ds}(\mathfrak{m}(t-s)u(s) + \mathfrak{n}(t-s)u'(s)) \\ &= -A\mathfrak{n}(t-s)u(s) + \mathfrak{m}(t-s)u'(s) \\ &- \mathfrak{m}(t-s)u'(s) + \mathfrak{n}(t-s)Au(s) = 0 , \end{aligned} \tag{2.19}$$

and (2.18) follows noting that $\mathfrak{m}(0) = R(\lambda;A)^3$, $\mathfrak{n}(0) = 0$. This completes the proof of Theorem 2.1.

REMARK 2.2. Theorem 2.1 shows in particular that we have the relation $\sigma(A) \subseteq \{\mu^2; \mathrm{Re}\mu \leq \omega\} = \{\lambda; \mathrm{Re}\lambda \leq \omega^2 - (\mathrm{Im}\lambda)^2/4\omega^2\}$, the region to the left of a parabola passing through the points ω^2, $\pm 2i\omega^2$. In particular, if $\omega = 0$, $\sigma(A)$ is contained in the negative real axis.

A notation similar to that for the first order case will be useful here. A closed, densely defined operator A <u>belongs to</u> $\mathfrak{C}^2(C_0,\omega)$ if the Cauchy problem for (1.1) is (uniformly) well posed in $-\infty < t < \infty$ and $C(t)$ satisfies (1.10). The class $\mathfrak{C}^2(\omega)$ is the union of all $\mathfrak{C}^2(C_0,\omega)$ for $C_0 \geq 1$ and $\mathfrak{C}^2$ is the union of all the $\mathfrak{C}^2(\omega)$ for $\omega \geq 0$ (note that $\mathfrak{C}^2(\omega)$ is empty for $\omega < 0$ by Theorem 2.1).

The following analogue of Theorem I.3.4 holds:

THEOREM 2.3. *Let A be an operator such that $R(\lambda^2;A)$ exists in the half plane $\mathrm{Re}\lambda > \lambda_0$, $C(\hat{t})$ an operator valued function strongly continuous in $t \geq 0$ and such that*

$$\|C(t)\| \leq C_0 e^{\omega t} \quad (t \geq 0). \tag{2.20}$$

Assume that, for each $u \in E$

$$\int_0^\infty e^{-\lambda t} C(t)u\, dt = \lambda R(\lambda^2;A)u \quad (\mathrm{Re}\lambda > \lambda_0). \tag{2.21}$$

Then $A \in \mathfrak{C}^2(C_0,\omega)$ and $C(|\hat{t}|)$ is the solution operator of (1.1).

The proof imitates that of Theorem I.3.4 and we omit it.

REMARK 2.4. Inequalities (2.1) follow from their real counterparts

$$\|(\lambda R(\lambda^2;A))^{(n)}\| \leq C_0 n!(\lambda - \omega)^{-(n+1)} \quad (\lambda > \omega, \quad n = 0,1,\ldots). \tag{2.22}$$

(see Remark I.3.5).

§II.3 Cosine function theory.

A (E)-valued function $C(\hat{t})$ defined in $-\infty < t < \infty$ is called a <u>cosine function</u> or <u>cosine operator function</u> if the cosine functional equations (1.9) hold, that is, if

$$C(0) = I, \quad C(s+t) + C(s-t) = 2C(s)C(t) \tag{3.1}$$

for all s,t in $(-\infty,\infty)$. The propagator $C(t)$ of a well posed second order Cauchy problem is a strongly continuous cosine function; we shall show in this section that, just as in the first order case, the converse is also true.

THEOREM 3.1. <u>Let</u> $C(\hat{t})$ <u>be a cosine function strongly continuous in</u> $-\infty < t < \infty$. <u>Then there exists a unique closed, densely defined operator</u> $A \in \mathfrak{C}^2$ <u>such that</u> $C(\hat{t})$ <u>is the evolution operator of</u>

$$u''(t) = Au(t). \tag{3.2}$$

<u>Proof</u>: The <u>infinitesimal generator</u> of $C(\hat{t})$ is defined by

$$Au = \lim_{h\to 0} \frac{1}{h^2}(C(h) - 2I + C(-h))u = \lim_{h\to 0} \frac{2}{h^2}(C(h) - I)u \tag{3.3}$$

(note that (3.1) yields $C(t) = C(-t)$ setting $s = 0$). The domain $D(A)$ of A is the set of all u such that the limit exists (roughly speaking, A is the "second derivative of C at the origin"). Let $u \in E$, $a > 0$. We define

$$u^a = \frac{2}{a^2}\int_0^a (a-t)C(t)u\,dt. \tag{3.4}$$

Using (3.1) we obtain

$$\frac{2}{h^2}(C(h) - I)u^a = \frac{2}{a^2h^2}\int_h^{a+h}(a+h-t)C(t)u$$

$$+ \frac{2}{a^2h^2}\int_{-h}^{a-h}(a-h-t)C(t)u\,dt - \frac{4}{a^2h^2}\int_0^a (a-t)C(t)u\,dt$$

$$= \frac{2}{a^2h^2}\int_a^{a+h}(a+h-t)C(t)u\,dt - \frac{2}{a^2h^2}\int_0^h (a+h-t)C(t)u\,dt$$

$$- \frac{2}{a^2h^2}\int_{a-h}^a (a-h-t)C(t)u\,dt + \frac{2}{a^2h^2}\int_{-h}^0 (a-h-t)C(t)u\,dt, \tag{3.5}$$

thus $u^a \in D(A)$ with

$$Au^a = \frac{2}{a^2}(C(a) - I)u. \tag{3.6}$$

Since $u^a \to u$ as $a \to 0$ it results that $D(A)$ is dense in E. To show that A is closed we note that (3.1) implies that $C(s)C(t) = C(t)C(s)$ for all s,t; hence

$$\frac{2}{h^2}(C(h) - I)C(t)u = C(t)\frac{2}{h^2}(C(h) - I)u$$

for $h \neq 0$. Accordingly, if $u \in D(A)$ then $C(t)u \in D(A)$ and

$$AC(t)u = C(t)Au \tag{3.7}$$

for any $u \in E$. Hence

$$(Au)^a = Au^a = \frac{2}{a^2}(C(a) - I)u \tag{3.8}$$

after (3.6). Let $\{u_n\}$ be a sequence in $D(A)$ such that $u_n \to u$, $Au_n \to v$. Then

$$\frac{2}{h^2}(C(h) - I)u = \lim_{n\to\infty}\frac{2}{h^2}(C(h) - I)u_n = \lim_{n\to\infty}(Au_n)^h = v^h,$$

so that taking limits as $h \to 0$ we deduce that $u \in D(A)$ and $Au = v$ as claimed.

If $u \in D(A)$ we have

$$\int_0^t (t - s)C(s)Au\, ds = \int_0^t (t - s)AC(s)u\, ds$$

$$= A\int_0^t (t - s)C(s)u\, ds = \frac{t^2}{2}Au^t = C(t)u - u. \tag{3.9}$$

Hence $C(\hat{t})u$ is twice continuously differentiable in $-\infty < t < \infty$ with

$$C''(t)u = C(t)Au = AC(t)u, \tag{3.10}$$

so that $C(\hat{t})$ is a solution of (3.2) in $-\infty < t < \infty$ and the existence postulate in the definition of well posed problem is verified. To show continuous dependence, we proceed as in Theorem I.4.1; first we prove with the help of the uniform boundedness theorem that $\|C(t)\|$ is bounded on compact subsets of $(-\infty, \infty)$ and then define

$$S(t)u = \int_0^t C(s)u\, ds, \tag{3.11}$$

so that $\|S(t)\|$ is as well bounded on compacts. If $u(\hat{t})$ is an arbitrary solution of (3.2) we can show essentially as in the end of the proof of Theorem 2.1 that the function $C(t - s)u(s) + S(t - s)u'(s)$ is constant, whence the formula

$$u'(t) = C(t)u(0) + S(t)u'(0)$$

results. This completes the proof of Theorem 3.1.

§II.4 The inhomogeneous equation.

We consider here the equation

$$u''(t) = Au(t) + f(t). \tag{4.1}$$

Solutions are defined in the same way as solutions of (1.1); again if $A \in \mathfrak{C}^2$ uniqueness of solutions of the initial value problem for (1.1) implies that there is at most one solution of (4.1) satisfying the initial conditions

$$u(0) = u_0, \ u'(0) = u_1. \tag{4.2}$$

Let $f(\hat{t})$ be a continuous function defined in $0 \leq t \leq T$, and let $u(\hat{t})$ be a solution of (4.1) satisfying (4.2). Assuming that $t > 0$ is fixed and that $\mathcal{C}(\hat{t})$, $\mathcal{S}(\hat{t})$ are the solution operators of (1.1) we obtain using the equation that the function $\mathcal{C}(t - \hat{s})u(\hat{s}) + \mathcal{S}(t - \hat{s})u'(\hat{s})$ is continuously differentiable in $0 \leq s \leq t$ with derivative $\mathcal{S}(t - s)f(s)$. Integrating in $0 \leq t \leq T$ we obtain

$$u(t) = \mathcal{C}(t)u_0 + \mathcal{S}(t)u_1 + \int_0^t \mathcal{S}(t - s)f(s)\, ds. \tag{4.3}$$

As in the first order case, (4.3) will be a genuine solution of (4.1) if additional conditions are imposed on $f(\hat{t})$.

LEMMA 4.1. *Assume that one of the following two conditions is satisfied:*

(a) $f(t) \in D(A)$ *and* $f(\hat{t})$, $Af(\hat{t})$ *are continuous in* $0 \leq t \leq T$,

or

(b) $f(\hat{t})$ *is continuously differentiable in* $0 \leq t \leq T$.

Suppose in addition that $u_0, u_1 \in D(A)$. *Then* (4.3) *is a genuine solution of* (4.1) *in* $0 \leq t \leq T$.

Proof: We have already noted that $D_0 = D(A)$, $D_1 \supseteq D(A)$ (see (2.8) and following comments) thus we only consider the last term in (4.3). Since $\mathcal{S}(\hat{t})$ is differentiable in the norm of (E) with derivative $\mathcal{C}(t)$, $u(t)$ is always continuously differentiable with derivative

$$u'(t) = \int_0^t \mathcal{C}(t - s)f(s)\, ds. \tag{4.4}$$

To establish the conclusion of Lemma 4.1 in either case it is enough to show that $u(t) \in D(A)$, $Au(t)$ is continuous and

$$u'(t) = \int_0^t (Au(s) + f(s))\, ds \quad (0 \le t \le T) . \tag{4.5}$$

If (a) holds we have

$$\int_0^t Au(s)\, ds = \int_0^t \left(\int_0^s \mathcal{S}(s-r)Af(r)dr \right) ds$$

$$= \int_0^t \left(\int_r^t A\mathcal{S}(s-r)f(r)ds \right) dr$$

$$= \int_0^t (C(t-r)f(r) - f(r))\, dr ,$$

thus proving (4.5). If we assume (b) instead we integrate (4.3) by parts:

$$u(t) = -\int_0^t \left(\frac{d}{ds} \int_0^{t-s} \mathcal{S}(r)f(s)dr \right) ds$$

$$= \int_0^t \mathcal{S}(s)f(0)\, ds + \int_0^t \left(\int_0^{t-s} \mathcal{S}(r)f'(s)dr \right) ds .$$

In view of (1.5),

$$Au(t) = C(t)f(0) - f(0) + \int_0^t (C(t-s) - I)f'(s)\, ds$$

$$= C(t)f(0) - f(t) + \int_0^t C(t-s)f'(s)\, ds .$$

We obtain now (4.5) integrating and reversing the order of integration (see the proof of Lemma I.5.1).

Just as in the first order case, we shall call (4.3) a <u>generalized solution</u> of (4.1) if u_0, u_1 are arbitrary elements of E and f is, say, a function in $L^1(0,T;E)$.

LEMMA 4.2. *Assume* $u_0, u_1 \in E$, $f \in L^1(0,T;E)$. *Then* $u(\hat{t})$ *is continuous in* $0 \le t \le T$ *with*

$$\|u(t)\| \le C_0 e^{\omega t}\|u_0\| + C_0\omega^{-1}(e^{\omega t} - 1)\|u_1\|$$

$$+ C_0\omega^{-1}(e^{\omega T} - 1)\|f\|_1 \quad (0 \le t \le T) . \tag{4.6}$$

($\omega^{-1}(e^{\omega t} - 1)$ *is replaced by* t *when* $\omega = 0$). *If* $u_0 = 0$ *then* $u(\hat{t})$ *is continuously differentiable and*

$$\|u'(t)\| \le C_0 e^{\omega t}\|u_1\| + C_0 e^{\omega T}\|f\|_1 \qquad (0 \le t \le T) . \tag{4.7}$$

The proof imitates that of Lemma I.5.2 and is therefore omitted.

§II.5 Estimations by hyperbolic functions.

It is sometimes convenient to replace the estimate (1.10) for a cosine function by

$$\|\mathfrak{C}(t)\| \le C_0 \cosh \omega t \qquad (-\infty < t < \infty) . \tag{5.1}$$

Obviously, both are equivalent, but the value of the constant C_0 will in general be different; precisely, (5.1) implies (1.10) with the same constant, whereas to pass from (1.10) to (5.1) the constant must be doubled. If we denote by $\tilde{\mathfrak{C}}^2(C_0,\omega)$ the class of all infinitesimal generators of strongly continuous cosine functions satisfying (5.1),

$$\tilde{\mathfrak{C}}^2(C_0,\omega) \subseteq \mathfrak{C}^2(C_0,\omega), \qquad \mathfrak{C}^2(C_0,\omega) \subseteq \tilde{\mathfrak{C}}^2(2C_0,\omega).$$

If $\omega > 0$, (5.1) implies the following estimate for $\mathfrak{S}(\hat{t})$:

$$\|\mathfrak{S}(t)\| \le C_0 \frac{\sinh \omega t}{\omega} \qquad (-\infty < t < \infty) . \tag{5.2}$$

If $\omega = 0$, the inequality is (1.15).

We can easily obtain a generation theorem based on (5.1) rather than on (1.10), although the counterparts of inequalities (2.1) are less simple.

THEOREM 5.1. <u>Let</u> A <u>be closed. The Cauchy problem for</u> (1.1) <u>is uniformly well posed in</u> $(-\infty,\infty)$ <u>with propagator</u> $\mathfrak{C}(\hat{t})$ <u>satisfying</u> (5.1) <u>if and only if</u> $R(\lambda^2;A)$ <u>exists in the half space</u> $\operatorname{Re}\lambda > \omega$ <u>and</u>

$$\|(\lambda R(\lambda^2;A))^{(n)}\| \le C_0(-1)^n n!(\operatorname{Re}\lambda((\operatorname{Re}\lambda)^2 - \omega^2)^{-1})^{(n)}$$

$$(\operatorname{Re}\lambda > \omega,\ n = 0,1,\dots\ ) , \tag{5.3}$$

<u>where the indicated derivatives on the right hand side are taken with respect to the variable</u> $\operatorname{Re}\lambda$.

<u>Proof</u>. Combining the basic formula (2.11) (which is obtained exactly as in Theorem 2.1) with inequality (5.1) we obtain

$$\|(\lambda R(\lambda^2;A))^{(n)}\| \le C_0 \int_0^\infty t^n e^{-(\mathrm{Re}\lambda)t} \cosh \omega t \, dt \,. \tag{5.4}$$

We use then again (2.11), this time for the scalar cosine function $\cosh \omega\hat{t}$ (whose infinitesimal generator is ω^2); the result is the sequence of inequalities (5.3).

To prove the converse, we only need to make a few minor changes in the proof of Theorem 2.1. Observe first that the first inequality (5.3) implies the first inequality (2.1). Thus the construction of the function $\tilde{u}(t;u)$ in (2.12) and the proof of its properties proceeds in exactly the same way. However, the estimation (2.15) is slightly different. We use again the Post inversion formula (I.3.14) obtaining

$$\|\tilde{u}(t,u)\| \le C_0 \lim_{n\to\infty} \frac{(-1)^n}{n!} \left(\frac{n}{t}\right)^{n+1} \left(\frac{n}{t}\left(\left(\frac{n}{t}\right)^2 - \omega^2\right)^{-1}\right)^{(n)} = C_0 \cosh \omega t \tag{5.5}$$

where we use on the right side (I.3.14) for the function $\cosh \omega t$ (whose Laplace transform is $\lambda(\lambda^2 - \omega^2)^{-1}$). The rest of the proof is just like that of Theorem 2.1 and we omit the details.

REMARK 5.2. Just as (2.1), inequalities (5.5) follow from their real counterparts

$$\|(\lambda R(\lambda^2;A))^{(n)}\| \le C_0 (-1)^n n! (\lambda(\lambda^2 - \omega^2)^{-1})^{(n)} \qquad (\lambda > \omega, n = 1,2,\ldots) \tag{5.6}$$

(see Remarks I.3.5 and 2.4). This can be again proved using Taylor series.

§II.6 Miscellaneous comments.

Strongly continuous cosine functions were introduced by SOVA [1966:1], who defined the infinitesimal generator and proved the generation theorem 2.1. Other proofs of Theorem 2.1 were given by DA PRATO-GIUSTI [1967:1] and the author [1969:3] in certain locally convex spaces. This last proof is the one we have employed here. Cosine functions continuous in the norm of (E) were considered earlier by KUREPA [1962:1] who treated as well the case where the cosine function takes values in a Banach algebra; the end result of this version of the theory is the representation $C(t) = \cos(tA^{1/2})$ (see Exercise 2 below). The definition of properly posed Cauchy problems for higher order equations (of which (1.1)) is a particular case) is due to the author [1969:2], as well as the relation between strongly continuous cosine functions and solution operators of second order equations. Theorem 1.1 is due to the author [1969:2]; a result of the same "measurability implies continuity" type was proved earlier by KUREPA [1962:1], where both measurability and continuity are understood in the norm of (E) (or, more generally, in the norm of a Banach algebra). Theorem 2.3 is due to the author [1969:2].

EXERCISE 1. Let A be a bounded operator in a Banach space E. Show that $A \in \mathfrak{C}^2$, that is, that the Cauchy problem for

$$u''(t) = Au(t) \tag{6.1}$$

is uniformly well posed in $-\infty < t < \infty$. The cosine function $\mathcal{C}(\hat{t})$ generated by A is given by

$$\mathcal{C}(t) = \cosh(tA^{1/2}) = \sum_{n=0}^{\infty} \frac{t^{2n}}{(2n)!} A^n, \tag{6.2}$$

the series (6.2) uniformly convergent on compact subsets of $-\infty < t < \infty$. We also have

$$\mathcal{S}(t) = A^{-1/2}\sinh(tA^{1/2}) = \sum_{n=0}^{\infty} \frac{t^{2n+1}}{(2n+1)!} A^n, \tag{6.3}$$

the series (6.3) converging in the same sense as (6.2). Note that the expressions $\cosh(tA^{1/2})$ and $A^{-1/2}\sinh(tA^{1/2})$ are purely symbolic and do not assume the existence of square roots of A; their actual meaning is given by the series (6.2) and (6.3). Prove that $\mathcal{C}(\hat{t})$, $\mathcal{S}(\hat{t})$ can also be expressed as

$$\mathcal{C}(t) = \frac{1}{2\pi i}\int_\Gamma \cosh(t\lambda^{1/2})R(\lambda;A)\, d\lambda , \tag{6.4}$$

$$\mathcal{S}(t) = \frac{1}{2\pi i}\int_\Gamma \lambda^{-1/2}\sinh(t\lambda^{1/2})R(\lambda;A)\, d\lambda , \tag{6.5}$$

where Γ is a simple closed curve (or the union of a finite number of these) oriented counterclockwise and enclosing $\sigma(A)$ in its (their) interior. (Note that the choice of square root $\lambda^{1/2}$ in the integrands of (6.4) and (6.5) is irrelevant, since $\cosh(t\lambda^{1/2}) = \sum t^{2n}\lambda^n/(2n)!$, $\lambda^{-1/2}\sinh(t\lambda^{1/2}) = \sum t^{2n+1}\lambda^n/(2n+1)!$). Show that (6.4) and (6.5) imply

$$\|\mathcal{C}(t)\| \le C \cosh \omega t , \tag{6.6}$$

$$\|\mathcal{S}(t)\| \le C \frac{\sinh \omega t}{\omega} , \tag{6.7}$$

in $-\infty < t < \infty$ if $\omega > \omega_0$, where

$$\omega_0 = \sup\{\operatorname{Re} \lambda^{1/2};\ \lambda \in \sigma(A)\} . \tag{6.8}$$

Alternately, ω_0 is the least positive number such that $\sigma(A)$ is contained in the closed region to the left of the parabola

$$\xi = \omega^2 - \eta^2/4\omega^2 , \tag{6.9}$$

passing through the points $\omega^2, \pm 2i\omega^2$. In particular, (6.6) and (6.7) hold with <u>any</u> $\omega > 0$ (with C naturally depending on ω) if $\sigma(A)$ is contained in the negative real axis. Produce an example to show that (6.6) and (6.7) do not necessarily hold with $\omega = \omega_0$. Show that $\mathcal{C}(\hat{t})$ and $\mathcal{S}(\hat{t})$ can be extended to functions $\mathcal{C}(\hat{\zeta})$ and $\mathcal{S}(\hat{\zeta})$ holomorphic (as (E)-valued functions) for all ζ; in particular, $\mathcal{C}(\hat{t})$ is continuous in the norm of (E) (of course, $\mathcal{S}(\hat{t})$ is <u>always</u> continuous in the norm of (E) due to formula (1.5) and boundedness of $\|\mathcal{C}(\hat{t})\|$ on compact subsets).

EXERCISE 2. Let $\mathcal{C}(\hat{t})$ be a cosine function continuous in the topology of (E) (continuity at $t = 0$ suffices). Show that the infinitesimal generator A of $\mathcal{C}(\hat{t})$ is a bounded operator. (Hint: use formula (2.11) for $n = 1$ and continuity of $\mathcal{C}(\hat{t})$ at $t = 0$ to show that $\|\lambda^2 R(\lambda^2;A) - I\| < 1$ for λ sufficiently large so that $\lambda^2 R(\lambda^2;A)$ has a bounded inverse). Show that $\mathcal{C}(\hat{t})$ admits the representations (6.2) and (6.4).

EXERCISE 3. Let $E = C_0(-\infty,\infty)$ be the Banach space of all (complex-valued) continuous functions $u(x)$ defined in $-\infty < x < \infty$ and such that $u(x) \to 0$ as $|x| \to \infty$ endowed with the norm (I.6.8).

Define

$$\mathcal{C}(t)u(x) = \frac{1}{2}\,(u(x + t) + u(x - t)) \qquad (-\infty < t < \infty). \tag{6.10}$$

Show that $\mathcal{C}(\hat{t})$ is a strongly continuous cosine function in E. Identify its infinitesimal generator. Show that

$$\mathcal{S}(t)u(x) = \frac{1}{2}\int_{x-t}^{x+t} u(\xi)\, d\xi \quad (-\infty < t < \infty) . \tag{6.11}$$

EXERCISE 4. Prove the result of Exercise 3 (and identify the infinitesimal generator) in the space $L^p(-\infty,\infty)$. Show that (6.11) is not strongly continuous (or even strongly measurable) in $L^\infty(-\infty,\infty)$.

EXERCISE 5. Let A be a normal operator in a Hilbert space H. Show that $A \in \mathfrak{C}^2$ (or, equivalently, to $\tilde{\mathfrak{C}}^2$) if and only if ω_0 (defined by 6.8) is finite. Prove the formulas

$$\mathcal{C}(t)u = \int_{\sigma(A)} \cosh(t\lambda^{1/2})P(d\lambda)u \qquad (u \in H,\ -\infty < t < \infty), \tag{6.12}$$

$$\mathcal{S}(t)u = \int_{\sigma(A)} \lambda^{-1/2}\sinh(t\lambda^{1/2})P(d\lambda)u \quad (u \in H,\ -\infty < t < \infty), \tag{6.13}$$

where $P(d\lambda)$ is the resolution of the identity associated with A. Using the functional calculus for normal operators show that

$$\|\mathcal{C}(t)\| = \cosh \omega_0 t , \tag{6.14}$$

$$\|\mathcal{S}(t)\| = \frac{1}{\omega_0} \sinh \omega_0 t . \tag{6.15}$$

EXERCISE 6. Let $\mathcal{C}(\hat{t})$ be a strongly continuous cosine function in a Hilbert space H, A its infinitesimal generator. Assume that each $\mathcal{C}(t)$ is normal. Show that A is normal (Hint: proceed as in Exercise I.9)

EXERCISE 7. Under the hypoteses of Exercise 6.6 assume that each $\mathcal{C}(t)$ is self adjoint. Then A is self adjoint. (Hint: proceed as in Exercise I.10)

EXERCISE 8. Show by means of an example, that there exists a strongly continuous cosine function $\mathcal{C}(\hat{t})$ in separable Hilbert space H with infinitesimal generator A such that $\omega > \omega_0$, where ω_0 is defined by (6.8) and

$$\omega = \limsup_{t \to \infty} \frac{1}{t} \log\|\mathcal{C}(t)\| . \tag{6.16}$$

More generally, we can construct $\mathcal{C}(\hat{t})$ in such a way that

$$\omega > \omega_0 + h ,$$

where $h > 0$ is arbitrarily preassigned (see Exercise I.13).

EXERCISE 9. Prove that the "real" inequalities (2.22) imply their complex counterparts (2.1). Likewise, show that inequalities (5.6) imply inequalities (5.3).

EXERCISE 10. Let $C(\hat{t})$ a strongly continuous cosine function, A its infinitesimal generator, ω the number defined by (6.16), $\omega' > \omega, 0$. Show that if $u \in D(A)$ then

$$S(t)u = \lim_{r \to \infty} \frac{1}{2\pi i} \int_{\omega'-ir}^{\omega'+ir} \lambda e^{\lambda t} R(\lambda^2; A)u \, d\lambda,$$

the limit being uniform on compact subsets of $|t| > 0$.

EXERCISE 11. Let $A \in \mathfrak{C}^2$. Show that $A \in \mathfrak{C}^1_+$, the propagator $S(\hat{t})$ of the equation

$$u'(t) = Au(t) \tag{6.17}$$

given by the <u>abstract Weierstrass formula</u>

$$S(t) = \frac{1}{(\pi t)^{1/2}} \int_0^\infty e^{-s^2/4t} C(s)u \, ds \qquad (t > 0), \tag{6.18}$$

where $C(\hat{t})$ is the propagator of

$$u''(t) = Au(t) . \tag{6.19}$$

FOOTNOTES TO CHAPTER II

(1) See inequalities (I.3.25).

(2) See footnote (2), p. 23; here u is meant to be the inverse Laplace transform of $\lambda R(\lambda; A)$ with the adjustments needed to produce convergence and tolerate two differentiations with respect to t.

CHAPTER III

REDUCTION OF A SECOND ORDER EQUATION TO A A FIRST ORDER SYSTEM. PHASE SPACES.

§III.1 Phase spaces.

It seems natural to ask whether the Cauchy problem for the equation

$$u''(t) = Au(t) \tag{1.1}$$

in the Banach space E can be reduced to the Cauchy problem for a first order equation: the obvious candidate is of course the system

$$u_0'(t) = u_1(t), \quad u_1'(t) = Au_0(t), \tag{1.2}$$

where $u_0(\hat{t}) = u(\hat{t})$. However, if (1.2) is considered in the space $E \times E$ it may give rise to an improperly posed Cauchy problem due to the fact that $u'(t)$, unlike $u(t)$, may fail to depend continuously on $u(0)$, $u'(0)$.

EXAMPLE 1.1. We can cast the one-dimensional wave equation $u_{tt} = u_{xx}$ in the form (1.1) in the following way (see Exercise II.3) let, say, $E = L^2(-\infty, \infty)$, $Au(x) = u''(x)$ where $D(A)$ is the set of all $u \in E$ such that u'' (understood in the sense of distributions) belongs to L^2. Then the Cauchy problem for (1.1) is well posed with propagators $\mathcal{C}(\hat{t})$, $\mathcal{S}(\hat{t})$ given by formulas (II. 6.10), (II. 6.11). Obviously, $\mathcal{C}'(t)u$ is given by

$$\mathcal{C}'(t)u(\hat{x}) = \frac{1}{2}\,(u'(\hat{x} + t) - u'(\hat{x} - t)),$$

thus it is not a bounded operator in E.

However, the reduction (1.2) will always suceed if $u_1(t)$ is measured in a different norm. Since more than one of these norms will be used in the future we give the following general definition, in which we do not postulate that the Cauchy problem for (1.1) is necessarily well posed in the sense of §II.1; we assume, however, that condition (a) in the definition of well posed problem holds in $t \geq 0$, that is, that there exists a dense subspace D of E such that for every $u_0, u_1 \in D$ there exists a solution of (1.1) satisfying the initial conditions

$$u(0) = u_0,\ u'(0) = u_1. \qquad (1.3)$$

A <u>phase space in</u> $t \geq 0$ for the equation (1.1) is a product space $\mathfrak{E} = E_0 \times E_1$ (endowed with any of its product norms) where E_0 and E_1 are Banach spaces satisfying the following assumptions:

(a) $E_0, E_1 \subseteq E$ <u>with bounded inclusion; moreover,</u> $D \cap E_0$ (<u>resp.</u> $D \cap E_1$) <u>is dense in</u> E_0 <u>in the topology of</u> E_0 (<u>resp. is dense in</u> E_1 <u>in the topology of</u> E_1).

(b) <u>There exists a strongly continuous semigroup</u> $\mathfrak{S}(\hat{t})$ <u>in</u> $\mathfrak{E} = E_0 \times E_1$ <u>such that</u>

$$\mathfrak{S}(t)\begin{bmatrix} u(0) \\ u'(0) \end{bmatrix} = \begin{bmatrix} u(t) \\ u'(t) \end{bmatrix}^{(1)} \qquad (1.4)$$

<u>in</u> $t \geq 0$ <u>for any solution</u> $u(\hat{t})$ <u>with</u> $u(0) \in E_0$, $u'(0) \in E_1$.

We note an obvious consequence of (b): if $u(\hat{t})$ is an arbitrary solution of (1.1) with $u(0) \in E_0$, $u'(0) \in E_1$ then $u(t) \in E_0$ and $u'(t) \in E_1$ for all $t \geq 0$, thus E_0 and E_1 are invariant subspaces for the values of a solution and for its derivative, respectively. Note also that there might well be solutions whose initial values do not belong to E_0, E_1; (b) does not apply to these solutions.

The notion of <u>phase space in</u> $-\infty < t < \infty$ is correspondingly defined; here $\mathfrak{S}(\hat{t})$ is required to be a <u>group</u> and (1.4) happens for all t if $u(\hat{t})$ is any solution of (1.1) in $-\infty < t < \infty$ with $u(0) \in E_0$, $u'(0) \in E_1$.

We show below that if the Cauchy problem for (1.1) is well posed in the sense of §II.1 then a phase space $\mathfrak{E}_m$ (which is maximal in an obvious sense) always exists (of course, the trivial choice $E_0 = E_1 = \{0\}$ also provides a phase space!) To construct $\mathfrak{E}_m$ we take $E_1 = E$ with its original norm, while $E_0 = \tilde{D}_1$, where $\tilde{D}_1$ is the space of all u such that $C(\hat{t})u$ is continuously differentiable in $-\infty < t < \infty$ (as we shall see later in Theorem 1.2, this space is none other than D_1, the space of admissible initial values for solutions of (1.1), defined in §II.1).

To define the norm in $E_0 = \tilde{D}_1$ we use the auxiliary result below.

LEMMA 1.2. <u>For every</u> $u \in E_0 = \tilde{D}_1$ <u>there exists a constant</u> $C = C(u)$ <u>such that</u>

$$\|C'(t)u\| \leq C(1 + |t|)e^{\omega|t|} \quad (-\infty < t < \infty) \tag{1.5}$$

(ω <u>the constant in</u> (II.1.10)).

<u>Proof</u>: We begin by renorming the space E as follows:

$$\|u\|' = \sup_{-\infty < s < \infty} e^{-\omega|s|}\|C(s)u\|.$$

Obviously, $\|u\| \leq \|u\|' \leq C_0\|u\|$, thus the new norm is equivalent to the original one and it is enough to prove (1.5) for $\|\cdot\|'$. We have

$$e^{-\omega|s|}\|C(s)C(t)u\| \leq \frac{1}{2} e^{\omega|t|}e^{-\omega|s+t|}\|C(s + t)u\|$$
$$+ \frac{1}{2} e^{\omega|t|}e^{-\omega|s-t|}\|C(s - t)u\|$$

by the cosine functional equation (II.1.9). Hence

$$\|C(t)u\|' \leq e^{\omega|t|}\|u\|' \quad (-\infty < t < \infty). \tag{1.6}$$

Let $u \in \tilde{D}_1$. Set $\eta(t) = \exp(-\omega t)\|C'(t)u\|'$ for $t \geq 0$. Since different values of $C(\hat{t})$ commute (a consequence of the cosine functional equation (II.1.9)) we can differentiate (II.1.7) with respect to s. Taking $\|\cdot\|'$ norms, multiplying by $\exp(-\omega(s + t))$ and using (II.1.10) we obtain

$$\eta(s + t) \leq \eta(s) + \eta(t) \quad (t \geq 0). \tag{1.7}$$

Obviously, (1.7) implies $\eta(m) \leq m\eta(1)$ for $m = 1,2,\ldots$ For $t \geq 0$ arbitrary let m be the greatest integer $\leq t$. Then $\eta(t) \leq \eta((t - m) + m) \leq \eta(t - m) + m\eta(1) \leq C(1 + t)$ where C is the supremum of $\eta(s)$ in $0 \leq s \leq 1$. Accordingly, (1.5) follows for $t \geq 0$; since $C'(\hat{t})u$ is an odd function, (1.5) holds for all t. This completes the proof.

We can now define a norm in $\tilde{D}_1$ by

$$\|u\|_0 = \max\{\|u\|, \sup_{t\geq 0} e^{-\omega t}(1+t)^{-1}\|C'(t)u\|\}. \tag{1.8}$$

A moment's consideration shows that $\tilde{D}_1$ is a Banach space equipped with $\|\cdot\|_0$.

THEOREM 1.3. *The space* $\mathfrak{E}_m = E_0 \times E_1$ *with* $E_1 = E$ *and* $E_0 = \tilde{D}_1$ *is a phase space for* (1.1) *in* $-\infty < t < \infty$; *moreover,*

$$\tilde{D}_1 = D_1. \tag{1.9}$$

The group $\mathfrak{S}(\hat{t})$ *is given by*

$$\mathfrak{S}(t) = \begin{bmatrix} C(t) & S(t) \\ C'(t) & S'(t) \end{bmatrix}, \tag{1.10}$$

and its infinitesimal generator is

$$\mathfrak{A} = \begin{bmatrix} 0 & I \\ A & 0 \end{bmatrix}, \tag{1.11}$$

with domain $D(\mathfrak{A}) = D(A) \times D_1$. *If, say, the norm* $\|u\|_{\mathfrak{E}_m} = \|[u,v]\|_{\mathfrak{E}_m} = \max(\|u\|, \|v\|)$ *is used in* $\mathfrak{E}_m$ *we have*

$$\|\mathfrak{S}(t)\|_{(\mathfrak{E}_m)} \leq (C_0+1)^2(1+|t|)e^{\omega|t|} \quad (-\infty < t < \infty). \tag{1.12}$$

where C_0 *and* ω *are the constants in* (II.1.10).

The proof that follows is by no means the shortest but can be adapted easily to a more general situation (for an alternate proof see Exercise III.15). Its basis are equalities (II.1.7) and (II.1.8). We have already noted that the second holds for every $u \in E$. The first can be extended from $u \in D$ to $u \in \tilde{D}_1$; for this we integrate with respect to t in $0 \leq t \leq \tau$, extend to all $u \in E$ by continuity, apply the equality thus obtained to any $u \in \tilde{D}_1$ and differentiate with respect to τ.

We prove now that each $\mathfrak{S}(t)$ is a bounded operator in $\mathfrak{E}_m$; since $C(\hat{t})$ is even and $C'(\hat{t})$, $S(\hat{t})$ are odd, it is obviously sufficient to work in $t \geq 0$. To show that $C(\hat{t})$ is a bounded operator from

E_1 into E_1 we use (II.1.7) differentiated with respect to s in the form $C'(s)C(t)u = C'(s+t)u - C(s)C'(t)u$. It implies that

$$\|C'(s)C(t)u\| \le \|C'(s+t)u\| + \|C(s)\|\,\|C'(t)u\|$$

$$\le (1+s+t)e^{\omega(s+t)}\|u\|_0 + C_0(1+t)e^{\omega(s+t)}\|u\|_0 .$$

Since $(1+s+t) \le (1+s)(1+t)$ we obtain.

$$\|C(t)u\|_0 \le (C_0+1)(1+t)e^{\omega t}\|u\|_0 \qquad (u_0 \in E_0) . \tag{1.13}$$

To handle $S(t)$ we use (II.1.8) differentiated with respect to s in the form $C'(s)S(t)u = C(s+t)u - C(s)C(t)u$; the result is

$$\|C'(s)S(t)u\| \le \|C(s+t)u\| + \|C(s)C(t)u\|$$

$$\le C_0e^{\omega(s+t)}\|u\| + C_0^2e^{\omega(s+t)}\|u\| ,$$

thus $S(t)$ is a bounded operator from E into E_0; we have

$$\|S(t)u\|_0 \le C_0(1+C_0)\, e^{\omega t}\|u\| \qquad (u \in E) . \tag{1.14}$$

On the other hand it is obvious that $C'(t)$ maps E_0 into E with

$$\|C'(t)u\| \le (1+t)e^{\omega t}\|u\|_0 \qquad (u \in E_0), \tag{1.15}$$

and that $S'(t) = C(t)$ maps E into E with

$$\|C(t)u\| \le C_0e^{\omega t}\|u\| \qquad (u \in E). \tag{1.16}$$

Estimates (1.13) to (1.16) clearly show that each $\mathfrak{S}(t)$ is a bounded operator from E into E with the norm indicated in (1.12). The fact that $\mathfrak{S}(\hat{t})$ satisfies the first of the semigroup equations is obvious; the second follows from (II.1.7) and (II.1.8).

The next task is to show that $\mathfrak{S}(\hat{t})$ is strongly continuous: since $\mathfrak{S}(t') - \mathfrak{S}(t) = \mathfrak{S}(t)(\mathfrak{S}(t'-t) - I)$ it is obviously sufficient to prove strong continuity at the origin and, due to the symmetries of the operators in the matrix of $\mathfrak{S}(t)$ we only need to show that

$$\|\mathfrak{S}(h)u - u\|_{\mathfrak{E}_m} \to 0 \qquad (h \to 0+) . \tag{1.17}$$

Again, this must be done in four steps. For the first we differentiate (II.1.7) with respect to s and recombine it in the form

$$\mathcal{C}'(s)(\mathcal{C}(h)u - u) = \mathcal{C}'(s + h)u - \mathcal{C}'(s)u - \mathcal{C}(s)\mathcal{C}'(h)u \qquad (u \in E_0). \tag{1.18}$$

We differentiate now the cosine functional equation $\mathcal{C}(s + t) + \mathcal{C}(s - t) = 2\mathcal{C}(s)\mathcal{C}(t)$ with respect to t, obtaining $\mathcal{C}'(s + t)u - \mathcal{C}'(s - t)u = 2\mathcal{C}(s)\mathcal{C}'(t)u$ for $u \in E_0$. Replacing s by $s + h/2$ and t by $h/2$ we obtain

$$\mathcal{C}'(s + h)u - \mathcal{C}'(s)u = 2\mathcal{C}(s + h/2)\mathcal{C}'(h/2). \tag{1.19}$$

Combining this equality with (1.18), we obtain

$$\|\mathcal{C}(h)u - u\|_0 \to 0 \qquad (h \to 0+) \tag{1.20}$$

for every $u \in E_0$. The second step is to prove that

$$\|\mathcal{S}(h)u\|_0 \to 0 \quad \text{as} \quad h \to 0+ \tag{1.21}$$

for every $u \in E$. To do this we use the differentiated version of (II.1.8)

$$\mathcal{C}'(s)\mathcal{S}(h)u = \mathcal{C}(s + h)u - \mathcal{C}(s)u + \mathcal{C}(s)(\mathcal{C}(h)u - u). \tag{1.22}$$

Finally, we must show that

$$\|\mathcal{C}'(h)u\| \to 0 \qquad (h \to 0+) \tag{1.23}$$

for every $u \in E_0$, and

$$\|\mathcal{S}'(h)u - u\| \to 0 \qquad (h \to 0+) \tag{1.24}$$

for every $u \in E$, both of which relations are obvious. Putting now (1.19) to (1.23) together, (1.17) follows, completing the proof of the strong continuity of $\mathfrak{S}(\hat{t})$.

It only remains to identify $\mathfrak{B}$, the infinitesimal generator of $\mathfrak{S}(\hat{t})$. To do this we show first that

$$\mathfrak{A} \subseteq \mathfrak{B}. \tag{1.25}$$

In fact, let $\mathfrak{u} = [u,v] \in D(\mathfrak{A})$, so that $u \in D(A)$ and $v \in \tilde{D}_1$. Divide (1.18) by h and use (1.19); the result is that $h^{-1}(\mathcal{C}(h)u - u) \to 0$ in the norm of E_0. We proceed in the same way with (1.22) applied to v, obtaining that $h^{-1}\mathcal{S}(h)v \to v$, also in the norm of E_0. Since $h^{-1}(C'(h)u - u) \to C''(0)u = Au$ and $h^{-1}(C(h)u - u) \to C'(0)u = 0$ in E, the proof of (1.25) is complete.

Two more auxiliary facts are necessary for the proof that $\mathfrak{A}$ and $\mathfrak{B}$ actually coincide. The first is

(i) $\mathfrak{A}$ is closed. To see this, let $\{u_n\} = \{[u_n, v_n]\}$ be a sequence in $D(\mathfrak{A})$ such that $u_n \to [u,v] \in \mathfrak{E}_m$ and $\mathfrak{A}u_n \to (w,z) \in \mathfrak{E}_m$ in the norm of $\mathfrak{E}_m$ so that $u_n \to u$ and $v_n \to w$ in E_0, $v_n \to v$ and $Au_n \to z$ in E. Then $w \in E_0$ and (by closedness of $\mathfrak{A}$), $u \in D(A)$ with $Au = z$, i.e. $[u,v] \in D(\mathfrak{A})$ and $\mathfrak{A}[u,v] = [v, Au] = [w,z]$.

(ii) For every $u \in E$ we have

$$u^h = \frac{1}{h}\int_0^h \mathfrak{S}(t)u \, dt \in D(\mathfrak{A}). \tag{1.26}$$

This inclusion relation reduces to the following four: $C(t)E_0 \subseteq E_0$, $S(t)E \subseteq E_0$,

$$\int_0^h C(t)u \, dt \in D(A) \qquad (u \in E_0), \tag{1.27}$$

and

$$\int_0^h S(t)u \, dt \in D(A) \qquad (u \in E). \tag{1.28}$$

The first two have already been established at the beginning of the proof. To prove (1.28) we note that for $u \in D(A)$ we have

$$A\int_0^h S(t)u \, dt = C(h)u - u, \tag{1.29}$$

which equality can be extended to all $u \in E$ approximating u by a sequence $\{u_n\}$ in $D(A)$. As for (1.27), write (1.29) for $u \in \tilde{D}_1$ and differentiate (using again closedness of A). The result is

$$A\int_0^h C(t)u \, dt = C'(h)u.$$

We show finally that

$$\mathfrak{A} = \mathfrak{B}. \tag{1.30}$$

Let u be an arbitrary element of $\mathfrak{E}_m$. Since elements of the form u^h are dense in $\mathfrak{E}_m$, if (1.25) holds we can select a sequence $\{u_n\}$

in $D(\mathfrak{A})$ such that $u_n \to u$ in $\mathfrak{E}_m$. Then $(u_n)^h \to u^h$ whereas $\mathfrak{A}u_n^h = (\mathfrak{A}u_n)^h = h^{-1}(\mathfrak{S}(h)u_n - u_n) \to h^{-1}(\mathfrak{S}(h)u - u)$ (see (I.4.5)) thus by closedness of $\mathfrak{A}$, $\mathfrak{A}u^h = h^{-1}(\mathfrak{S}(h)u - u)$. If u in addition belongs to $D(\mathfrak{B})$ we let $h \to 0$ and obtain again from closedness of $\mathfrak{A}$ that $u = \lim u^h$ belongs to $D(\mathfrak{A})$ with $\mathfrak{A}u = \lim h^{-1}(\mathfrak{S}(h)u - u) = \mathfrak{B}u$. This completes the proof of (1.30).

Since $D(\mathfrak{A})$ must be dense in $\mathfrak{E}_m$ (Theorem I.4.1) we obtain in particular that $D(\mathfrak{A})$ is dense in E_0 in the topology of E_0. Let u be an arbitrary element of $E_0 = \tilde{D}_1$, $\{u_n\}$ a sequence in $D(A)$ such that $u_n \to u$ in E_0. Then we have $\mathcal{S}(t)u_n \to \mathcal{S}(t)u$ and $A\mathcal{S}(t)u_n = \mathcal{S}''(t)u_n = \mathcal{C}'(t)u_n \to \mathcal{C}'(t)u$ for all t whence it follows that $\mathcal{S}(t)u \in D(A)$ and $A\mathcal{S}(t)u = \mathcal{S}''(t)u$, so that $\mathcal{S}(\hat{t})u$ is a solution of (1.1) in $-\infty < t < \infty$, proving that $\tilde{D}_1 \subseteq D_1$. The opposite inclusion is obvious, thus (1.9) holds and the proof of Theorem 1.3 is complete.

Theorem 1.2 might be taken to imply that "every well posed second order Cauchy problem can be reduced to an equally well posed first order system", thus the theory of (1.1) could be reduced from the word go to semigroup theory. In reality, the situation is not so simple; on the one hand, the growth of $\mathfrak{S}(\hat{t})$ at infinity (see 1.11) does not match accurately the growth of $\mathcal{C}(\hat{t})$ or even that of $\mathcal{S}(\hat{t})$, while on the other hand the norm of E_0 is not easy to identify (it depends on complete knowledge of the solutions of (1.1)). Taking a clue from the wave and Klein-Gordon equations we may expect that the reduction (1.2) to first order can be made to work if one measures $u(t)$ in the graph norm of a square root of A. This prescription will be seen to work, albeit with some restrictions, in the next sections.

§III.2 Fractional powers of closed operators.

A densely defined operator A is said to belong to the class $\mathfrak{Z}(C)$ $(C > 0)$ if and only if $R(\lambda;A)$ exists for $\lambda > 0$ and

$$\|R(\lambda;A)\| \le C/\lambda \quad (\lambda > 0) . \tag{2.1}$$

Obviously we have $\mathfrak{Z}(1) = \mathfrak{C}^1(1,0)$ since (2.1) implies all the inequalities (I.3.24) with $C_0 = 1$, $\omega = 0$. If $C > 1$, however, we only can assure that $\|R(\lambda;A)^n\| \le C^n/\lambda^n$ $(n = 0,1,\ldots)$ which does not imply that $A \in \mathfrak{C}^1$ (for a counterexample see HILLE-PHILLIPS [1957:1, p. 371]).

We show below that fractional powers of A (or rather of $(-A)$) can be constructed employing the formula

$$(-A)^{\alpha} = \frac{\sin\alpha\pi}{\pi}\int_0^{\infty} \lambda^{\alpha-1}R(\lambda;A)(-A)\, d\lambda , \tag{2.2}$$

which is the vector counterpart of a well known scalar formula (GRADSTEIN-RIDZYK [1963:1, p. 303]). Although complex values of α are handled just as easily, we shall only utilize α real, mostly in the range $0 \leq \alpha \leq 1$. We restrict the treatment accordingly.

Let $0 < \alpha < 1$. The operator K_{α} (with domain $D(K_{\alpha}) = D(A)$) is defined by the right hand side of (2.2), that is, by

$$K_{\alpha}u = \frac{\sin\alpha\pi}{\pi}\int_0^{\infty} \lambda^{\alpha-1}R(\lambda;A)(-A)u\, d\lambda , \tag{2.3}$$

where the branch of $\lambda^{\alpha-1}$ is that which is real when $\lambda \geq 0$. That the integral (2.3) is convergent at infinity is clear from (2.1); near zero we use the second resolvent equation in the form

$$R(\lambda;A)(-A)u = u - \lambda R(\lambda;A)u. \tag{2.4}$$

LEMMA 2.1 <u>The operator</u> K_{α} <u>is closable for each</u> α, $0 < \alpha < 1$.

<u>Proof</u>: Let $0 < \alpha < 1$, $\{u_n\}$ a sequence in $D(A)$ with $u_n \to 0$, $K_{\alpha}u_n \to v \in E$. Select $\mu > 0$ fixed and write

$$v_n = R(\mu;A)K_{\alpha}u_n . \tag{2.5}$$

It follows from its definition that K_{α} commutes with $R(\mu;A)$ in the sense that

$$R(\mu;A)K_{\alpha}u = K_{\alpha}R(\mu;A)u \tag{2.6}$$

for $u \in D(K_{\alpha}) = D(A)$. Moreover, $K_{\alpha}R(\mu;A)$ is everywhere defined and bounded since

$$K_{\alpha}R(\mu;A)u = \frac{\sin\alpha\pi}{\pi} R(\mu;A)\int_0^{1} \lambda^{\alpha-1}(I - \lambda R(\lambda;A))u\, d\lambda$$
$$- \frac{\sin\alpha\pi}{\pi} AR(\mu;A)\int_1^{\infty} \lambda^{\alpha-1}R(\lambda;A)u\, d\lambda . \tag{2.7}$$

Hence we obtain, taking limits in (2.5) that $v_n \to 0$ so that $R(\mu;A)v = 0$, thus $v = 0$, ending the proof.

Lemma 2.1 makes possible the following definition:

$$(-A)^{\alpha} = \bar{K}_{\alpha} \quad (0 < \alpha < 1), \tag{2.8}$$

where the bar indicates closure. Of course, we define $(-A)^0 = I$, $(-A)^1 = -A$. The following results imply that $(-A)^{\alpha}$ exhibits, at least in part, the behavior expected of a fractional power.

LEMMA 2.2. *Let* $u \in D(A)$. *Then*

$$\lim_{a \to 1-} K_{\alpha} u = -Au. \tag{2.9}$$

Proof: Using the scalar counterpart of (2.2) we obtain

$$K_{\alpha}u - (-A)u = \frac{\sin \alpha\pi}{\pi} \int_0^{\infty} \lambda^{\alpha-1}(R(\lambda;A) - \frac{1}{\lambda+1} I)(-A)u \, d\lambda = I_1 + I_2, \tag{2.10}$$

where I_1 and I_2 correspond to the division of the domain of integration at $\lambda = a > 0$. For the second integral we make use again of the scalar version of (2.2); the result is

$$\|I_2\| \leq \sup_{\lambda \geq a} \|((\lambda + 1)R(\lambda;A) - I)Au\|. \tag{2.11}$$

To estimate the right hand side of (2.11) we return to (2.1). Since $\lambda R(\lambda;A)u = u + R(\lambda;A)Au$ we have

$$\lambda R(\lambda;A)u \to u \quad \text{as} \quad \lambda \to \infty \tag{2.12}$$

for $u \in D(A)$; by virtue of the fact that $\|\lambda R(\lambda;A)\| \leq C$, (2.12) can be extended to all $u \in E$ and shows that the right hand side of (2.11) can be made arbitrarily small for a large enough. Obviously the estimate is independent of α. To deal with I_1 we only have to observe that the integrand can be bounded uniformly with respect to α for α near 1 (see (2.4)) thus the factor $\sin \alpha\pi$ drags $\|I_1\|$ to zero as $\alpha \to 1$. This completes the proof.

The corresponding result in the neighborhood of $\alpha = 0$ is

LEMMA 2.3 *Let* $u \in D(A)$ *be such that*

$$\lambda R(\lambda;A)u \to 0 \quad \text{as} \quad \lambda \to 0+. \tag{2.13}$$

Then

$$\lim_{\alpha \to 0+} K_\alpha u = u. \tag{2.14}$$

Proof: Proceeding in a way similar to that used in Lemma 2.2 but making use of (2.4) we obtain

$$\begin{aligned} K_\alpha u - u &= -\frac{\sin\alpha\pi}{\pi}\int_0^\infty \lambda^{\alpha-1}(R(\lambda;A)Au + \frac{1}{\lambda+1}u)\,d\lambda \\ &= -\frac{\sin\alpha\pi}{\pi}\int_0^\infty \frac{\lambda^\alpha}{\lambda+1}((\lambda+1)R(\lambda;A)u - u)\,d\lambda = J_1 + J_2 \end{aligned}$$

where J_1 and J_2 arise once again from division of the interval of integration at $\lambda = a > 0$. To estimate the second integral we observe that $(\lambda + 1)R(\lambda;A)u - u = R(\lambda;A)(Au + u)$ and use (2.1); it follows that $\|J_2\| \to 0$ as $\alpha \to 0$. The corresponding bound for J_1 follows from the inequality

$$\|(\lambda + 1)R(\lambda;A)u - u\| \le C + \|R(\lambda;A)u\|.$$

In fact, the portion of the bound corresponding to the constant C obviously tends to zero when $\alpha \to 0$. On the other hand,

$$\begin{aligned} &\frac{\sin\alpha\pi}{\pi}\int_0^a \frac{\lambda^\alpha}{\lambda+1}\|R(\lambda;A)u\|\,d\lambda \\ &= \frac{\sin\alpha\pi}{\pi}\int_0^a \frac{\lambda^{\alpha-1}}{\lambda+1}\|\lambda R(\lambda;A)u\|\,d\lambda \le \max_{0\le\lambda\le a}\|\lambda R(\lambda;A)u\|, \end{aligned} \tag{2.15}$$

thus we can make $\|J_1\|$ arbitrarily small by taking a small enough. This completes the proof.

An obviously desirable property of any definition of fractional power is the additivity relation $(-A)^{\alpha+\beta} = (-A)^\alpha(-A)^\beta$. In the present level of generality only a slightly weaker statement can be proved (Theorem 2.5 below). Its key ingredient is the following result.

LEMMA 2.4. Let $\alpha,\beta > 0$, $\alpha + \beta < 1$. Then

$$K_\alpha K_\beta u = K_{\alpha+\beta}u \quad (u \in D(A^2)). \tag{2.16}$$

Proof: Let $u \in D(A^2)$. Since $K_\beta u \in D(A) = D(K_\alpha u)$ we may write

$$K_\alpha K_\beta u = \frac{\sin\alpha\pi}{\pi}\frac{\sin\beta\pi}{\pi}\int_0^\infty\int_0^\infty \lambda^{\alpha-1}\mu^{\beta-1}R(\lambda;A)R(\mu;A)A^2u\,d\lambda d\mu\,. \tag{2.17}$$

(that the integral is absolutely convergent in the quadrant $\lambda,\mu \geq 0$ is shown in the same way as for (2.3)). Divide the domain of integration into the two triangles $\mu \leq \lambda$ and $\lambda \leq \mu$ and perform the change of variable $\mu = \lambda\sigma$ in the first, $\lambda = \mu\sigma$ in the second; interchanging λ by μ in the second integral we obtain.

$$K_\alpha K_\beta u = \frac{\sin\alpha\pi}{\pi}\frac{\sin\beta\pi}{\pi}\int_0^\infty\int_0^1 (\sigma^{\alpha-1} + \sigma^{\beta-1})\lambda^{\alpha+\beta-1}R(\lambda\sigma;A)R(\lambda;A)A^2u\,d\sigma\,d\lambda\,. \tag{2.18}$$

Using the second resolvent equation we deduce that

$$R(\lambda\sigma;A)R(\lambda;A)A^2u = \frac{1}{1-\sigma}\{R(\lambda;A) - \sigma R(\lambda\sigma;A)\}(-A)u \tag{2.19}$$

and replace this expression in the integrand of (2.18). On account of the factor $(1-\sigma)^{-1}$ the two integrals resulting from (2.19) cannot be separated, but we can write (2.18) as the limit when $\rho \to 1-$ of

$$\frac{\sin\alpha\pi}{\pi}\frac{\sin\beta\pi}{\pi}\int_0^\infty\int_0^\rho \frac{\sigma^{\alpha-1} + \sigma^{\beta-1}}{1-\sigma}\lambda^{\alpha+\beta-1}R(\lambda;A)(-A)u\,d\sigma\,d\lambda$$

$$-\frac{\sin\alpha\pi}{\pi}\frac{\sin\beta\pi}{\pi}\int_0^\infty\int_0^\rho \frac{\sigma^{\alpha} + \sigma^{\beta}}{1-\sigma}\lambda^{\alpha+\beta-1}R(\lambda\sigma;A)(-A)u\,d\sigma\,d\lambda\,. \tag{2.20}$$

We compute the second term integrating first in λ, making the change of variable $\lambda\sigma = \mu$ and then changing again μ by λ. We obtain

$$K_\alpha K_\beta u = C_{\alpha,\beta}\int_0^\infty \lambda^{\alpha+\beta-1}R(\lambda;A)(-Au)\,d\lambda, \tag{2.21}$$

where

$$C_{\alpha,\beta} = \frac{\sin\alpha\pi}{\pi}\frac{\sin\beta\pi}{\pi}\int_0^1 \frac{\sigma^{\alpha-1} + \sigma^{\beta-1} - \sigma^{-\alpha} - \sigma^{-\beta}}{1-\sigma}\,d\sigma.$$

To evaluate the constant $C_{\alpha,\beta}$ we apply the argument above to $E = \mathbb{C}$, $A = 1$, so that $K_\alpha = 1$ for all α and it follows from (2.21) that

$$C_{\alpha,\beta} = \frac{\sin(\alpha+\beta)\pi}{\pi}\,.$$

This completes the proof of (2.16).

THEOREM 2.5. Let $\alpha,\beta \geq 0$, $\alpha+\beta \leq 1$. Then $(-A)^{\alpha}(-A)^{\beta}$ is closable and

$$\overline{(-A)^{\alpha}(-A)^{\beta}} = (-A)^{\alpha+\beta}\ , \tag{2.22}$$

where the bar indicates closure.

Proof: (2.22) (in an improved version) is obvious if either α or β are zero, thus we may suppose that $\alpha,\beta > 0$. Assume first that $\alpha+\beta < 1$ and let $u \in D((-A)^{\alpha}(-A)^{\beta})$ and $\mu > 0$. In view of Lemma 2.4 and of the fact that $(-A)^{\alpha}$ commutes with $R(\mu;A)$ for any α (which in turn follows from commutativity of K_{α} and $R(\mu;A)$) we have

$$\begin{aligned} K_{\alpha+\beta}(\mu R(\mu;A))^2u &= K_{\alpha}K_{\beta}(\mu R(\mu;A))^2u = (-A)^{\alpha}(-A)^{\beta}(\mu R(\mu;A))^2u = \\ &= (\mu R(\mu;A))^2(-A)^{\alpha}(-A)^{\beta}u \to (-A)^{\alpha}(-A)^{\beta}u \quad \text{as} \quad \mu\to\infty. \end{aligned}$$

Since $(\mu R(\mu;A))^2u \to u$ as $\mu\to\infty$ we deduce that $u \in D((-A)^{\alpha+\beta})$ with $(-A)^{\alpha+\beta}u = (-A)^{\alpha}(-A)^{\beta}u$, thus $(-A)^{\alpha}(-A)^{\beta} \subseteq (-A)^{\alpha+\beta}$. Since $(-A)^{\alpha+\beta}$ is closed, $(-A)^{\alpha}(-A)^{\beta}$ is closable and $\overline{(-A)^{\alpha}(-A)^{\beta}} \subseteq (-A)^{\alpha+\beta}$.

We prove next the opposite inclusion. To this end, denote by K'_{γ} the operator K_{γ} restricted to $D(A^2)$. We propose to show that for $0 < \gamma < 1$ we have

$$\overline{K'_{\gamma}} = \overline{K_{\gamma}} = (-A)^{\gamma}\ . \tag{2.23}$$

Obviously we have $K'_{\gamma} \subseteq K_{\gamma}$, thus $\overline{K'_{\gamma}} \subseteq \overline{K_{\gamma}}$. Assume that $u \in D(\overline{K_{\gamma}}) = D((-A)^{\gamma})$ so that there exists a sequence $\{u_n\} \subseteq D(A)$ with $u_n \to u$, $K_{\gamma}u_n \to (-A)^{\gamma}u$. If $\{\mu_n\}$ is a sequence of positive numbers with $\mu_n \to \infty$ then $v_n = \mu_n R(\mu_n;A)u_n \in D(A^2) = D(K'_{\gamma})$; on the other hand, $v_n \to u$ and $K'_{\gamma}v_n = \mu_n R(\mu_n;A)K_{\gamma}u_n \to (-A)^{\gamma}u$ so that $u \in D(\overline{K'_{\gamma}})$ with $\overline{K'_{\gamma}}u = \overline{K_{\gamma}}u$ which shows that $\overline{K_{\gamma}} \subseteq \overline{K'_{\gamma}}$ and completes the proof of (2.23). Since Lemma 2.4 implies that $\overline{(-A)^{\alpha}(-A)^{\beta}}u = (-A)^{\alpha}(-A)^{\beta}u = K_{\alpha}K_{\beta}u = K_{\alpha}K'_{\beta}u = K'_{\alpha+\beta}u$ for $u \in D(K'_{\alpha+\beta}) = D(A^2)$ we have $\overline{(-A)^{\alpha}(-A)^{\beta}} \supseteq K'_{\alpha+\beta}$; taking closures, $\overline{(-A)^{\alpha}(-A)^{\beta}} \supseteq (-A)^{\alpha+\beta}$ by virtue of (2.23) for $\gamma = \alpha+\beta$. This completes the proof of (2.22) for $\alpha+\beta < 1$.

Assume now that $\alpha+\beta = 1$. Let $u \in D(A)$ and consider the function

$$\gamma \to K_{\gamma}u \tag{2.24}$$

in $0 < \gamma \leq 1$. Continuity of (2.24) in $0 < \gamma < 1$ is obvious from the definition and continuity at $\gamma = 1$ has been proved in Lemma 2.2. Accordingly, if $u \in D(A^2)$ we have

$$K_\alpha K_\beta u = \lim_{\gamma \to \alpha -} K_\gamma K_\beta u = \lim_{\gamma \to \alpha -} K_{\gamma+\beta} u = \lim_{\gamma \to 1-} K_\gamma u = -Au.$$

The proof that $-A \supseteq \overline{(-A)^\alpha (-A)^\beta}$ is exactly the same as in the previous case; the opposite inclusion depends on the fact that if A' is the restriction of A to $D(A^2)$ then $\overline{A'} = A$ which is shown using the argument employed above for K'_γ. We omit the details.

§III.3 Resolvents of fractional powers.

As seen in the next result, condition (2.1) implies existence of $R(\lambda;A)$ in a sector containing the positive real axis.

We introduce some notations. Given φ, $0 < \varphi < \pi$, we write

$$\Sigma(\varphi) = \{\lambda; |\arg \lambda| \leq \varphi\}.$$

The sector $\Sigma(\varphi-)$ is defined in the same way with $\leq$ replaced by $<$. Obviously, $\lambda = 0$ belongs to any sector $\Sigma(\varphi)$ or $\Sigma(\varphi-)$. The subindex + indicates exclusion of 0; $\Sigma_+(\varphi)$ (resp. $\Sigma_-(\varphi-)$) is the set of all $\lambda \neq 0$ with $|\arg \lambda| \leq \varphi$ (resp. $|\arg \lambda| < \varphi$).

THEOREM 3.1. <u>Assume that</u> $A \in \mathfrak{C}(C_1)$ <u>for some</u> $C_1 > 0$. <u>Then</u> $R(\lambda;A)$ <u>exists in the sector</u> $\Sigma_+(\varphi-)$ <u>with</u> $\varphi = \arcsin(1/C_1)$ <u>and for every</u> φ', $0 < \varphi' < \varphi$, <u>there exists a constant</u> $C = C_{\varphi'}$ <u>such that</u>

$$\|R(\lambda;A)\| \leq C/|\lambda| \qquad (\lambda \in \Sigma_+(\varphi')). \tag{3.1}$$

<u>Proof</u>: Let $\lambda_0 > 0$. Since $\|R(\lambda_0;A)\| \leq C_1/\lambda_0$ it follows that $R(\lambda;A)$ exists in $|\lambda - \lambda_0| < \lambda_0/C_1 \leq 1/\|R(\lambda_0;A)\|^{-1}$ and can be expressed there by the power series

$$R(\lambda;A) = \sum_{j=0}^{\infty} (\lambda_0 - \lambda)^j R(\lambda_0;A)^{j+1}. \tag{3.2}$$

Since

$$\Sigma_+(\varphi-) = \bigcup_{\lambda_0 > 0} \{\lambda; |\lambda - \lambda_0| < \lambda_0/C_1\}$$

with $\varphi = \arcsin(1/C_1)$, the first statement stands proved. Let $0 < \varphi' < \varphi$, $\lambda \in \Sigma_+(\varphi')$, $\lambda_0 = |\lambda|/\cos\arg\lambda$. Then $|\lambda - \lambda_0| \le \lambda_0 \sin|\arg\lambda| \le \lambda_0 \sin\varphi'$ and we can estimate (3.2) as follows:

$$\|R(\lambda;A)\| \le \sum_{j=0}^{\infty} |\lambda - \lambda_0|^j \left(\frac{C_1}{\lambda_0}\right)^{j+1} = \frac{C_1}{\lambda_0(1 - C_1 \sin\varphi')} = \frac{C'}{\lambda_0} \le \frac{C}{|\lambda|} .$$

This completes the proof.

The next step is that of establishing existence and estimates similar to (3.1) for the resolvent of the fractional powers; we restrict ourselves here to the case $\alpha = 1/2$, since other values of α are of less interest in connection with second order equations. Let μ be a nonzero complex number in the sector defined by

$$(\pi - \varphi)/2 \le \varphi < \pi/2, \tag{3.3}$$

so that $-\mu^2$ belongs to the sector $\Sigma_+(\varphi-)$ where the resolvent of A exists. Define

$$Q(\mu) = -(\mu I - (-A)^{1/2})R(-\mu^2;A). \tag{3.4}$$

Since $D((-A)^{1/2}) \supseteq D(A)$, $Q(\mu)$ is everywhere defined; on the other hand, we observed in the previous section (see (2.7)) that $K_\alpha R(\lambda;A)$ is bounded for $0 < \alpha < 1$, thus $(-A)^\alpha R(\lambda;A)$ - a fortiori $Q(\mu)$- is bounded. Since $(-A)^\alpha$ and $R(\lambda;A)$ commute, we have

$$Q(\mu)u = -R(\mu^2;A)(\mu I - (-A)^{1/2})u \tag{3.5}$$

for $u \in D(A^{1/2})$. It also follows from commutativity that if $u \in D(((-A)^{1/2})^2)$ then $R(-\mu^2)u \in D(((-A)^{1/2})^2)$ and

$$(\mu I + (-A)^{1/2})Q(\mu)u = -(\mu^2 I + A)R(-\mu^2;A)u = u \tag{3.6}$$

since $((-A)^{1/2})^2 \subseteq -A$ (Theorem 2.5). On the other hand, $D(((-A)^{1/2})^2) \supseteq D(A^2)$ is dense in E; we can then use (3.6) and the facts that $Q(\mu)$ is bounded and that $(-A)^{1/2}$ is closed to deduce that (3.6) holds for every u, that is, $Q(\mu)E \subseteq D((-A)^{1/2})$ and

$$(\mu I + (-A)^{1/2})Q(\mu)u = u . \tag{3.7}$$

In particular, $\mu I + (-A)^{1/2} : D((-A)^{1/2}) \to E$ is <u>onto</u>. On the other hand, $\mu I - (-A)^{1/2}$ is <u>one-to-one</u>; for if $u \in D((-A)^{1/2})$ is such that

$(\mu I-(-A)^{1/2})u = 0$ then it results from (3.5) that $Q(\mu)u = 0$ and then from (3.7) that $u = 0$.

We run now the entire argument replacing $\mu I-(-A)^{1/2}$ in (3.4) by $\mu I+(-A)^{1/2}$. This time we obtain that $\mu I+(-A)^{1/2} : D((-A)^{1/2} \to E$ is onto, while $\mu I-(-A)^{1/2}$ is one-to-one. The conclusion is that both $\mu I-(-A)^{1/2}$ and $\mu I+(-A)^{1/2}$ are invertible, that is, μ and $-\mu$ belong to $\rho((-A)^{1/2}$; in particular, we have

$$R(\mu;-(-A)^{1/2}) = Q(\mu) = -(\mu I-(-A)^{1/2})R(-\mu^2;A). \qquad (3.8)$$

Although the mere fact that $\rho((-A)^{1/2})$ is nonempty is all what is needed for the moment, significant information on the resolvent can be obtained with a little more work. We start from (3.8) using the basic formula (2.3), obtaining

$$R(\mu;-(-A)^{1/2}) = -\mu R(-\mu^2;A) - \frac{1}{\pi}\int_0^\infty \lambda^{-1/2}R(\lambda;A)AR(-\mu^2;A)\,d\lambda. \qquad (3.9)$$

Using the second resolvent equation we deduce that

$$R(\lambda;A)AR(-\mu^2;A) = -(\lambda+\mu^2)^{-1}(\lambda R(\lambda;A) + \mu^2R(-\mu^2;A)).$$

Replacing into (3.9) we obtain after a simple calculation the formula

$$R(\mu;-(-A)^{1/2}) = \frac{1}{\pi}\int_0^\infty \frac{\lambda^{1/2}}{\lambda+\mu^2}R(\lambda;A)\,d\lambda \qquad (3.10)$$

(which has a well known scalar counterpart; see GRADSTEIN-RIDZYK [1963:1, p. 303]). Up to this point, μ has been a complex number in the sector defined by (3.3). However, (3.10) defines a (E) - valued analytic function in the entire plane minus the imaginary axis, thus it provides an analytic extension of $R(\mu;-(-A)^{1/2})$ to the right half plane $\operatorname{Re}\mu > 0$. Since the norm of the resolvent $R(\mu;B)$ of an arbitrary operator B must tend to infinity as μ approaches the boundary of the resolvent set $\rho(B)$ it follows that every μ with $\operatorname{Re}\mu > 0$ belongs to $\rho(-(-A)^{1/2})$ with $R(\mu;-(-A)^{1/2})$ given by (3.10). This statement can be improved as follows:

THEOREM 3.2. <u>Let</u> $\varphi = \arcsin(1/C_1)$ <u>as in Theorem</u> 3.1. <u>Then</u> $R(\mu;-(-A)^{1/2})$ <u>exists in the sector</u> $\Sigma_+(\frac{1}{2}(\varphi+\pi)-)$; <u>moreover, if</u> $0 < \varphi' < \varphi$ <u>there exists a constant</u> $C = C_{\varphi'}$ <u>such that</u>

$$\|R(\mu;-(-A)^{1/2})\| \le C/|\mu| \quad (\mu \in \Sigma_+(\tfrac{1}{2}(\varphi'+\pi)) \qquad (3.11)$$

Proof: Let $\varphi' < \varphi'' < \varphi$. Using Theorem 3.1 we can deform the path of integration in (3.10) to either of the rays $\lambda e^{\pm i\varphi''}$, $0 \leq \lambda < \infty$ so that if, say, $\mu > 0$ we have

$$R(\mu;-(-A)^{1/2}) = \frac{1}{\pi} \int_0^\infty \frac{\lambda^{1/2} e^{i\varphi''/2}}{\lambda + e^{-i\varphi''}\mu^2} R(\lambda e^{i\varphi''};A)d\lambda \tag{3.12}$$

and

$$R(\mu;-(-A)^{1/2}) = \frac{1}{\pi} \int_0^\infty \frac{\lambda^{1/2} e^{-i\varphi''/2}}{\lambda + e^{i\varphi''}\mu^2} R(\lambda e^{-\varphi''};A)d\lambda\ . \tag{3.13}$$

The first integral formula defines an analytic function in the whole plane minus the line $\arg \mu = \frac{1}{2}(\varphi' + \pi)$, thus provides an analytic extension of $R(\lambda;-(-A)^{1/2})$ to the half plane $\operatorname{Re} e^{-i\varphi''/2}\mu > 0$. In a symmetric fashion, (3.13) yields an analytic extension to the half plane $\operatorname{Re} e^{i\varphi''/2}\mu > 0$, thus $R(\lambda;-(-A)^{1/2})$ exists in $\Sigma_+(\frac{1}{2}(\varphi' + \pi))$.
It remains to estimate the resolvent. Let $\mu \in \Sigma_+(\frac{1}{2}(\varphi' + \pi))$, $\operatorname{Im} \mu \geq 0$ so that $\mu = |\mu|e^{i\psi}$ with $0 \leq \psi \leq \frac{1}{2}(\pi + \varphi')$. Then

$$|\lambda + e^{-i\varphi''}\mu^2| \geq c(\lambda + |\mu|^2),$$

with $c > 0$, so that

$$\|R(\mu;-(-A)^{1/2})\| \leq C' \int_0^\infty \frac{d\lambda}{\lambda^{1/2}(\lambda + |\mu|^2)} = \frac{C}{|\mu|}\ .$$

For $\mu \in \Sigma_+(\frac{1}{2}(\varphi' + \pi))$, $\operatorname{Im} \mu \leq 0$ we use (3.13) instead obtaining the same estimate. This completes the proof of Theorem 3.2.

An important consequence of the mere fact that $\rho((-A)^{1/2} \neq \emptyset$ is the following improvement of a (particular case) of Theorem 2.5:

COROLLARY 3.3.

$$((-A)^{1/2})^2 = (-A)^{1/2}(-A)^{1/2} = -A\ . \tag{3.14}$$

Proof: Let μ be a complex number such that $-\mu^2 \in \rho(A)$. Since both μ and $-\mu$ belong to $\rho(-(-A)^{1/2})$ we have $D(((-A)^{1/2})^2) = R(\mu;-(-A)^{1/2})R(-\mu;-(-A)^{1/2})E = R(-\mu^2;A)E = D(A)$. Since (2.22) implies that $((-A)^{1/2})^2 \subseteq -A$, (3.14) stands proved.

§III.4 Translation of generators of cosine functions.

If A is the infinitesimal generator of a strongly continuous semigroup $S(\hat{t})$ then $A - bI$ (b any complex number) is the infinitesimal generator of the strongly continuous semigroup $S_b(\hat{t}) = e^{-b\hat{t}}S(\hat{t})$. The corresponding result for cosine functions is somewhat harder to prove.

LEMMA 4.1. Let A be the infinitesimal generator of a strongly continuous cosine function $C(\hat{t})$ satisfying

$$\|C(t)\| \leq C_0 e^{\omega|t|} \qquad (-\infty < t < \infty) \tag{4.1}$$

and let b be an arbitrary complex number. Then $A_b = A - b^2 I$ is the infinitesimal generator of a strongly continuous cosine function $C_b(\hat{t})$ satisfying

$$\|C_b(t)\| \leq C_0 e^{(\omega+|b|)t} {}^{(2)} \qquad (-\infty < t < \infty) \tag{4.2}$$

Proof: Define, inductively,

$$C_0(t) = C(t), \quad C_n(t)u = (S * C_{n-1})(t)u = \int_0^t S(t-s)C_{n-1}(s)u\, ds \tag{4.3}$$

$$(u \in E, \ -\infty < t < \infty, \ n = 1,2,\ldots).$$

Obviously $C_0(\hat{t}), C_1(\hat{t}),\ldots$ are all strongly continuous operator valued functions in $-\infty < t < \infty$. Denote by $W^{(n)}$ the set of all n-tuples $(e_1,e_2,\ldots,e_n)$ with $e_j = \pm 1$; $W^{(n)}$ has 2^n elements. It is easy to see by induction starting from the cosine functional equation (II.3.1) that for arbitrary real numbers $t_0, s_1,\ldots,s_n$ we have

$$C(s_n)C(s_{n-1}) \ldots C(s_1)C(t_0) = \frac{1}{2^n}\sum C(t_0 + e_1 s_1 + \cdots + e_n s_n), \tag{4.4}$$

where the sum is extended over all $(e_1,\ldots,e_n) \in W^{(n)}$. Let $0 \leq t_0 \leq t_1 \leq \ldots \leq t_n = t > 0$, $0 \leq s_j \leq t_j - t_{j-1}$ ($j = 1,\ldots,n$; $t_n = t$). Since $|t_0 + e_1 s_1 + \ldots + e_n s_n| \leq t$ we obtain from (4.4) that

$$\|C(s_n)C(s_{n-1}) \ldots C(s_1)C(t_0)\| \leq C_0 e^{\omega t}. \tag{4.5}$$

We integrate now the left hand side of (4.4) in the n-dimensional parallelepipedon $0 \leq s_j \leq t_j - t_{j-1}$ ($j = 1,\ldots,n$; $t_n = t$). Making use of (4.5) we get

$$\|S(t - t_{n-1})S(t_{n-1} - t_{n-2}) \ldots S(t_1 - t_0)C(t_0)\|$$
$$\leq C_0(t - t_{n-1})(t_{n-1} - t_{n-2}) \ldots (t_1 - t_0)e^{\omega t}.$$

We note next that

$$\mathcal{C}_n(t)u = \int \mathcal{S}(t - t_{n-1}) \ldots \mathcal{S}(t_1 - t_0)\mathcal{C}(t_0)u \, dt_0 \ldots dt_{n-1},$$

the integral taken on the region $0 \leq t_0 \leq t_1 \leq \ldots \leq t_{n-1} \leq t$, thus it follows from (4.5) that

$$\|\mathcal{C}_n(t)\| \leq C_0 \frac{t^{2n}}{(2n)!} e^{\omega t} . \tag{4.6}$$

Consequently, the series

$$\mathcal{C}_b(t) = \sum_{n=0}^{\infty} (-b^2)^n \mathcal{C}_n(t) \tag{4.7}$$

converges uniformly on compacts of $t \geq 0$. This plainly implies that $\mathcal{C}_b(\hat{t})$ is a strongly continuous (E)-valued function with $\mathcal{C}_b(0) = I$. Moreover

$$\|\mathcal{C}_b(t)\| \leq C_0 e^{\omega t} \sum_{n=0}^{\infty} \frac{|bt|^{2n}}{(2n)!} \leq C_0 e^{(\omega + |b|)t} . \tag{4.8}$$

We obtain from (II.2.11), (4.3) and (a vector valued variant) of the convolution formula for Laplace transforms that

$$\int_0^{\infty} e^{-\lambda t} \mathcal{C}_n(t)u \, dt = \lambda R(\lambda^2; A)^{n+1} u \tag{4.9}$$

for $\operatorname{Re}\lambda > \omega$. Hence, after a clearly permissible term-by-term integration,

$$\int_0^{\infty} e^{-\lambda t} \mathcal{C}_b(t)u \, dt = \lambda \sum_{n=0}^{\infty} (-b^2)^n R(\lambda^2; A)^{n+1} u$$

$$= \lambda R(\lambda^2 + b^2; A)u = \lambda R(\lambda^2; A_b)u \quad (u \in E) \tag{4.10}$$

for $\operatorname{Re}\lambda > \omega + |b|$. It follows then from Lemma II.2.3 that $\mathcal{C}_b(|\hat{t}|)$ is a cosine function with A_b as infinitesimal generator, completing the proof of Lemma 4.1.

In the sequel, b will be a real number with $b \geq \omega$. For those values of b we have, in view of the first inequality (II.2.11),

$$\|R(\lambda^2; A_b)\| = \|R(\lambda^2 + b^2; A)\|$$

$$\leq \frac{C_0}{(\lambda^2 + b^2)^{1/2}((\lambda^2 + b^2)^{1/2} - \omega)} \leq \frac{C_1}{\lambda^2} \quad (\lambda > 0) \tag{4.11}$$

(where C_1 depends of course on b). Accordingly, A_b belongs to the class $\mathfrak{G}(C_1)$ defined in §III.2 and the fractional powers $(-A_b)^\alpha = (b^2 I - A)^\alpha$ can be defined.

LEMMA 4.2. Let $0 < \alpha < 1$, $b, b' \geq \omega$. Then

$$(-A_{b'})^\alpha - (-A_b)^\alpha = (b'^2 I - A)^\alpha - (b^2 I - A)^\alpha \tag{4.12}$$

is a bounded operator in $D(A)$; as a consequence,

$$D((-A_{b'})^\alpha) = D((-A_b)^\alpha). \tag{4.13}$$

Proof: Denote by $K_{\alpha,b}$ the operator (2.3) corresponding to the operator A_b. If $u \in D(A)$ we have

$$\begin{aligned}(-A_{b'})^\alpha - (-A_b)^\alpha &= (K_{\alpha,b'} - K_{\alpha,b})u = \\ &- \frac{\sin\alpha\pi}{\pi}\int_0^\infty \lambda^{\alpha-1}(R(\lambda;A_{b'})A_{b'} - R(\lambda;A_b)A_b)u\, d\lambda \\ &= - \frac{\sin\alpha\pi}{\pi}\int_0^1 \lambda^{\alpha}(R(\lambda;A_{b'}) - R(\lambda;A_b))u\, d\lambda \\ &+ (b'^2 - b^2)\frac{\sin\alpha\pi}{\pi}\int_1^\infty \lambda^{\alpha}R(\lambda;A_{b'})R(\lambda;A_b)u\, d\lambda .\end{aligned}$$

Both integrals define bounded operators, thus boundedness of (4.12) follows. If $u \in D((-A_b)^\alpha)$ then there exists a sequence $\{u_n\} \in D(A)$ such that $u_n \to u$, $(-A_b)^\alpha u_n \to (-A_b)^\alpha u$. By virtue of (4.12), $(-A_{b'})^\alpha u_n$ is convergent as well thus $u \in D((-A_{b'})^\alpha)$. Since the argument works equally well with the roles of b and b' reversed, (4.13) follows.

It remains to define the fractional powers of A_b itself. We select arbitrarily one of the possible values:

$$A_b^\alpha = e^{i\alpha\pi}(-A_b)^\alpha . \tag{4.14}$$

In particular, $A_b^{1/2} = i(-A_b)^{1/2}$.

§III.5 The principal value square root reduction.

Throughout the rest of this chapter A will be the infinitesimal generator of a strongly continuous cosine function $C(\hat{t})$ satisfying (4.1), that is, $A \in \mathfrak{C}^2(C_0, \omega)$. As seen in Section II.3, $A \in \mathfrak{C}^2$ is equivalent to the fact that the Cauchy problem for

$$u''(t) = Au(t) \tag{5.1}$$

is (uniformly) well posed in $t \geq 0$ (equivalently, it is uniformly well posed in $-\infty < t < \infty$). We shall also require the following ad hoc postulate, which will be shown to hold automatically in certain Banach spaces in the following section.

ASSUMPTION 5.1. *Let* $b \geq \omega$. *Then* $\mathcal{S}(t)E \in D(A_b^{1/2})$ *and* $A_b^{1/2}\mathcal{S}(t)$ *is a strongly continuous function of* t *for* $-\infty < t < \infty$.

It follows from Lemma 4.2 that Assumption 5.1 is independent of the particular $b \geq \omega$ used. The following result establishes another translation invariance property.

LEMMA 5.2. *Let* $b \geq \omega$, A_b, $\mathcal{C}_b(\hat{t})$ *as in Lemma* 4.1, $\mathcal{S}_b(t)u = \int_0^t \mathcal{C}_b(s)u\,ds$. *Then if Assumption* 5.1 *is satisfied,* $\mathcal{S}_b(t)E \subseteq D(A_b^{1/2})$ *and* $A_b^{1/2}\mathcal{S}_b(t)$ *is a strongly continuous function of* t *in* $-\infty < t < \infty$.

Proof: Term-by-term integration of the series (4.7) shows that

$$\mathcal{S}_b(t) = \sum_{n=0}^{\infty} (-b^2)^n \mathcal{S}_n(t), \tag{5.2}$$

where

$$\mathcal{S}_0(t) = \mathcal{S}(t), \quad \mathcal{S}_n(t)u = (\mathcal{S} * \mathcal{S}_{n-1})(t)u = \int_0^t \mathcal{S}(t-s)\mathcal{S}_{n-1}(s)u\,ds$$

$$(u \in E,\ -\infty < t < \infty,\ n = 1,2,\dots)\ .$$

On the other hand, consider the series

$$\mathcal{D}_b(t) = \sum_{n=0}^{\infty} (-b^2)^n \mathcal{D}_n(t)\ , \tag{5.3}$$

where

$$\mathcal{D}_0(t) = A_b^{1/2}\mathcal{S}(t), \quad \mathcal{D}_n(t)u = (\mathcal{S} * \mathcal{D}_{n-1})(t)u = \int_0^t \mathcal{S}(t-s)\mathcal{D}_{n-1}(s)u\,ds$$

$$(u \in E,\ -\infty < t < \infty,\ n = 1,2,\dots)\ .$$

An argument very similar to that used in Lemma 4.1 shows that each term in (5.2) and (5.3) is strongly continuous and that both series are uniformly convergent on compact subsets of $-\infty < t < \infty$, thus the limit

$\mathfrak{D}_b(\hat{t})$ of (5.3) is a strongly continuous function. On the other hand, $A_b^{1/2}$ commutes with $\mathcal{S}(t)$ (this follows from the fact that $R(\lambda;A)$ commutes with $\mathcal{S}(t)$ and from the definition of $A_b^{1/2}$ in terms of $R(\lambda;A)$), hence $\mathfrak{D}_n(t) = A_b^{1/2}\mathcal{S}_n(t)$; since $A_b^{1/2}$ is closed, $\mathcal{S}_b(t)E \subseteq D(A_b^{1/2})$ and $A_b^{1/2}\mathcal{S}_b(t)u = \mathfrak{D}_b(t)$, which completes the argument.

THEOREM 5.3. <u>Let Assumption</u> 5.1 <u>hold. Then for each</u> $b \geq \omega$, $A_b^{1/2}$ <u>is the infinitesimal generator of the strongly continuous group</u>

$$\mathcal{U}_b(t) = \mathcal{C}_b(t) + A_b^{1/2}\mathcal{S}_b(t) \tag{5.4}$$

<u>If</u> ω <u>is the constant in</u> (4.1) <u>then there exists</u> $C > 0$ <u>such that</u>

$$\|\mathcal{U}_b(t)\| \leq C(1 + |t|)e^{(\omega+b)|t|} \quad (-\infty < t < \infty). \tag{5.5}$$

(3)

To prove (5.5) we renorm the space E in the same way as in Lemma 1.2, so that (1.6) holds in the new norm; then Theorem 4.1 implies that

$$\|\mathcal{C}_b(t)\|' \leq e^{(\omega+b)|t|} \quad (-\infty < t < \infty). \tag{5.6}$$

Define $\eta(t) = \exp(-(\omega + b)t)\|A_b^{1/2}\mathcal{S}_b(t)\|$ for $t \geq 0$ and use (II.1.8) for $\mathcal{C}_b$, $\mathcal{S}_b$ after multiplication by $\exp(-(\omega + b)(s + t))A_b^{1/2}$. The result is (1.7), which has been seen to imply that $\eta(t) \leq C(1 + t)$. This shows (5.5) for $t \geq 0$ and it follows for all t by symmetry.

The group equations are an immediate consequence of (II.1.7) and (II.1.8).

Finally, if $\mathrm{Re}\lambda > \omega$ we have

$$\int_0^\infty e^{-\lambda t}\mathcal{U}_b(t)u\,dt = (\lambda I + A_b^{1/2})R(\lambda^2;A_b)u$$

$$= R(\lambda;A_b^{1/2})u \quad (\mathrm{Re}\lambda > \omega + b), \tag{5.7}$$

thus it results from Theorem I.3.4 that $A_b^{1/2}$ is the infinitesimal generator of $\mathcal{U}_b(\hat{t})$ as claimed.

Assumption 5.1 and the previous results provide a phase space for the equation (5.1), where $E_1 = E$ and $E_0 = D(A_b^{1/2})$ endowed with its graph norm (note that, by virtue of Lemma 4.2 the graph norms of any two $A_b^{1/2}$ are equivalent for $b, b' \geq \omega$). This reduction of (5.1) to a first order Cauchy problem will be restated in a different form at the end of the section.

THEOREM 5.4. <u>The space</u> $\mathfrak{Z} = D(A_b^{1/2}) \times E$ <u>($D(A_b^{1/2})$ equipped with its graph norm) is a phase space for</u> (5.1) <u>in</u> $-\infty < t < \infty$ <u>which is contained in the state space</u> $\mathfrak{E}_m$ <u>of Theorem</u> 1.2: <u>precisely,</u>

$$D(A_b^{1/2}) \subseteq \tilde{D}_1 = D_1, \tag{5.8}$$

<u>the inclusion being bounded. The group</u> $\mathfrak{S}(\hat{t})$ <u>and its infinitesimal generator</u> $\mathfrak{A}$ <u>are given by</u> (1.10) <u>and</u> (1.11) <u>respectively and are the restrictions of the corresponding operators in Theorem</u> 2.1 <u>to</u> $D(A_b^{1/2}) \times E$ (resp. $D(A) \times D(A_b^{1/2})$). <u>The group</u> $\mathfrak{S}(\hat{t})$ <u>satisfies</u>

$$\|\mathfrak{S}(t)\|_{(\mathfrak{Z})} \leq C(1 + |t|)e^{\omega|t|} \qquad (-\infty < t < \infty) \tag{5.9}$$

<u>for an adequate constant</u> $C > 0$.

<u>Proof</u>: We show first (5.8). Note that $C'(s)u = AS(s)u$ for $u \in D(A)$; integrating in $0 \leq s \leq t$,

$$C(t)u = A\int_0^t S(s)u\,ds . \tag{5.10}$$

Since the left hand side is a bounded operator of u it follows that the integral of $S(\hat{s})$ maps E into $D(A)$ and (5.10) holds for all $u \in E$. Assume now that $u \in D(A_b^{1/2})$: using Assumption 5.1 we can write

$$\begin{aligned} C(t)u &= (A_b + b^2 I)\int_0^t S(s)u\,ds \\ &= \int_0^t A_b^{1/2} S(s) A_b^{1/2} u\,ds + b^2\int_0^t S(s)u\,ds , \end{aligned}$$

which obviously implies that $C(\hat{t})u$ is continuously differentiable with derivative

$$C'(t)u = A_b^{1/2} S(t) A_b^{1/2} u + b^2 S(t)u . \tag{5.11}$$

This establishes the inclusion relation (5.8). To show that the inclusion is bounded it is sufficient to show that

$$\|A_b^{1/2} S(t)\| \leq C(1 + |t|)e^{\omega|t|} \qquad (-\infty < t < \infty) \tag{5.12}$$

for an adequate constant C. This is proved just as the corresponding estimate for $A_b^{1/2} S_b(t)$; after renorming of the space we write (II.1.8) for $\mathcal{C}$ and $\mathcal{S}$, premultiply by $A_b^{1/2}$ and work with the inequality (1.7) for $\eta(t) = \exp(-\omega t)\|A_b^{1/2}\mathcal{S}(t)\|$. Once in possession of (5.12) we obtain from (5.11) that

$$\|\mathcal{C}'(t)u\| \leq C\|A_b^{1/2}u\|(1 + |t|)e^{\omega|t|} + C'\|u\|e^{\omega|t|} \tag{5.13}$$

(with the obvious modification if $\omega = 0$) which shows the identity operator from $D(A_b^{1/2})$ into D_1 to be bounded.

The fact that $\mathfrak{S}(t)$ is a group has already been proved in Theorem 1.2. To show that each $\mathfrak{S}(t) : \mathfrak{F} \to \mathfrak{F}$ is a bounded operator and $\mathfrak{S}(\hat{t})$ is a strongly continuous function in $\mathfrak{F}$ we only have to show that $\mathcal{C}(t) : D(A_b^{1/2}) \to D(A_b^{1/2})$, $\mathcal{S}(t) : E \to D(A_b^{1/2})$, $\mathcal{C}'(t) : D(A_b^{1/2}) \to E$ and $\mathcal{S}'(t) : E \to E$ are all bounded operators strongly continuous in t and obeying estimates of the form (5.12). This is obvious for $\mathcal{S}'(t) = \mathcal{C}(t)$ as an operator in E; in $D(A_b^{1/2})$ we use the fact that $\mathcal{C}(t)$ commutes with $A_b^{1/2}$. The treatment of $\mathcal{C}'(t)$ is based on (5.11) and (5.13) and the desired property of $\mathcal{S}(t)$ is just Assumption 1.5 complemented by inequality (5.12). It only remains to identify the infinitesimal generator of $\mathfrak{S}(t)$ in the space $\mathfrak{F}$. This is done as follows: if $\operatorname{Re}\lambda > \omega$ and $\mathfrak{u} = [u,v] \in \mathfrak{F}$ then we have

$$\int_0^\infty e^{-\lambda t}\mathfrak{S}(t)\mathfrak{u}\, dt = \begin{bmatrix} \lambda R(\lambda^2;A) & R(\lambda^2;A) \\ \lambda^2 R(\lambda^2;A) - I & \lambda R(\lambda^2;A) \end{bmatrix} \begin{bmatrix} u \\ v \end{bmatrix} = \mathfrak{R}(\lambda)\mathfrak{u} \tag{5.14}$$

(implicit in (5.14) is the easily verifiable fact that the Laplace transforms of $\mathcal{C}(t)u$ and $\mathcal{C}'(t)u$ exist in the norm of $D(A_b^{1/2})$). A routine computation shows that

$$\mathfrak{R}(\lambda) = R(\lambda;\mathfrak{A}), \tag{5.15}$$

thus $\mathfrak{A}$ is the infinitesimal generator of $\mathfrak{S}(\hat{t})$ by virtue of Theorem I.3.4.

Let $\mathfrak{D} = E \times E$ endowed with any of its product norms. The following result is essentially equivalent to Theorem 5.4; we still assume that the Cauchy problem for (5.1) is uniformly well posed in $(-\infty,\infty)$ and that Assumption 5.1 holds.

THEOREM 5.5. Let $b \geq \omega$. Consider the operator

$$\mathfrak{A}_b = \begin{bmatrix} 0 & A_b^{1/2} + ibI \\ A_b^{1/2} - ibI & 0 \end{bmatrix} \tag{5.16}$$

in $\mathfrak{D} = E \times E$ with domain $D(A_b) = D(A_b^{1/2}) \times D(A_b^{1/2})$. Then A is the infinitesimal generator of a strongly continuous group $\mathfrak{U}_b(t)$ such that (4)

$$\|\mathfrak{U}_b(t)\|_{(\mathfrak{D})} \leq C(1 + |t|)^2 e^{(\omega+2b)|t|} \qquad (-\infty < t < \infty). \tag{5.17}$$

There is a one-to-one correspondence between solutions $u(\hat{t})$ of (5.1) with $u'(0) \in D(A_b^{1/2})$ and solutions $\mathfrak{u}(\hat{t})$ of

$$\mathfrak{u}'(t) = \mathfrak{A}_b \mathfrak{u}(t) \tag{5.18}$$

given by

$$u(\hat{t}) \longleftrightarrow \begin{bmatrix} (A_b^{1/2} + ibI)u(\hat{t}) \\ u'(\hat{t}) \end{bmatrix} \tag{5.19}$$

Proof: Let

$$\mathfrak{B}_b = \begin{bmatrix} 0 & A_b^{1/2} \\ A_b^{1/2} & 0 \end{bmatrix}$$

with domain $D(\mathfrak{B}_b) = D(A_b^{1/2}) \times D(A_b^{1/2}) = D(\mathfrak{A}_b^{1/2})$, so that $\mathfrak{A}_b = \mathfrak{B}_b + b\mathfrak{P}$ with

$$\mathfrak{P} = \begin{bmatrix} 0 & iI \\ -iI & 0 \end{bmatrix}$$

Finally, let

$$\mathfrak{N}_b(t) = \begin{bmatrix} C_b(t) & A_b^{1/2}\mathfrak{S}_b(t) \\ A_b^{1/2}\mathfrak{S}_b(t) & C_b(t) \end{bmatrix}$$

In view of Theorem 5.3 we have

$$\|\mathfrak{R}_b(t)\|_{(\mathfrak{D})} \le C(1+|t|)e^{(\omega+b)|t|} \qquad (-\infty < t < \infty); \tag{5.20}$$

moreover, it follows again from (II.1.7), (II.1.8) and Assumption 1.5 that $\mathfrak{R}_b(\hat{t})$ is a strongly continuous group. If $\operatorname{Re}\lambda > \omega + b$ and $u = [u,v] \in \mathfrak{D}$ we have

$$\int_0^\infty e^{-\lambda t}\mathfrak{R}_b(t)u\,dt =$$

$$= \begin{bmatrix} \lambda R(\lambda^2;A_b) & A_b^{1/2}R(\lambda^2;A_b) \\ A_b^{1/2}R(\lambda^2;A_b) & \lambda R(\lambda^2;A_b) \end{bmatrix} \begin{bmatrix} u \\ v \end{bmatrix} = \mathfrak{R}(\lambda)u \tag{5.21}$$

and we check easily enough that $\mathfrak{R}(\lambda) = R(\lambda\,;\mathfrak{B}_b)$ so it follows again from Theorem I.3.4 that $\mathfrak{B}_b$ is the infinitesimal generator of $\mathfrak{R}_b(\hat{t})$. The fact that $\mathfrak{A}_b = \mathfrak{B}_b + b\mathfrak{P}$ generates a group $\mathfrak{U}_b(\hat{t})$ is (a particular case of) a classical result in perturbation theory (see HILLE-PHILLIPS [1957: 1, p. 389]) but the estimate (5.17) is slightly nonstandard, thus we start the proof from scratch. The main ingredient will be the following result, which is a sort of complement to the Hille-Yosida Theorem I.3.1. We state and prove it in a rather general version.

LEMMA 5.6. <u>Let</u> $S(\hat{t})$ <u>be a strongly continuous semigroup in the Banach space</u> E, A <u>its infinitesimal generator. Then</u> $S(\hat{t})$ <u>satisfies</u>

$$\|S(t)\| \le C_0(1+t^m)e^{\omega t} \qquad (t \ge 0) \tag{5.22}$$

<u>for an integer</u> $m \ge 1$ <u>if and only if</u>

$$\|R(\lambda;A)^n\| \le C_0\Big((\operatorname{Re}\lambda-\omega)^{-n} + n(n+m)\,\dots\,(n+m-1)(\operatorname{Re}\lambda-\omega)^{-(n+m)}\Big)$$

$$(\operatorname{Re}\lambda > \omega,\ n = 0,1,\dots\ ). \tag{5.23}$$

<u>If</u> $S(\hat{t})$ <u>is a strongly continuous group, then</u>

$$\|S(t)\| \le C_0(1+|t|^m)e^{\omega|t|} \qquad (-\infty < t < \infty) \tag{5.24}$$

<u>if and only if</u>

$$\|R(\lambda;A)^n\| \le C_0\Big((|\operatorname{Re}\lambda|-\omega)^{-n} + n(n+m)\,\dots\,(n+m-1)(|\operatorname{Re}\lambda|-\omega)^{-(n+m)}\Big)$$

$$(|\operatorname{Re}\lambda| > \omega,\ n = 0,1,\dots\ ). \tag{5.25}$$

Proof: Using (5.22) in formula (I.3.8) all inequalities (5.23) result instantly. The corresponding formula for $-A$ takes care of estimates (5.25) when $S(\hat{t})$ is a group. The opposite implications are a consequence of Lemma I.3.2. In fact, since $R(\lambda;A)^{(n)} = (-1)^n n!\, R(\lambda;A)^n$ we obtain from formula (I.3.14) (see also (I.3.15)) that

$$\|S(t)\| \le C_0 \lim \left(1 - \frac{\omega t}{n}\right)^{-(n+1)} +$$

$$C_0 \lim t^m n^{-m}(n+1) \ldots (n+m)\left(1 - \frac{\omega t}{n}\right)^{-(n+m+1)} = C_0(1+t^m)e^{\omega t}. \quad (t \ge 0)$$

The corresponding estimate in $t \le 0$ for groups follows in the same way. We observe in passing that (5.23) and (5.25) need only be assumed for $\lambda > \omega$ real.

End of proof of Theorem 5.5. The operator $\mathfrak{A}_b$ generates a group $\mathfrak{A}_b(\cdot)$ satisfying (5.20). Applying Lemma 5.6 we obtain

$$\|R(\lambda;\mathfrak{A}_b)^n\|_{(\mathfrak{D})} \le C((|\lambda| - \omega - b)^{-n} + n(|\lambda| - \omega - b)^{-(n+1)})$$

$$(|\lambda| > \omega + b,\ n = 0, 1, \ldots). \tag{5.26}$$

Consider the series

$$\sum R(\lambda;\mathfrak{A}_b)(R(\lambda;\mathfrak{A}_b)b\mathfrak{P})^{k_1} \ldots R(\lambda;\mathfrak{A}_b)(R(\lambda;\mathfrak{A}_b)b\mathfrak{P})^{k_n} \tag{5.27}$$

for $k_1, k_2, \ldots = 0,1,2,\ldots, |\lambda| > \omega + 2b$. It is easy to see that

$$R(\lambda;\mathfrak{A}_b)\mathfrak{P} = -\mathfrak{P}R(-\lambda;\mathfrak{A}_b),$$

thus each term of (5.27) can be written in the form

$$\pm (b\mathfrak{P})^k R(-\lambda;\mathfrak{A}_b)^p R(\lambda;\mathfrak{A}_b)^q, \tag{5.28}$$

where $k = \sum k_j$ and $p + q = k + n$. We make use of this relation and of (5.26) to deduce that the generic term in the series (5.27) is bounded in norm by an expression of the form

$$Cb^k\left(\frac{1}{(|\lambda| - \omega - b)^p} + \frac{p}{(|\lambda| - \omega - b)^{p+1}}\right)\left(\frac{1}{(|\lambda| - \omega - b)^q} + \frac{q}{(|\lambda| - \omega - b)^{q+1}}\right). \tag{5.29}$$

We can estimate (5.29) by

$$Cb^k \frac{1}{(|\lambda| - \omega - b)^{k+n}} + Cb^k \frac{k+n}{(|\lambda| - \omega - b)^{k+n+1}} +$$

$$+ Cb^k \frac{(k+n)(k+n+1)}{(|\lambda| - \omega - b)^{k+n+2}} . \tag{5.30}$$

We observe next that

$$\sum \frac{b^k}{(|\lambda| - \omega - b)^{k+n}} = \frac{1}{(|\lambda| - \omega - b)^n} \sum \left(\frac{b}{|\lambda| - \omega - b} \right)^k =$$

$$= \frac{1}{(|\lambda| - \omega - b)^n} \left\{ \sum \left(\frac{b}{|\lambda| - \omega - b} \right)^j \right\}^n = \frac{1}{(|\lambda| - \omega - 2b)^n} \tag{5.31}$$

for $|\lambda| > \omega + 2b$, where it must be remembered that $k = \sum k_i$ and that $k_1, \ldots, k_n$ assume independently all the values $0, 1, \ldots$ We differentiate next (5.31) repeatedly with respect to $|\lambda|$ obtaining the equalities

$$\sum \frac{(k+n)b^k}{(|\lambda| - \omega - b)^{k+n+1}} = \frac{n}{(|\lambda| - \omega - 2b)^{n+1}} , \tag{5.32}$$

$$\sum \frac{(k+n)(k+n+1)b^k}{(|\lambda| - \omega - b)^{k+n+2}} = \frac{n(n+1)}{(|\lambda| - \omega - 2b)^{n+2}} . \tag{5.33}$$

Accordingly, the series (5.27) is convergent in the norm of the space $(\mathfrak{D})$ and we can estimate the norm of the sum by (a constant times) the sum of (5.31), (5.32) and (5.33). We observe finally that (2.57) is nothing but

$$\{R(\lambda;\mathfrak{A}_b) \sum (R(\lambda;\mathfrak{A}_b)b\mathfrak{P})^j\}^n$$

and check (by direct application of the definition) that

$$R(\lambda;\mathfrak{A}_b) \sum (R(\lambda;\mathfrak{A}_b)b\mathfrak{P})^j = R(\lambda;\mathfrak{A}_b + b\mathfrak{P}) = R(\lambda;\mathfrak{A}_b). \tag{5.34}$$

The end result is the sequence of inequalities

$$\|R(\lambda;\mathfrak{A}_b)^n\|_{(\mathfrak{D})} \le C(|\lambda| - \omega - 2b)^{-n} + Cn(|\lambda| - \omega - 2b)^{-(n+1)}$$

$$+ Cn(n+1)(|\lambda| - \omega - 2b)^{-(n+2)} \quad (|\lambda| > \omega + 2b,\ n = 0, 1, \ldots) \tag{5.35}$$

Using (a slight extension of) Lemma 5.6 for $m = 2$ we deduce that $\|\mathfrak{U}_b(t)\| \le C(1 + |t| + |t|^2)e^{(\omega+2b)|t|} \le C(1 + |t|)^2 e^{(\omega+2b)|t|}$ in

$-\infty < t < \infty$, which completes the proof of (5.17).

We attend finally to the last statement in the proof of Theorem 5.5. Let $u(\hat{t})$ be a solution of the second order equation (5.1) with $u'(0) \in D(A_b^{1/2})$. Since $u(t) = \mathcal{C}(t)u(0) + \mathcal{S}(t)u'(0)$ with $u(0) \in D(A)$ it follows that $u'(t) \in D(A_b^{1/2})$ for all t and we check easily that $u(t)$, given by (5.19), is a solution of (5.18). That the correspondence is one-to-one is obvious.

§III.6 Second order equations in L^p spaces.

All the results in the previous section (in particular the principal value square root reduction in Theorem 5.5) depend on the validity of Assumption 5.1. As seen in this section, Assumption 5.1 is automatically verified in a class of spaces identified below.

A triple (X, Σ, μ) where X is an arbitrary set is a measure space if Σ is a σ-field of subsets of X and μ is a countably additive measure in Σ. All measure spaces considered here will be positive, i.e. $\mu(e) \geq 0$ for any $e \in \Sigma$ (for details on this definition and connected facts see DUNFORD-SCHWARTZ [1958:1, Chapter III]). The (complex) Lebesgue spaces $L^r(X, \Sigma, \mu) = L^r(X)$ $(1 \leq r \leq \infty)$ are defined in the customary manner. All results below are based on the following theorem on Hilbert transform of vector valued functions.

THEOREM 6.1. *Let (X, Σ, μ) be a measure space and let*

$$E = L^r(X, \Sigma, \mu), \tag{6.1}$$

where $1 < r < \infty$. For $\varepsilon > 0$, define

$$(H_\varepsilon f)(t) = \int_{|s-t| \geq \varepsilon} \frac{f(s)}{s-t}\, ds \tag{6.2}$$

for $f \in L^p(-\infty, \infty; E)$, $1 < p < \infty$. Then there exists a constant C depending on p but not on f or ε such that

$$\|H_\varepsilon f\|_p \leq C\|f\|_p, \tag{6.3}$$

where $\|\cdot\|_p$ indicates the norm of $L^p(-\infty, \infty; E)$. Moreover,

$$Hf = \lim_{\varepsilon \to 0} H_\varepsilon f \tag{6.4}$$

exists (in the norm of $L^p(-\infty, \infty; E)$) for every $f \in L^p(-\infty, \infty; E)$ so that,

<u>in view of</u> (6.3), H <u>is a bounded operator in</u> $L^p(-\infty,\infty;E)$ <u>with norm</u> $\leq C$.

For a proof see DUNFORD-SCHWARTZ [1963:1, p. 1173]. Actually Theorem 6.1 will be only needed for an arbitrarily fixed value of p; for $p = r$, the result is nothing but an integrated form of M. Riesz's well known result on L^p boundedness of the ordinary Hilbert transform (DUNFORD-SCHWARTZ [1963:1, p. 1059]).

We shall need in the sequel a corrollary of Theorem 6.1. For $\omega \geq 0$ denote by $\mathcal{L}_\omega(E)$ the space of all strongly measurable E - valued functions $f(\hat{t})$ defined in $-\infty < t < \infty$ such that

$$\|f\|_{\infty,\omega} = \operatorname*{ess.\,sup}_{-\infty<t<\infty} e^{-\omega|t|}\|f(t)\|, \tag{6.5}$$

endowed with the norm $\|\cdot\|_{\infty,\omega}$. On the other hand, for $1 \leq p < \infty$, $\mathfrak{m}_p(E)$ denotes the space of all E-valued strongly measurable functions $f(\hat{t})$ defined in $-\infty < t < \infty$ such that

$$\|f\|_{p,\tau} = \left(\int_{-\tau}^{\tau} \|f(s)\|^p \, ds\right)^{1/p} < \infty \tag{6.6}$$

for all $\tau > 0$; $\mathfrak{m}_p(E)$ is a locally convex space endowed with the family of semi-norms $\{\|\cdot\|_{p,\tau}; \tau > 0\}$.

THEOREM 6.2. <u>Let</u> $c > \omega$, $1 < p < \infty$, E <u>a Banach space satisfying</u> (6.1) <u>for some</u> r, $1 < r < \infty$. <u>Then</u>

$$(H^c f)(t) = \lim_{\varepsilon\to 0} (H^c_\varepsilon f)(t) = \lim_{\varepsilon\to 0} \int_{|s|\geq\varepsilon} \frac{e^{-c|t-s|}}{t-s} f(s)\,ds \tag{6.7}$$

<u>exists in the topology of</u> $\mathfrak{m}_p(E)$ <u>and defines a continuous operator</u> $H^c : \mathcal{L}_\omega \to \mathfrak{m}_p(E)$.

<u>Proof</u>: For any $a > 0$ let χ_a be the characteristic function of the interval $|t| < a$. Let $\tau \geq 1$, $0 < \varepsilon \leq 1$, $t \leq |\tau|$. We have

$$(H^c_\varepsilon f)(t) = \int_{|s-t|\geq\varepsilon} \frac{1}{t-s} \chi_{\tau+1}(s) f(s)\,ds$$

$$+ \int_{|s-t|\geq\varepsilon} \frac{1}{t-s}(e^{-c|t-s|}-1)\chi_{\tau+1}(s)f(s)\,ds + \int_{|s|\geq\tau+1} \frac{e^{-c|t-s|}}{t-s} f(s)\,ds$$

$$= (H^{c,1}_\varepsilon f)(t) + (H^{c,2}_\varepsilon f)(t) + (H^{c,3}_\varepsilon f)(t)\,. \tag{6.8}$$

It is obvious that $H_\varepsilon^{c,3} = H^{c,3}$ is independent of ε and that

$$\|H_\varepsilon^{c,3}f(t)\| \le C\|f\|_{\infty,\omega} \qquad (|t| \le \tau) . \tag{6.9}$$

On the other hand, $H_\varepsilon^{c,2}f(t)$ tends uniformly on compacts to the convolution of $(e^{-\varepsilon|\hat{t}|}-1)/\hat{t}$ by $\chi_{\tau+1}(\hat{t})f(\hat{t})$ thus an estimate of the type of (6.9) holds for $H^{c,2}f(t) = \lim H_\varepsilon^{c,2}f(t)$. Finally it results from Theorem 6.1 that $H^{c,1}f = \lim H_\varepsilon^{c,1}f$ exists in the norm of $L^p(-\infty,\infty;E)$ and

$$\|H^{c,1}f\|_{p,\tau} \le \|H^{c,1}f\|_p \le C'\|\chi_{\tau+1}f\|_p \le C\|f\|_{\infty,\omega} , \tag{6.10}$$

where again C depends only on τ. This ends the proof of Theorem 6.2.

THEOREM 6.3. _Let_ E _be a Banach space obeying_ (6.1) _for some_ r, $1 < r < \infty$, _and let_ $A \in \mathfrak{C}^2(\omega)$. _Then Assumption_ 5.1 _holds: equivalently,_

$$\mathfrak{U}_b(t) = \mathfrak{C}_b(t) + A_b^{1/2}\mathfrak{S}_b(t) \tag{6.11}$$

is a strongly continuous group (with infinitesimal generator $A_b^{1/2}$) _satisfying_ (5.5)) _for_ $b \ge \omega$.

Proof: Let $b \ge \omega$. If $u \in D(A)$ we have

$$A_b^{1/2}u = \frac{i}{\pi}\int_0^\infty \lambda^{-1/2}R(\lambda;A_b)(-A_b)u\,d\lambda . \tag{6.12}$$

We have already proved that A_b generates a cosine function $\mathfrak{C}_b(\hat{t})$ satisfying

$$\|\mathfrak{C}_b(t)\| \le C_0 e^{(\omega+b)|t|} \qquad (-\infty < t < \infty) . \tag{6.13}$$

Let $c > \omega + b$. We divide the interval of integration in (6.12) at $\lambda = c^2$ and use in the second integral the integrated-by-parts version of (II.2.11) for $n = 1$:

$$R(\lambda^2;A_b)u = \int_0^\infty e^{-\lambda s}\mathfrak{S}_b(s)u\,ds , \tag{6.14}$$

valid for $\operatorname{Re}\lambda > \omega + b$. The result is

$$A_b^{1/2}\mathfrak{S}_b(t)u = \frac{i}{\pi}\mathfrak{S}_b(t)\int_0^{c^2} \lambda^{-1/2}R(\lambda;A_b)(-A_b)u\,d\lambda +$$

$$+\frac{i}{\pi}\int_{c^2}^{\infty}\lambda^{-1/2}\left(\int_0^{\infty}e^{-\lambda^{1/2}}(-\mathcal{S}_b(s)\mathcal{S}_b(t)A_bu)ds\right)d\lambda = I_1 + I_2 \ . \quad (6.15)$$

The integrand in the second integral is now transformed as follows. Write (II.1.7) for $\mathcal{C}_b$, $\mathcal{S}_b$, $u \in D(A)$ for both t and $-t$. Using the fact that $C'(t)u = \mathcal{S}(t)Au$ and subtracting the equalities so obtained we prove that

$$-\mathcal{S}_b(s)\mathcal{S}_b(t)A_bu = \frac{1}{2}(\mathcal{C}_b(s-t) - \mathcal{C}_b(s+t))u \ . \quad (6.16)$$

Since $u \in D(A)$, $(\mathcal{C}_b(s-t) - \mathcal{C}_b(s+t))u = O(|s|)$ as $s \to 0$ so that there exists a constant C such that

$$\|(\mathcal{C}_b(s-t) - \mathcal{C}_b(s+t))u\| \le Cse^{(\omega+b)s} \qquad (s \ge 0)$$

(observe that C will depend on u and t). Accordingly, we deduce that

$$\int_0^{\infty}e^{-\lambda^{1/2}s}\|(\mathcal{C}_b(s-t) - \mathcal{C}_b(s+t))u\|\,ds \le C\int_0^{\infty}se^{-(\lambda^{1/2}-\omega-b)s}ds =$$
$$= C(\lambda^{1/2} - \omega - b)^{-2} \ .$$

Since $\lambda^{-1/2}(\lambda^{1/2} - \omega - b)^{-2}$ is summable in (c^2,∞) it follows from the Lebesgue dominated convergence theorem that

$$I_2 = \lim_{\varepsilon\to 0}\frac{i}{2\pi}\int_{c^2}^{\infty}\left(\lambda^{-1/2}\int_{\varepsilon}^{\infty}e^{-\lambda^{1/2}s}(\mathcal{C}_b(s-t) - \mathcal{C}_b(s+t))u\,ds\right)d\lambda \ . \quad (6.17)$$

By (a vector-valued version of) Tonelli's theorem the order of integration in the integral on the right hand side of (6.17) can be reversed. Once this is done, the attractive formula

$$A_b^{1/2}\mathcal{S}_b(t)u = \frac{i}{\pi}\mathcal{S}_b(t)\int_0^{c^2}\lambda^{-1/2}R(\lambda;A_b)(-A_b)u\,d\lambda$$
$$+ \lim_{\varepsilon\to 0}\frac{i}{\pi}\int_{\varepsilon}^{\infty}\frac{e^{-cs}}{s}(\mathcal{C}_b(s-t) - \mathcal{C}_b(s+t))u\,ds$$
$$= \frac{i}{\pi}\mathcal{S}_b(t)\int_0^{c^2}\lambda^{-1/2}(I - \lambda R(\lambda;A_b))u\,d\lambda$$
$$+ \lim_{\varepsilon\to 0}\frac{i}{\pi}\int_{|s|\ge\varepsilon}\frac{e^{-c|s|}}{s}\mathcal{C}_b(t-s)u\,ds$$
$$= \frac{i}{\pi}\mathcal{S}_b(t)\int_0^{c^2}\lambda^{-1/2}(I - \lambda R(\lambda;A_b))u\,d\lambda +$$

$$+ \lim_{\varepsilon \to 0} \frac{i}{\pi} \int_{|s-t| \geq \varepsilon} \frac{e^{-c|t-s|}}{t-s} \mathcal{C}_b(s)u \, ds \tag{6.18}$$

results. So far, so good: but (6.18) (specifically, the integral with respect to s) still only makes sense for $u \in D(A)$ and thus does not provide enough information on $A_b^{1/2}\mathcal{S}_b(t)$. It is in this connection, of course, that Theorem 6.2 proves useful. Assume that u is an arbitrary element of E and let $\{u_n\}$ be a sequence in $D(A)$ with $u_n \to u$. Then $\mathcal{C}_b(s)u_n \to \mathcal{C}_b(s)u$ uniformly on compact subsets of $(-\infty,\infty)$. In view of (6.13), $\mathcal{C}_b(\hat{s})u_n \to \mathcal{C}_b(\hat{s})u$ in $\mathcal{L}_{\omega+b}(E)$ and it then results from (6.18) and Theorem 6.2 that, for $1 < p < \infty$, $A_b^{1/2}\mathcal{S}_b(\hat{t})u_n$ converges in $\mathfrak{M}_p(E)$, that is, converges in $L^p((-\tau,\tau);E)$ for every $\tau > 0$. Passing if necessary to a subsequence we can then insure that $A_b^{1/2}\mathcal{S}_b(t)u_n \to g(t)$ for t outside of a set d of full measure in $(-\infty,\infty)$ depending of course on u) where $g(t)$ is a E-valued function (belonging to $L^p((-\tau,\tau);E)$ for every $\tau > 0$). Since $A_b^{1/2}$ is a closed operator, it follows that if $t \in d$, $\mathcal{S}_b(t)u \in D(A_b^{1/2})$ and

$$A_b^{1/2}\mathcal{S}_b(t)u = g(t) \qquad (t \in d). \tag{6.19}$$

Let

$$e = e(u) = \{t; u \in D(A_b^{1/2}\mathcal{S}_b(t))\},$$

so that e has a complement of null measure in $(-\infty,\infty)$. We have

$$e = -e, \qquad e + e \subseteq e. \tag{6.20}$$

The first equality follows from the fact that $A_b^{1/2}\mathcal{S}_b(t)$ is an even function; to show the second we apply $A_b^{1/2}$ to both sides of (II.1.8) obtaining

$$A_b^{1/2}\mathcal{S}(s+t)u = \mathcal{C}(s)A_b^{1/2}\mathcal{S}_b(t)u + \mathcal{C}(t)A_b^{1/2}\mathcal{S}_b(s)u.$$

The two equalities (6.20) combined imply that $e - e \subseteq e$, thus by a classical result in measure theory, e contains an interval $(-a,a)$, $a > 0$. Applying then repeatedly the second relation (6.20) we deduce that $e = (-\infty,\infty)$, so that $\mathcal{S}_b(t)E \subseteq D(A_b^{1/2})$ for all t; by the closed graph theorem $A_b^{1/2}\mathcal{S}_b(t)$ is a bounded operator.

Consider now the group $\mathcal{U}_b(\hat{t})$. It follows from (6.19) that

$\mathcal{U}_b(\hat{t})u$ is strongly measurable for each u so that the argument yielding Theorem I.2.1. (iii) shows that $\mathcal{U}_b(\hat{t})$ is strongly continuous. This ends the proof of Theorem 6.3.

Theorem 6.3 assures that all the results in §III.5 - including the construction of the state space in Theorem 5.4 and the principal value square root reduction in Theorem 5.5-hold. More can be proved; for instance, as the following result shows, the phase spaces constructed in Theorems 1.2 and 5.4 are actually the same (except for a readjustment of the norm).

THEOREM 6.4. Let E be as in Theorem 6.3. Then if $b \geq \omega$,

$$\tilde{D}_1 = D(A_b^{1/2}) , \tag{6.21}$$

where, as in Theorem 1.2, $\tilde{D}_1$ is the set of all u such that $C(\hat{t})u$ is continuously differentiable in $-\infty < t < \infty$. Moreover the norm (1.8) of D_1 is equivalent to the graph norm of $D(A_b^{1/2})$.

Proof: We have already shown in Theorem 5.4 that $D(A_b^{1/2}) \subseteq \tilde{D}_1$ with bounded inclusion, thus as soon as we prove that

$$\tilde{D}_1 \subseteq D(A_b^{1/2}) \tag{6.22}$$

the closed graph theorem will take care of equivalence of the norms. To show (6.22) we start with formula (6.18) in its second version, always for $u \in D(A)$ and check that the principal value integral can be differentiated with respect to t under the integral sign. This is obvious enough if we note that (always for $c > \omega + b$) we have

$$\lim_{\varepsilon \to 0} \int_{|s| \geq \varepsilon} \frac{e^{-c|s|}}{s} C_b(t-s)u \, ds$$

$$= \int_{-\infty}^{\infty} e^{-c|s|} \left\{ \frac{1}{s} (C_b(t-s) - C_b(t))u \right\} ds ,$$

which, as an ordinary integral, can be differentiated under the integral sign. The derivative equals

$$\int_{-\infty}^{\infty} e^{-c|s|} \left\{ \frac{1}{s} (C_b'(t-s) - C_b'(t))u \right\} ds$$

$$= \lim_{\varepsilon \to 0} \int_{|s| \geq \varepsilon} \frac{e^{-c|s|}}{s} C_b'(t-s)u \, ds .$$

Hence

$$A_b^{1/2}\mathcal{C}_b(t)u = \frac{i}{\pi}\,\mathcal{C}_b(t)\int_0^{c^2} \lambda^{-1/2}(I - \lambda R(\lambda;A_b))u\,d\lambda$$

$$+ \lim_{\varepsilon\to 0}\frac{i}{\pi}\int_{|s|\geq\varepsilon} \frac{e^{-c|s|}}{s}\,\mathcal{C}_b'(t-s)u\,ds\,. \tag{6.23}$$

Let $u \in \tilde{D}_1$, so that $\mathcal{C}_b(\hat{t})u$ is continuously differentiable in $(-\infty,\infty)$. As a particular case of Theorem 1.3 we know that $D(\mathfrak{A})$ is dense in $\mathfrak{E}_m$, thus $D(A)$ is dense in $\tilde{D}_1$ and we may select a sequence $\{u_n\} \subseteq D(A)$ with $u_n \to u$ in the norm of $\tilde{D}_1$: in particular $\mathcal{C}_b'(t)u_n \to \mathcal{C}_b'(t)u$ uniformly on compact subsets of $(-\infty,\infty)$ thus, by virtue of Theorem 6.2, $\mathcal{C}_b'(\hat{t})u_n \to \mathcal{C}_b'(\hat{t})u$ in $\mathcal{L}_c(E)$. Applying again Theorem 6.2 we deduce that there exists $h(\hat{t}) \in \mathfrak{m}_p(E)$ such that $A_b^{1/2}\mathcal{C}_b(\hat{t})u_n \to h(\hat{t})$ in $\mathfrak{m}_p(E)$, and and argument similar to the one used in Theorem 6.3 shows that $\mathcal{C}_b(t)u \in D(A_b^{1/2})$ for t in a set e of full measure in $(-\infty,\infty)$.

We make use now of (II.1.7). Originally, this equality was proved for $u \in D(A)$, but we can extend it easily to $u \in D_1$ using the fact that $D(A)$ is dense in D_1 in the topology of D_1 (see the proof of Theorem 1.3). Setting $s = -t$ we get

$$u = C(t)C(t)u + \mathcal{S}(t)C'(t)u\,.$$

On account of the facts that $u \in D_1$ and that $\mathcal{S}(t)E \subseteq D(A_b^{1/2})$ we deduce that $u \in D(A_b^{1/2})$. This completes the proof of (6.21) and thus of Theorem 6.4.

We reformulate below some of the results hitherto obtained. A strongly continuous cosine function $C(t)$ is said to possess a <u>group decomposition</u> if there exists a strongly continuous group $\mathcal{U}(\hat{t})$ such that

$$C(t) = \frac{1}{2}\,(\mathcal{U}(t) + \mathcal{U}(-t)) \qquad (-\infty < t < \infty). \tag{6.24}$$

Let B be the infinitesimal generator of the group $\mathcal{U}(\hat{t})$. If $u \in D(B^2)$ then $\mathcal{U}(\hat{t})u$ is twice continuously differentiable with $(\mathcal{U}(t)u)'' = B^2\mathcal{U}(t)u$, thus u belongs to the domain $D(A)$ of the infinitesimal generator A of the cosine function $C(t)$ and $Au = B^2u$; in other words,

$$B^2 \subseteq A \tag{6.25}$$

Let λ be so large that $\lambda, -\lambda \in \rho(B)$ and $\lambda^2 \in \rho(A)$. Then $\lambda^2 \in \rho(B^2)$ (with $R(\lambda^2;B^2) = -R(\lambda;B)R(-\lambda;B)$. The equality $\lambda^2 I - B^2 \subseteq \lambda^2 I - A$ (that follows from (6.25)) and the existence of inverses are then easily seen to imply that

$$B^2 = A. \tag{6.26}$$

Thus, to exhibit an example of a cosine function not having any group decomposition it is sufficient to take

$$C(t) = \cos(tA) , \tag{6.27}$$

where A is a bounded operator having no square root whatsoever. An operator like that can be found in HALMOS [1967:1, p. 115]:

EXAMPLE 6.5. Let $E = \ell^2$ be the Hilbert space of all sequences $\xi = (\xi_0, \xi_1, \dots)$ such that $\|\xi\| = (\sum |\xi_j|^2)^{1/2} < \infty$. Consider the <u>unilateral shift</u>

$$A(\xi_0, \xi_1, \dots) = (\xi_1, \xi_2, \dots) .$$

Then there exists no (bounded or unbounded) operator B such that $B^2 = A$. In fact, assume such a B exists. Then, since $D(A) = E$ we must have as well $D(B) = E$; moreover, B must be onto since A is. Let $N(A)$ (resp. $N(B)$) be the nullspace of A (resp. of B). Since $\dim N(A) = 1$ and $N(B) \subseteq N(A)$ we must have

$$\dim N(B) \leq 1 .$$

Assume $\dim N(B) = 1$. Then there exists $\xi \in N(B)$, $\xi \neq 0$. Take η with $B\eta = \xi$. If $\eta = \mu\xi$ then $\xi = 0$, a contradiction. Hence η and ξ are linearly independent. But both η and ξ belong to $N(A)$, which belies the fact that $\dim N(A) = 1$. We deduce then that $N(B) = \{0\}$, so that B is one-to-one. But this is impossible since A is not one-to one.

In view of the failure of the group decomposition (6.24) in general and of the results in the last three sections one may surmise that a formula like (6.24) may be available only for a convenient translate of $C_b(\hat{t})$. This is of course the case under the assumptions in vigor here.

THEOREM 6.6. *Let* E *be a Banach space satisfying the assumptions of Theorem* 6.3 *and let* $C(\hat{t})$ *be a strongly continuous cosine function in* E *such that*

$$\|C(t)\| \le C_0 e^{\omega|t|} \qquad (-\infty < t < \infty) . \tag{6.28}$$

Finally, let $b \ge \omega$. *Then the strongly continuous cosine function* $C_b(t)$ *generated by* $A - b^2 I$ (*which satisfies*[5]

$$\|C_b(t)\| \le C_0 e^{(\omega+b)|t|} \qquad (-\infty < t < \infty) \tag{6.29}$$

possesses a group decomposition

$$C_b(t) = \frac{1}{2}(\mathcal{U}_b(t) + \mathcal{U}_b(-t)) \qquad (-\infty < t < \infty) \tag{6.30}$$

where the strongly continuous group $\mathcal{U}_b(\hat{t})$ *satisfies*[6]

$$\|\mathcal{U}_b(t)\| \le C(1 + |t|)e^{(\omega+b)|t|} \qquad (-\infty < t < \infty) \tag{6.31}$$

for some constant $C > 0$.

Theorem 6.6 is of course true in an arbitrary Banach space if Assumption 5.1 holds: the only thing to be proved is the estimate (6.31).

The most significant particular case of Theorem (6.6) is undoubtely $\omega = 0$, where $C(\hat{t})$ is uniformly bounded:

$$\|C(t)\| \le C_0 \qquad (-\infty < t < \infty). \tag{6.32}$$

Here $C(\hat{t})$ itself admits the group decomposition (6.24) with

$$\|\mathcal{U}(t)\| \le C(1 + |t|) \qquad (-\infty < t < \infty) . \tag{6.33}$$

It is natural to ask whether the growth of $\mathcal{U}(\hat{t})$ at $|t| = \infty$ can be better matched to that of $C(\hat{t})$, that is whether the factor $(1 + |t|)$ can be eliminated in (6.33). The answer is in the affirmative when E is a Hilbert space; this will be shown in Chapter 5 by completely different methods. In a general Banach space, however, the answer is not known.

§III.7 Analyticity properties of $\mathcal{U}_b(\hat{t})$.

There is more than meets the eye in the semigroup $\mathcal{U}_b(\hat{t})$: as we shall see in this section, each $\mathcal{U}_b(t)$ is the boundary value of a semigroup analytic in a half plane. The proof of this result will require a short digression into the theory of analytic semigroups.

Let $S(\hat{t})$ be an arbitrary strongly continuous semigroup. We say that $S(\hat{t})$ is analytic if there exists φ, $0 < \varphi < \pi/2$ such that

(a) $S(\hat{t})$ admits an analytic extension $S(\hat{\zeta})$ to the sector $\Sigma_+(\varphi) = \{\zeta;\ |\arg \zeta| \leq \varphi,\ \zeta \neq 0\}$.

(b) For every $u \in E$,

$$\lim_{|\zeta| \to 0,\ \zeta \in \Sigma_+(\varphi)} S(\zeta)u = u. \tag{7.1}$$

An argument very similar to that leading to (I.1.9) shows the existence of constants C, β such that

$$\|S(\zeta)\| \leq C\, e^{\beta|\zeta|}. \tag{7.2}$$

If A is the infinitesimal generator of $S(\hat{t})$ we write $A \in \mathcal{Q}(\varphi)$ to indicate that $S(\hat{t})$ satisfies (a) and (b).

The following result provides a fairly complete characterization of generators of analytic semigroups.

THEOREM 7.1 (i) *Let* $A \in \mathcal{Q}(\varphi)$. *Then there exists a real constant* α *such that* $R(\lambda;A)$ *exists in the sector*

$$\{\lambda;\ |\arg(\lambda - \alpha)| < \varphi + \pi/2,\ \lambda \neq \alpha\} \tag{7.3}$$

and for every φ', $0 < \varphi' < \varphi$ *there exists a constant* $C = C_{\varphi'}$ *such that*

$$\|R(\lambda;A)\| \leq C_{\varphi'}/|\lambda - \alpha| \quad (|\arg(\lambda - \alpha)| \leq \varphi',\ \lambda \neq \alpha). \tag{7.4}$$

(ii) *Assume that* $R(\lambda;A)$ *exists in a sector*

$$\{\lambda;\ |\arg(\lambda - \alpha)| \leq \varphi + \pi/2,\ \lambda \neq \alpha\} \tag{7.5}$$

(*where* α *is real and* $0 < \varphi < \pi/2$) *and satisfies*

$$\|R(\lambda;A)\| \leq C/|\lambda - \alpha| \tag{7.6}$$

there for some constant $C > 0$. Then $A \in \mathcal{a}(\varphi')$ if $0 < \varphi' < \varphi$.

Proof: Assume that $A \in \mathcal{a}(\varphi')$. Consider formula I.3.8 $(n = 1)$ for $R(\lambda;A)$. In view of the fact that $S(\hat{\zeta})$ is analytic in $\Sigma_+(\varphi)$ and of (7.2), if $\mathrm{Re}\lambda$ is large enough, the path of integration can be deformed to either of the rays $\Gamma_+(\varphi)$ or $\Gamma_-(\varphi)$, respectively the upper and lower boundaries of the sector $\Sigma_+(\varphi)$. The first choice yields the formula

$$R(\lambda;A)u = \int_{\Gamma_+} e^{-\lambda\zeta} S(\zeta)u \, d\zeta = e^{i\varphi} \int_{\Gamma_+} e^{-\lambda t e^{i\varphi}} S(te^{i\varphi})u \, dt. \tag{7.7}$$

Since

$$\|e^{-\lambda t e^{i\varphi}} S(te^{i\varphi})\| \le Ce^{-(\mathrm{Re}\,\lambda e^{i\varphi} - \beta)t}, \tag{7.8}$$

the integral (7.7) provides an analytic extension of $R(\lambda;A)$ to the tilted half-plane $\mathrm{Re}\,\lambda e^{i\varphi} > \beta$ and the estimate

$$\|R(\lambda;A)\| \le C/|\mathrm{Re}\,\lambda e^{i\varphi} - \beta| \quad (\mathrm{Re}\,\lambda e^{i\varphi} > \beta). \tag{7.9}$$

Using Γ_- instead of Γ_+ the resolvent can be extended to $\mathrm{Re}\,\lambda e^{-i\varphi} > \beta$ and the twin estimate

$$\|R(\lambda;A)\| \le C/|\mathrm{Re}\,\lambda e^{-i\varphi} - \beta| \quad (\mathrm{Re}\,\lambda e^{-i\varphi} > \beta) \tag{7.10}$$

results. Obviously, the union of the two slanted half planes is a sector of the type of (7.3) and the estimate, (7.4) can be easily assembled on the basis of (7.8) and (7.10).

The proof of (ii) is less trivial. Assume that $R(\lambda;A)$ exists in a sector of the form (7.5) and that (7.6) holds. Since $A \in \mathcal{a}(\psi)$ if and only if $A + cI \in \mathcal{a}(\psi)$ for any c we may assume that $\alpha < 0$. Let $0 < \varphi_1, \varphi_2 < \varphi$ and let $\Gamma(\varphi_1, \varphi_2)$ be the contour consisting of the two rays $\arg \lambda = \varphi_1 + \pi/2$, $\mathrm{Im}\lambda \ge 0$ and $\arg \lambda = -(\varphi_2 + \pi/2)$, $\mathrm{Im}\lambda \le 0$, the entire contour oriented clockwise with respect to the right half-plane. Define

$$S(\zeta) = \frac{1}{2\pi i} \int_{\Gamma(\varphi_1, \varphi_2)} e^{\lambda\zeta} R(\lambda;A) \, d\lambda \,. \tag{7.11}$$

Obviously the integral does not depend on the particular choice of φ_1, φ_2 and defines an infinitely differentiable E-valued function in $\zeta \ge 0$. Let now φ' such that $0 < \varphi' < \varphi$ and φ'' with

$\varphi' < \varphi'' < \varphi$. If we use in the integral the contour $\Gamma(\varphi''-\varphi',\varphi'')$ it is clear that $S(\hat{\zeta})$ can be extended as an (E)-valued analytic function to the half sector $0 \le \arg \zeta \le \varphi'$, $\zeta \ne 0$. Symetrically, the contour $\Gamma(\varphi'',\varphi''-\varphi')$ provides a (E)-valued analytic extension to $-\varphi' \le \arg \zeta \le 0$, $\zeta \ne 0$ thus taking care of (a) in the definition of analytic semigroups. It remains to show (b). To do this we take first $0 \le \arg \zeta \le \varphi'$ and use $\Gamma(\varphi''-\varphi',\varphi'')$ as the contour of integration in (7.11); making later the change of variable $z = \lambda\zeta$ we obtain

$$S(\zeta) = \frac{1}{2\pi i} \int_{\zeta\Gamma(\varphi''-\varphi',\varphi'')} \frac{e^{z}}{\zeta} R(\frac{z}{\zeta};A)\, dz . \tag{7.12}$$

Obviously, we can deform further the contour of integration to, say, $\Gamma(\varphi'',\varphi''-\varphi')$ and then modify this contour in such a way that z, as it travels upwards, does not hit the origin and passes to it right. Call Γ' the contour thus obtained. To estimate the integrand we use (7.6):

$$\|R(\frac{z}{\zeta};A)\| \le C\left|\frac{z}{\zeta} - \alpha\right|^{-1} \le C'\left|\frac{\zeta}{z}\right| \quad (z \in \Gamma') . \tag{7.13}$$

We obtain then from (7.12) that

$$\|S(\zeta)\| \le C' \int_{\Gamma'} \left|\frac{e^{z}}{z}\right| |dz| .$$

A symmetric argument takes care of the range $-\varphi' \le \mathrm{Re}\, \zeta \le 0$. Putting together the two estimations we obtain

$$\|S(\varphi)\| \le C_{\varphi'} \qquad (\zeta \in \Sigma_+(\varphi')) . \tag{7.14}$$

In view of this inequality, it is enough to show (b) for u in a dense subset of E, say, for $u \in D(A)$; as customary, we treat separately the cases $0 \le \mathrm{Re}\, \zeta \le \varphi'$ and $-\varphi' \le \mathrm{Re}\, \zeta \le 0$. In the first we use the contour $\Gamma(\varphi''-\varphi,\varphi'')$ modified near the origin in the same way as Γ'; we call Γ'' the resulting contour. If $u \in D(A)$ we have $R(\lambda;A)u = \lambda^{-1}R(\lambda;A)Au + \lambda^{-1}u$ so that

$$S(\zeta)u = \frac{1}{2\pi i}\int_{\Gamma'} e^{\lambda\zeta}R(\lambda;A)u\, d\lambda$$

$$= \frac{1}{2\pi i}\int_{\Gamma'} \frac{e^{\lambda\zeta}}{\lambda} R(\lambda;A)Au\, d\lambda + \left(\frac{1}{2\pi i}\int_{\Gamma'} \frac{e^{\lambda\zeta}}{\lambda}\, d\lambda\right)u .$$

In view of (7.6) the first integral is a continuous function of ζ in $0 \leq \arg \zeta \leq \varphi'$; for $\zeta = 0$ the integral vanishes, as can be easily seen translating to the right the domain of integration. The second integral evaluates to 1. This shows that (b) holds for $u \in D(A)$ and, in view of (7.4), ends the proof of Theorem 7.1.

Through the rest of this section A is the infinitesimal generator of a strongly continuous cosine function $C(\hat{t})$ satisfying (4.1). We require Assumption 5.1 to hold so that, if $b \geq \omega$,

$$\mathcal{U}_b(t) = \mathcal{C}_b(t) + A_b^{1/2}\mathcal{S}_b(t) \tag{7.15}$$

is a strongly continuous group.

THEOREM 7.2. <u>The group</u> $\mathcal{U}_b(\hat{t})$ <u>admits an extension</u> $\mathcal{U}_b(\hat{\zeta})$ <u>to the upper half plane</u> $\operatorname{Im} \zeta \geq 0$ <u>such that</u> (i) $\mathcal{U}_b(\hat{\zeta})$ <u>is strongly continuous in</u> $\operatorname{Im} \zeta \geq 0$ (ii) $\mathcal{U}_b(\hat{\zeta})$ <u>is an</u> (E)-<u>valued analytic function in</u> $\operatorname{Re} \zeta > 0$. (iii) <u>There exist a constant</u> $C > 0$ <u>such that</u>

$$\|\mathcal{U}_b(\zeta)\| \leq C(1 + |\operatorname{Re} \zeta|)(1 + |\operatorname{Im} \zeta|)\, e^{(\omega+b)|\operatorname{Re} \zeta|^{(7)}} \quad (\operatorname{Im} \zeta > 0). \tag{7.16}$$

<u>Proof</u>: If $b \geq \omega$ then the operator $-A_b$ satisfies (2.1) (see (4.11)). It follows then from Theorem 3.2 that there exists $\psi > 0$ such that $R(\lambda; -(-A_b)^{1/2})$ exists in $|\arg \lambda| \leq \psi + \pi/2$ and satisfies

$$\|R(\lambda; -(-A_b)^{1/2})\| \leq C/|\lambda| \quad (|\arg \lambda| \leq \psi + \pi/2). \tag{7.17}$$

Thus Theorem 7.1 applies to show that $-(-A_b)^{1/2}$ is the infinitesimal generator of a strongly continuous semigroup $\mathcal{V}_b(\zeta)$ analytic in $|\arg \zeta| < \psi$, $\zeta \neq 0$. Since Theorem 3.2 does not provide direct information on the growth of $\mathcal{V}_b(t)$ for t real we shall obtain this information by means of an explicit representation for $\mathcal{V}_b(t)$. Define

$$\mathfrak{J}_b(t) = \lim_{a \to \infty} \int_0^a \sin t\, \lambda^{1/2} R(\lambda; -A_b)\, d\lambda \quad (t > 0). \tag{7.18}$$

To show that the limit exists we perform an integration by parts, obtaining the equivalent expression

$$\mathfrak{J}_b(t) = \int_0^\infty h(t,\lambda) R(\lambda; -A_b)^2\, d\lambda , \tag{7.19}$$

where

$$h(t,\lambda) = \frac{1}{\pi}\int_0^{\lambda} \sin t\sigma^{1/2} d\sigma = \frac{2}{\pi t^2}(\sin t\lambda^{1/2} - t\lambda^{1/2}\cos t\lambda^{1/2}),$$

and we check that if $\delta > 0$ there exists a constant C_δ such that

$$|h(t,\lambda)| \leq C_\delta \lambda^{1/2} \quad (\lambda \geq 0), \tag{7.20}$$

$$|h(t,\lambda)| \leq C_\delta t\lambda^{3/2} \quad (0 \leq \lambda \leq 1), \tag{7.21}$$

for $t \geq \delta$. Dividing the domain of integration in (7.19) at $\lambda = 1$ and using (7.20) and (7.21) we obtain the estimate

$$\|\mathcal{J}_b(t)\| \leq C + C't \quad (t \geq \delta > 0), \tag{7.22}$$

where C, C' may in principle depend on δ. (if A_b^{-1} exists we need only use (7.20) in $\lambda \geq 0$ so that $C' = 0$ in (7.22): this is certainly the case if $b > \omega$). We prove easily that $\mathcal{J}_b(\hat{t})$ is continuous in $t > 0$ in the norm of (E). On the other hand, if $u \in D(A)$ we have

$$\begin{aligned}\mathcal{J}_b(t)u - u &= \lim_{a\to\infty} \frac{1}{\pi}\int_0^a \sin t\lambda^{1/2}(R(\lambda;-A_b)u - \frac{1}{\lambda}u)\, d\lambda \\ &= \frac{1}{\pi}\int_0^1 \sin t\lambda^{1/2}(R(\lambda;-A_b)u - \frac{1}{\lambda}u)\, d\lambda \\ &= \frac{1}{\pi}\int_1^\infty \frac{1}{\lambda}\sin t\lambda^{1/2}R(\lambda;-A_b)(-A_b u)\, d\lambda ,\end{aligned} \tag{7.23}$$

and this expression tends to zero when $t \to 0$. Although $\mathcal{J}_b(\hat{t})$ is actually strongly continuous at the origin (i.e. $\mathcal{J}_b(t)u \to u$ as $t \to 0$ for every $u \in E$) we need not prove this directly, as a far stronger result will be obtained below. If we take $u \in D(A)$ then the previous steps show that the following computation is justified:

$$\begin{aligned}\int_0^\infty e^{-\mu t}\mathcal{J}_b(t)u\, dt &= \frac{1}{\mu}u + \frac{1}{\pi}\int_0^\infty \int_0^\infty e^{-\mu t}\sin t\lambda^{1/2}dt\,(R(\lambda;-A_b)u - \frac{1}{\lambda}u)\, d\lambda \\ &= \frac{1}{\pi}\int_0^\infty \frac{\lambda^{1/2}}{\lambda+\mu^2}R(\lambda;-A_b)u\, d\lambda = R(\mu;-(-A_b)^{1/2})u.\end{aligned} \tag{7.24}$$

Since the same Laplace transform relation must of needs hold for $\mathcal{U}_b(\hat{t})u$, where $\mathcal{U}_b(\hat{t})$ is the analytic semigroup generated by $-(-A_b)^{1/2}$, we have $\mathcal{J}_b(t)u = \mathcal{U}_b(t)u$ (by uniqueness of Laplace transforms) for $u \in D(A)$, a fortiori for $u \in E$.

We extend $\mathcal{U}_b(\hat{t})$ to the upper half plane $\tau \geq 0$ by means of the formula

$$\mathcal{U}_b(\zeta) = \mathcal{U}_b(t + i\tau) = \mathcal{U}_b(t)\mathcal{V}_b(\tau) \qquad (-\infty < t < \infty,\ \tau \geq 0). \tag{7.25}$$

Since $\mathcal{U}_b(\hat{t})$ is strongly continuous in $-\infty < t < \infty$ and $\mathcal{V}_b(\hat{\tau})$ is strongly continuous in $\tau \geq 0$, $\mathcal{U}_b(\zeta)$ is strongly continuous in the upper half plane $\operatorname{Im} \zeta \geq 0$. We observe next that it follows from formula (7.11), the fact that A is closed, the resolvent equation for $R(\lambda;A)$ and Cauchy's formula that if $S(\hat{t})$ is an analytic semigroup and A its infinitesimal generator then $S(t)E \subseteq D(A^m)$ for each $m = 1,2,\ldots$ and $t > 0$ and

$$A^m S(t) = \frac{1}{2\pi i}\int_{\Gamma(\varphi_1,\varphi_2)} \lambda^m e^{\lambda t} R(\lambda;A)\, d\lambda \tag{7.26}$$

(so that, incidentally, $A^m S(t)$ is (E)-continuous in $t > 0$; obviously, (7.26) can be extended to t complex as well). We apply this observation (for $m = 1$) to the analytic semigroup generated by $-(-A_b)^{1/2}$. Let u be an arbitrary element of E. Then, if $\tau > 0$, $\mathcal{V}_b(\tau)u \in D(-(-A_b)^{1/2}) = D(i(-A_b)^{1/2}) = D(A_b^{1/2})$. Since $A_b^{1/2}$ is the infinitesimal generator of $\mathcal{U}_b(\hat{t})$, $t \to \mathcal{U}_b(t)\mathcal{V}_b(\tau)u$ is differentiable with respect to t with

$$\begin{aligned} D_t\mathcal{U}_b(t)\mathcal{V}_b(\tau)u &= \mathcal{U}_b(t)A_b^{1/2}\mathcal{V}_b(\tau)u = -i\,\mathcal{U}_b(t)(-(-A_b)^{1/2})\mathcal{V}_b(\tau)u \\ &= -i\,D_\tau\mathcal{U}_b(t)\mathcal{V}_b(\tau)u \qquad (-\infty < t < \infty,\ \tau > 0). \end{aligned} \tag{7.27}$$

But this equality is nothing but the Cauchy-Riemann equation for the (continuously differentiable) function $\mathcal{U}_b(\hat{\zeta})u$, thus $\mathcal{U}_b(\hat{\zeta})u$ is analytic in $\operatorname{Im} \zeta > 0$ as a E-valued function; it follows then from a well known result (HILLE-PHILLIPS [1957:1, p. 93]) that $\mathcal{U}_b(\hat{\zeta})$ itself is analytic in $\operatorname{Im} \zeta > 0$ as a (E)-valued function.

To complete the proof of Theorem 7.2 we only have to observe that the fact that $\mathcal{U}_b(\hat{\zeta})$ is an extension of $\mathcal{U}_b(\hat{t})$ is obvious since $\mathcal{V}_b(0) = I$. The estimate (7.16) follows from (7.22) for $\mathcal{V}_b(\hat{\tau})$ and from (5.5) for $\mathcal{U}_b(\hat{t})$. Note, incidentally, that the factor $1 + |\operatorname{Im} \zeta|$ can be altogether eliminated if A_b has a bounded inverse: this is certainly the case if $b > \omega$. Finally, it is of interest to observe that $\mathcal{U}_b(t)$ and $\mathcal{V}_b(\tau)$ commute for arbitrary values of t and τ; the easiest way to see this is by means of formula (7.18) for $\mathcal{V}_b(\hat{t})$, using the fact that $\mathcal{U}_b(t)$ commutes with $R(\lambda;-A_b)$. As a consequence, we obtain

$$\mathcal{U}_b(z+\zeta) = \mathcal{U}_b(z)\mathcal{U}_b(\zeta) \qquad (\mathrm{Im}\ z,\ \mathrm{Im}\ \zeta \geq 0). \tag{7.28}$$

For the sake of completeness, we restate Theorem 7.2 using the group decomposition language of Theorem 6.6.

THEOREM 7.3. <u>Let the assumptions of Theorem</u> 6.6 <u>be satisfied for the strongly continuous cosine function</u> $C(\hat{t})$. <u>Then, if</u> $b \geq \omega$, $C_b(\hat{t})$ <u>admits a group decomposition of the form</u> (6.30), <u>where the strongly continuous group</u> $\mathcal{U}_b(\hat{t})$ <u>admits a strongly continuous extension to</u> $\mathrm{Im}\ \zeta \geq 0$ <u>which is analytic in</u> $\mathrm{Im}\ \zeta > 0$ <u>and satisfies</u> (7.28) <u>and the estimate</u> (7.16). <u>If</u> $\omega = 0$, $C(\hat{t})$ <u>itself admits the group decomposition</u> (6.24), <u>where</u> $\mathcal{U}(\hat{\zeta})$ <u>is analytic in</u> $\mathrm{Im}\ \zeta \geq 0$, <u>strongly continuous in</u> $\mathrm{Im}\ \zeta > 0$ <u>and satisfies</u>

$$\|\mathcal{U}_b(\zeta)\| \leq C(1+|\mathrm{Re}\ \zeta|)(1+|\mathrm{Im}\ \zeta|) \qquad (\mathrm{Im}\ \zeta > 0). \tag{7.29}$$

§III.8 <u>Other square root reductions.</u>

The theory developed in the last six sections lacks subtlety in one important respect, namely in the rigid choice of the principal value square root of A_b in the reduction of (1.1) to a first order system. Although this has been seen to work satisfactorily in L^r spaces, the following example illustrates the drawbacks of the choice in the general case.

EXAMPLE 8.1. Let $E = C_{2\pi}(-\infty,\infty)$ be the space of all complex valued 2π-periodic continuous functions defined in $(-\infty,\infty)$ endowed with its usual supremum norm. Consider the strongly continuous cosine function

$$C(t)u(x) = \frac{1}{2}(u(x+t)+u(x-t)) \qquad (-\infty < t < \infty). \tag{8.1}$$

Its infinitesimal generator is $A = D^2 = d^2/ds^2$ with domain $D(A)$ consisting of all $u \in E$ such that u is twice continuously differentiable. Since $\mathcal{C}(t)$ is uniformly bounded, we can apply the theory in §5 with $b \geq \omega = 0$. An explicit formula for $A^{1/2}$ can be obtained setting $t = 0$ in (6.23) for any $c > 0$; a more attractive expression results if we take limits as $c \to 0$ (see Exercise 11 in this Chapter). In the particular case under consideration the formula reads

$$A^{1/2}u(x) = \lim_{a \to \infty} \frac{i}{\pi} \int_0^a \frac{u'(x+s) - u'(x-s)}{s}\, ds. \tag{8.2}$$

Similar manipulations with (6.18) (see Exercise 10) yield the formula

$$A^{1/2}\mathcal{S}(t)u(x) = A^{1/2}\frac{1}{2}\int_{x-t}^{x+t} u(\xi)\, d\xi = \frac{it}{\pi}\int_0^\infty \frac{u(x+s) + u(x-s)}{t^2 - s^2}\, ds, \tag{8.3}$$

where the integral is understood as the limit when $\varepsilon \to 0$ of the integral over $|s - t| \geq \varepsilon, s \geq 0$. However, the singular integral operator on the right hand side of (8.3) is not bounded in $C_{2\pi}(-\infty,\infty)$ so that Assumption 5.1 does not hold and the principal value square root reduction fails utterly. Nevertheless, a far more elementary reduction works. Let $B = d/dx$ with domain $D(B)$ consisting of all continuously differentiable functions in E. Then $B^2 = A$ and we obtain a group decomposition of the form (6.24) with

$$\mathfrak{U}(t)u(x) = u(x+t). \tag{8.4}$$

In other words, the second order equation $u_{tt} = u_{xx}$ can be reduced to the system $(u_1)_t = (u_2)_x, (u_2)_t = (u_1)_x$ as we learn in elementary courses on partial differential equations.

In view of Example 8.1 one may wonder whether a different choice of square root may make the square root reduction work for general cosine functions in general Banach spaces. The following example shows that the answer to this question is in general negative.

EXAMPLE 8.2. Let $E = C^o_{2\pi}(-\infty,\infty)$ be the subspace of $C_{2\pi}(-\infty,\infty)$ consisting of <u>odd</u> functions, $\hat{\mathcal{C}}(t)$ the cosine function defined by (8.1) (note that $\mathcal{C}(t)$ maps $C^o_{2\pi}(-\infty,\infty)$ into itself). However, the translation group (8.4) does not provide a group decomposition like (6.24), since $\mathfrak{U}(t)$ does not map $C^o_{2\pi}(-\infty,\infty)$ into itself. As we shall see below, there exists no strongly continuous group in $C^{o}_{2\pi}(-\infty,\infty)$ making (6.24) happen. In fact, much more than this is true: if $b \in \mathbb{C}$, $A_b = A - b^2 I$ and $\mathcal{C}_b(t)$ is the cosine function generated by A_b there exists no strongly continuous group $\mathcal{U}_b(\hat{t})$ such that

$$\mathcal{C}_b(t) = \frac{1}{2}(\mathcal{U}_b(t) + \mathcal{U}_b(-t)). \tag{8.5}$$

In view of the remarks in §III.6, this statement is equivalent to: if $b \in \mathbb{C}$, A_b has no square root $A_b^{1/2}$ whatsoever that generates a strongly continuous group.

Our claim is proved as follows. Assume the representation (8.5) holds for some strongly continuous group $\mathcal{U}_b(\hat{t})$. For $n = 1,2,\ldots$ let E_n be the one-dimensional subspace of $E = C^{o}_{2\pi}(-\infty,\infty)$ generated by $\sin n\hat{x}$. Then $\mathcal{C}(t)$ leaves each E_n invariant since $\mathcal{C}(t)\sin nx = \cos nt \sin nx$. We deduce from this and the perturbation series representation (4.3) that $\mathcal{C}_b$ leaves each E_n invariant as well; precisely,

$$\mathcal{C}_b(t)\sin nx = \cos (n^2 + b^2)^{1/2} t \sin nx \tag{8.6}$$

for $b \geq 0$ (formula (8.6) can also be obtained noting that $u(t,x) = \mathcal{C}_b(t)\sin nx$ is a solution of $u_{tt}(t,x) = u_{xx}(t,x) - b^2 u(t,x)$ with initial conditions $u(0,x) = \sin nx$, $u_t(0,x) = 0$). We show next that $\mathcal{U}_b$ also leaves each subspace E_n invariant. To prove this we observe first that $\mathcal{P}_n$, the projection of E into E_n is given by

$$\mathcal{P}_n u = \frac{1}{\pi}\int_{-\pi}^{\pi} \cos ns\, \mathcal{C}(s)u\, ds. \tag{8.7}$$

Since $\mathcal{U}_b(t)$ commutes with $\mathcal{C}_b(s)$ it also commutes with $\mathcal{C}(s)$ (again because of the perturbation series (4.3)) and a fortiori commutes with $\mathcal{P}_n$; this shows that $\mathcal{U}_b(t)$ maps E_n into itself. Making use of the fact that every semigroup in a finite dimensional space must be an exponential, equation (8.7) reduces to

$$\cos (n^2 + b^2)^{1/2} t = \frac{1}{2}\left(e^{i\tau_n(b)t} + e^{-i\tau_n(b)t}\right) \quad (-\infty < t < \infty) \tag{8.8}$$

where $\tau_n(b)$ is a complex number; obviously the only choice is $\tau_n(b) = \sigma_n(n^2 + b^2)^{1/2}$, where, for each $n \geq 1$ σ_n equals 1 or -1 and does not depend on t. We conclude that $\mathcal{U}_b(t)$ must be a multiplier operator of the form

$$\mathcal{U}_b(t)u \sim \sum_{n=1}^{\infty} e^{i\tau_n(b)t} a_n \sin nx, \tag{8.9}$$

where $u \sim \sum a_n \sin nx$. Notice now that, for fixed b, $\tau_n(b) - n = (n^2 + b^2)^{1/2} - n = O(n^{-1})$ as $n \to \infty$ so that if $\mathcal{U}_0(t)$ is the operator in (8.9) there exist coefficients $c_n(t)$, $|c_n(t)| \leq C|t|n^{-1}$ $(n = 1,2,\dots)$ such that

$$\|(\mathcal{U}_b(t) - \mathcal{U}_0(t))u(\hat{x})\| \leq \|\sum c_n(t)a_n \sin n\hat{x}\|$$

$$\leq (\sum |c_n(t)|^2)^{1/2}(\sum |a_n|^2)^{1/2} \leq C|t|\,\|u\|. \tag{8.10}$$

It follows that $\mathcal{U}_b(\hat{t})$ is a strongly continuous group in E if and only if $\mathcal{U}_0(\hat{t})$ is; each $\mathcal{U}_0(t)$ is given by

$$\mathcal{U}_0(t)u(x) \sim \sum_{n=1}^{\infty} e^{in\sigma_n t} a_n \sin nx. \tag{8.11}$$

Consider now the following operator in $C_{2\pi}$: if $u \sim \sum c_n e^{inx}$ then

$$Qu(x) \sim \sum_{n=-\infty}^{\infty} \sigma_n c_n e^{inx} \tag{8.12}$$

where we have set

$$\sigma_0 = 0,\ \sigma_{-n} = -\sigma_n \quad (n = 1,2,\dots).$$

The domain $D(Q)$ of Q consists of all $u \in C_{2\pi}$ such that $Qu \in C_{2\pi}$ and includes the subspace $P_{2\pi}$ of all trigonometric polynomials in $C_{2\pi}$. Since $Q \sin nx = -i\sigma_n \cos nx$ we have

$$\sin nx - Q \sin nx = i\sigma_n(\cos nx - i\sigma_n \sin nx)$$

$$= i\sigma_n e^{-in\sigma_n x}. \tag{8.13}$$

On the other hand, if $\mathcal{U}(t)$ is the translation operator (8.4) we have

$$\frac{2}{\pi}\int_0^{\pi} \mathcal{U}(t)\mathcal{U}_0(t)\sin nx\, dt$$

$$= \frac{2}{\pi}\int_0^{\pi} e^{in\sigma_n t} \sin n(x + t)\, dt = i\sigma_n e^{-in\sigma_n x}. \tag{8.14}$$

Hence, if $u(x)$ is a trigonometric polynomial in $C_{2\pi}^{o}$ we have, combining (8.13) and (8.14),

$$Qu(x) = u(x) - \frac{2}{\pi}\int_0^{\pi} \mathcal{U}(t)\mathcal{U}_0(t)u(x)\, dt, \tag{8.15}$$

so that Q is a bounded operator in the space $P^{o}_{2\pi}$ of all odd trigonometric polynomials. We define now a functional Φ taking the value of $Qu(\hat{x})$ at 0:

$$\Phi u = Qu(0). \tag{8.16}$$

Again, the domain of Φ includes $P_{2\pi}$. Moreover, in view of (8.12), $\Phi u = 0$ for any even trigonometric polynomial. Hence

$$\Phi u(\hat{x}) = \Phi\left\{\frac{1}{2}(u(\hat{x}) - u(-\hat{x}))\right\}, \tag{8.17}$$

so that, by continuity of Q in $P^{o}_{2\pi}$, Φ is a continuous linear functional in $P_{2\pi}$. We can then extend Φ to $C_{2\pi}$ preserving continuity and by the theorem of F. Riesz there exists a periodic finite Borel measure μ such that

$$\mu \sim \sum \sigma_n e^{inx}. \tag{8.18}$$

Since $\sigma_n^2 = \sigma_n$ for all n we have

$$\mu * \mu = \mu,$$

and a theorem of HELSON [1953:1] applies to show that, after eventual modification of a finite number of its Fourier coefficients μ must reduce to a finite linear combination of point masses, that is,

$$\mu = \sum_{j=1}^{m} a_j \delta(\hat{x} - x_j), \tag{8.19}$$

(in a somewhat ad hoc notation). Hence

$$\tilde{\sigma}_n = \frac{1}{2\pi} \sum_{j=1}^{m} a_j e^{-inx_j}, \tag{8.20}$$

where $\{\tilde{\sigma}_n\}$ is the sequence $\{\sigma_n\}$, possibly after modification of a finite number of its elements. Obviously, a contradiction will be obtained if we show that any sequence of the form (8.20) <u>must either be periodic or take an infinite number of values</u>, since the sequence $\{\tilde{\sigma}_n\}$ takes only a finite number of values (obvious) and cannot be periodic (since $\sigma_{-n} = -\sigma_n$, to make $\{\sigma_n\}$ periodic would imply an infinite number of changes). The underlined statement on (8.20) is proved by induction on m and we omit the details.

The absence of any group decomposition in Example 8.2 must of course be traced in part to the lack of desirable properties of the space $E = C^0_{2\pi}$ (for instance, E is not reflexive). One may raise the question of whether some group decomposition can be guaranteed under assumptions on E (say, reflexivity or uniform convexity) that fall short of the very stringent ones in §III.5. As far as reflexivity goes the answer is in the negative, as we shall see below.

EXAMPLE 8.3. There exists a reflexive space F and and a closed, densely defined operator A in F such that (i) A generates a strongly continuous cosine function (ii) For any complex number b the operator $A_b = A - b^2 I$ possesses no square root whatsoever generating a strongly continuous group. This is easily seen by modifying the space in Example 8.2 in a way suggested by F. LAUFBURSCHEN [private communication]. Let $E = C^0_{2\pi}$, and $m \geq 1$ a fixed integer. Consider (a) the subspace E^0_m generated by the functions $\{\sin nx;\ m \leq n\}$ (b) for $k \geq m$ the (finite dimensional) subspace $E^0_{m,k}$ of E generated by the functions $\{\sin nx;\ m \leq n \leq k\}$. Let $C(\hat{t})$ be the cosine function (8.1) and denote by $C_{b,m}(\hat{t})$ the restriction of $C_b(\hat{t})$ to E^0_m (obviously, E^0_m is invariant by $C_b(t)$). Assume there exists a strongly continuous group $\mathcal{U}_{b,m}(t)$ in E^0_m such that

$$C_{b,m}(t) = \frac{1}{2}(\mathcal{U}_{b,m}(t) + \mathcal{U}_{b,m}(t)) \qquad (-\infty < t < \infty). \tag{8.21}$$

The argument used for $C_b(t)$ in E applies equally well here: the decomposition (8.21) for $C_{b,m}(\hat{t})$ implies a similar decomposition for $C_m(\hat{t})$, the restriction of $C(\hat{t})$ to E_m with a strongly continuous group $\mathcal{U}_{0,m}(\hat{t})$ admitting the representation

$$\mathcal{U}_{0,m}(t)u(x) \sim \sum_{n=m}^{\infty} e^{in\sigma_n t} a_n \sin nx \tag{8.22}$$

for each t, where each σ_n equals 1 or -1 and is independent of t. The role of the operator Q is now played by

$$Q_m u(x) = \sum_{|n| \geq m} \sigma_n c_n e^{inx} \tag{8.23}$$

in the subspace E_m generated by all the functions $\{e^{inx};\ m \leq n\}$. The operator Q_m admits the representation

$$Q_m u(x) = u(x) - \frac{2}{\pi}\int_0^{\pi} \mathcal{U}(t)\mathcal{U}_{0,m}(t)u(x)\,dt \tag{8.24}$$

for every trigonometric polynomial in E^0_m, thus it is bounded there and we can define a continuous linear functional Φ_m in E_m^0 by (8.16) (with $Q = Q_m$) and (8.17). After extending Φ to $E_1^0 = C_{2\pi}$ by the Hahn-Banach theorem we deduce the existence of the measure μ in (8.18) as before; since $\sigma_n = \pm 1$ for $|n| \geq m$ and the argument applied to μ is unmindful of modification of a finite number of Fourier coefficients, the conclusion is identical: Q_m cannot be a bounded operator in E_m^0, thus $\mathcal{U}_{0,m}(\hat{t})$ cannot be a strongly continuous group in E_m^0. This implies that <u>there exists some</u> t_0 in $0 \leq t_0 \leq \pi$ <u>such that</u> $\mathcal{U}_{0,m}(t_0)$ <u>is unbounded</u>; for if $\mathcal{U}_{0,m}(t)$ is bounded for all t and $u \in E_m^0$ is arbitrary, let $\{u_n\}$ be a sequence of trigonometric polynomials in E_m^0 such that $u_n \to u$. Then $\mathcal{U}_{0,m}(\hat{t})u_n$ is continuous for all n and $\mathcal{U}_{0,m}(\hat{t})u_n \to \mathcal{U}_{0,m}(\hat{t})u$ pointwise, thus $\mathcal{U}_{0,m}(\hat{t})u$ is strongly measurable for all u and applying Theorem I.2.1 we obtain that $\mathcal{U}_{0,m}(\hat{t})$ is a strongly continuous group, a contradiction. We note that t_0 will depend, of course, on the choice of the sequence $\{\sigma_n;\ n \geq m\}$ in (8.22).

For arbitrary t and $k \geq m$ define

$$\gamma_{m,k}(t) = \inf \|\mathcal{U}_{b,m}(t)\|_{m,k} ,$$

where the norm is that of $\mathcal{U}_{b,m}(t)$ as an operator in $E_{m,k}^0$ and the infimum is taken over <u>all possible choices</u> of $\{\sigma_n;\ m \leq n \leq k\}$ with $\sigma_n = \pm 1$. We claim that, for any m, there exists a sequence $\{t_k\}$ (say, in $|t| \leq 1$) such that

$$\gamma_{m,k}(t_k) \to \infty \quad \text{as} \quad k \to \infty . \tag{8.25}$$

In fact, assume this is not true for a certain m. Then for each k there exists a finite sequence $\sigma_{k,m}, \sigma_{k,m+1}, \ldots, \sigma_{k,k}$ with $\sigma_{k,n} = \pm 1$ and such that

$$\|\mathcal{U}_{0,m}(t)\|_{m,k} \leq C \qquad (|t| \leq 1,\ k \geq m) , \tag{8.26}$$

where C is a constant independent of k and the sequence $\{\sigma_{k,n}; m \leq n \leq k\}$ has been used in the definition (8.22) of $\mathcal{U}_{0,m}(t)$. Note that the (possibly unbounded) operator $\mathcal{U}_{0,m}(t)$ also depends on the σ_n with $n \geq k$ but these are irrelevant when obtaining $\gamma_{m,k}(t)$. It is now easy to see that there exists a subsequence $k(1) < k(2) < \ldots$ such that $\sigma_{k(j),m} = \sigma_m$ $(j \geq 1)$, $\sigma_{k(j),m+1} = \sigma_{m+1}$

$(j \geq 2), \ldots$ i.e. such that each sequence $\underset{\sim}{\sigma}_{k(j),\ell}$ $(\ell \geq m)$ becomes stationary after a while. Accordingly, if $\tilde{u}_{0,m}(t)$ is the group corresponding to the sequence $\sigma_m, \sigma_{m+1}, \ldots$ in (8.22) we obtain

$$\|\tilde{u}_{0,m}(t)\|_{m,k} \leq C \qquad (|t| \leq 1,\ k \geq m) \tag{8.27}$$

(note that $\|\cdot\|_{m,k}$ is an increasing function of k). But $\bigcup_{k \geq m} E_{m,k}$ is dense in E_m, thus (8.27) implies that the operators $u_{0,m}(t)$ are bounded in E_m for $|t| \leq 1$, thus for all t, which has already seen to be impossible. This completes the proof of (8.25).

With this relation in our hands we construct the space F and the operator A as follows. Using (8.25) we can produce a sequence $m(1) < m(2) < \ldots$ of positive integers and a sequence $\{t_j\}$, $|t_j| \leq 1$ such that

$$\gamma_{m(j),m(j+1)}(t_j) \to \infty \quad \text{as} \quad j \to \infty \ . \tag{8.28}$$

The space F is the Hilbert sum of all the finite dimensional spaces $F_j = E_{m(j),m(j+1)}$, thus is a reflexive Banach space; since the L^2 norm in $|x| \leq \pi$ is dominated by $\sqrt{2\pi}$ times the supremum norm in the same interval, F is a subspace of the space $L^{o}_{2\pi}(-\infty,\infty)$ of all odd, 2π-periodic functions which are square integrable in $|x| \leq \pi$ endowed with the corresponding norm. The cosine function $C(\hat{t})$ is defined by (8.1); its infinitesimal generator A is d^2/dx^2 with maximal domain. Assume that for some b we can find a group decomposition (8.21) for $C_b(\hat{t})$. An argument very similar to that pertaining to the space $C_{2\pi}$ shows that $u_b(\hat{t})$ must obey (8.9) and that in case $\|u_b(t)\|_F$ is bounded in, say, $|t| \leq 1$ the same must be true of $\|u_0(t)\|$, $u_0(t)$ defined by (8.9) with $b = 0$. But if $\{t_j\}$ is the sequence in (8.28) then

$$\|u_0(t_k)\|_{(F)} = \sup_{j \geq 1} \|u_0(t_k)\|_{m(j),m(j+1)}$$

$$\geq \sup_{j \geq 1} \gamma_{m(j),m(j+1)}(t_k) \geq \gamma_{m(k),m(k+1)}(t_k), \tag{8.29}$$

thus $\|u_0(t_k)\| \to \infty$ by virtue of (8.28), and a contradiction is obtained.

The following result shows that the problem of finding a group decomposition of a cosine function becomes radically simpler if one is allowed to enlarge the underlying space.

THEOREM 8.4. *Let* $C(\hat{t})$ *be a strongly continuous cosine function in the Banach space* E *satisfying*

$$\|C(t)\|_{(E)} \le Ce^{\omega|t|} \qquad (-\infty < t < \infty). \tag{8.30}$$

Then there exists a Banach space F *such that* $E \subseteq F$,

$$\|u\|_E \le \|u\|_F \le C\|u\|_E \qquad (u \in E) , \tag{8.31}$$

(C *the constant in* (8.30)) *and a strongly continuous group* $U(\hat{t})$ *such that*

$$\|U(t)\|_{(F)} \le e^{\omega|t|} \qquad (-\infty < t < \infty) \tag{8.32}$$

and

$$C(t)u = \frac{1}{2}(U(t) + U(-t))u \qquad (-\infty < t < \infty,\ u \in E). \tag{8.33}$$

Moreover, F *is minimal in the sense that finite linear combinations of elements of the form* $U(t)u$ $(-\infty < t < \infty,\ u \in E)$ *are dense in* F.

Proof: Let G be the space of all continuous E-valued functions $u(\hat{t})$ defined in $-\infty < t < \infty$ such that $\|u(s)\| = O(e^{\omega|s|})$ as $|s| \to \infty$ (ω the constant in (8.30)). The space G becomes a Banach space equipped with the norm

$$\|u(\hat{s})\| = \sup_{-\infty < s < \infty} e^{-\omega|s|}\|u(s)\|. \tag{8.34}$$

The space E can be identified with a subspace $\hat{E}$ of F through the map

$$u \longleftrightarrow u(\hat{s}) = C(\hat{s})u , \tag{8.35}$$

and we obviously have

$$\|u\|_E \le \|C(\hat{s})u\|_G \le C\|u\|_E . \tag{8.36}$$

If $u(\hat{s}) \in \hat{E}$ we have

$$C(t)u(\hat{s}) = C(t)C(\hat{s})u = \frac{1}{2}(C(t+\hat{s})u + C(t-\hat{s})u) =$$

$$= \frac{1}{2}\,(u(\hat{s} + t) + u(\hat{s} - t)) , \tag{8.37}$$

thus $C(t)$ acts on elements of $\hat{E}$ in the same way as the "translation cosine functions" appearing in Examples (8.1), (8.2) and (8.3). The space F is defined as the closed subspace of G generated by all elements of the form $u(\hat{s} + t)$ with $-\infty < t < \infty$. The semigroup $U(\hat{t})$ in F is defined by

$$U(t)u(\hat{s}) = u(\hat{s} + t) . \tag{8.38}$$

Identifying E with $\hat{E}$ (as we may), (8.33) follows; obviously (8.36) is nothing but (8.30). To check (8.31) we note that if $u(\hat{s})$ belongs to G,

$$\|U(t)u(\hat{s})\| = \sup e^{-\omega|s|}\|u(s + t)\|$$

$$\leq e^{-\omega|t|}\sup e^{-\omega|\hat{s}+t|}\|u(\hat{s} + t)\| = e^{-\omega|t|}\|u(\hat{s})\|.$$

§III.9 <u>Miscellaneous comments</u>.

Phase spaces for abstract higher order equations more general than (1.1) were introduced by WEISS [1967:1] (see Chapter 8 for additional information). The maximal phase space $\mathfrak{E}_m$ in §III.1 was introduced by KISYŃSKI [1970:1]. The theory of fractional powers of closed operators in §III.2 and §III.3 was developed by BALAKRISHNAN [1960:1] although somewhat less general definitions were around before. The results in §III.4, §III.5 and §III.6 are slightly improved versions of the author [1969:2] and [1969:3]. Some further improvements regarding the growth of the cosine function $C_b(\hat{t})$ generated by $A - b^2I$ are possible: see Chapter VI, Exercises 3 to 8. Theorem 7.1 on analytic semigroups and their infinitesimal generators is a classical result of HILLE in the first edition (published in 1948) of his treatise HILLE-PHILLIPS [1957:1]. The application to second order equations (Theorem 7.2) is due to the author [1981;2], although the fact that fractional roots of $-A$ generate analytic semigroups was proved much before by BALAKRISHNAN [1960:1]; the estimation of $u(\hat{t})$ based on (7.19) is there. The problem of finding general square root reductions for the equation (1.1) was systematically examined by KISYŃSKI [1972:1], who provided Example 8.2 for $b = 0$. The reflexive extension (Example 8.3) is in the author [1981:2] and follows an unpublished suggestion of

F. LAUFBURSCHEN.

EXERCISE 1. Let $A \in \mathfrak{C}^1(C,\omega)$ (that is, let A generate a strongly continuous semigroup $S(t)$ such that

$$\|S(t)\| \le Ce^{\omega t} \qquad (t \ge 0)). \tag{9.1}$$

Then $A - bI \in \mathfrak{J}(C)$ for $b \ge \omega$ (that is, $R(\lambda; A - bI)$ exists for $\lambda > 0$ and satisfies

$$\|R(\lambda; A - bI)\| \le C/\lambda \qquad (\lambda > 0)).$$

For $b > \omega$, $(bI - A)^\alpha$ $(\alpha > 0)$ is invertible and

$$(bI - A)^{-\alpha} v = \frac{1}{\Gamma(\alpha)} \int_0^\infty t^{\alpha - 1} e^{-bt} S(t) v \, dt \tag{9.2}$$

(see YOSIDA [1978:1, p. 260]). A scalar counterpart of (9.2) is in GRADSTEIN-RIDZYK [1963:1, p. 331].

EXERCISE 2. (the author, [1983:3]). Let $A \in \mathfrak{C}^2(C,\omega)$ (that is, let A generate a strongly continuous cosine function $C(\hat{t})$ such that

$$\|C(t)\| \le Ce^{\omega|t|} \qquad (-\infty < t < \infty)). \tag{9.3}$$

Then $A - b^2 I \in \mathfrak{J}(C)$ for $b \ge \omega$. For $b > \omega$, $(b^2 I - A)^\alpha$ $(\alpha > 0)$ is invertible and

$$(b^2 I - A)^{-\alpha} u = \frac{2^{3/2 - \alpha} b^{1/2 - \alpha}}{\pi^{1/2} \Gamma(\alpha)} \int_0^\infty s^{\alpha - 1/2} K_{\alpha - 1/2}(bs) \, C(s) u \, ds, \tag{9.4}$$

where K_ν denotes the Macdonald function defined by

$$K_\nu(t) = \frac{\pi}{2} \, \frac{I_{-\nu}(t) - I_\nu(t)}{\sin \nu\pi} \tag{9.5}$$

for $\nu \ne \pm 1, \pm 2, \ldots$ and extended by continuity to all values of ν (WATSON [1944:1, p. 78]). We note that (9.4) is a vector-valued analogue of a well known integral formula (GRADSTEIN-RIDZYK [1963:1, p. 763]).

EXERCISE 3. Let A, $S(t)$ be as in Exercise 1. Given $u \in E$ we say that $S(\hat{t})u$ is <u>continuously differentiable of order</u> $\alpha \geq 0$ <u>in</u> $t \geq 0$ if and only if there exists $\beta > \omega$ and a function $f_\beta(\hat{s})$ continuous, with $s^\alpha \|f_\beta(s)\|$ integrable in $s \geq 0$ and such that

$$e^{-\beta t} S(t)u = \frac{e^{i\pi\alpha}}{\Gamma(\alpha)} \int_t^\infty (s - t)^{\alpha-1} f_\beta(s)\, ds \qquad (t \geq 0). \quad (9.6)$$

The function $f_\beta(\hat{t})$ is the <u>derivative of order</u> α of $e^{-\beta\hat{t}}u$. Show that $S(\hat{t})u$ is continuously differentiable of order α if and only if

$$u \in E_\alpha = D((bI - A)^\alpha) \qquad (9.7)$$

(note that by Lemma 4.2, $D((bI - A)^\alpha)$ does not depend on b). The result shows, in particular that the definition of continuous differentiability of order α does not depend on the $\beta > \omega$ chosen. See KOMATSU [1966:1] or the author [1983:3] for additional details.

EXERCISE 4. (the author, [1983:3]). Let A, $C(\hat{t})$ be as in Exercise 2. Given $u \in E$ we say that $C(\hat{t})\, u$ is <u>continuously differentiable of order</u> α <u>in</u> $-\infty < t < \infty$ if and only if there exists $\beta > \omega$ and a function $f_\beta(\hat{s})$ continuous in $-\infty < t < \infty$, with $s^\alpha \|f_\beta(s)\|$ integrable at $+\infty$ and such that

$$e^{-\beta t}\, C(t)u = \frac{e^{i\pi\alpha}}{\Gamma(\alpha)} \int_t^\infty (s - t)^{\alpha-1} f_\beta(s) ds$$

$$(-\infty < t < \infty). \qquad (9.8)$$

The function $f_\beta(\hat{s})$ is the derivative of order α of $e^{-\beta\hat{t}}C(\hat{t})u$. Show that if $C(t)u$ is continuously differentiable of order 2α with $\alpha \neq n + 1/2$, $n = 0,1,\ldots$ then

$$u \in E_\alpha = D((b^2 I - A)^\alpha)\,. \qquad (9.9)$$

For the exceptional values of α the implication is in general false.

EXERCISE 5. (the author, [1983:3]). Let A, $C(\hat{t})$ be as in Exercise 2. Show that if $\alpha > 0$, $\delta > 0$ and $u \in E_{\alpha+\delta}$, then $C(t)u$ is continuously differentiable of order 2α

EXERCISE 6. (the author, [1983:3]). Let E be a space satisfying

(6.1) with $1 < r < \infty$, A, $C(\hat{t})$ as in Exercise 2. Then $C(\hat{t})u$ is continuously differentiable of order 2α with $\alpha > 0$ if and only if (9.9) holds.

EXERCISE 7. Let $S(\hat{t})$ be a strongly continuous semigroup, $0 < \alpha \leq 1$. Assume that $S(\hat{t})u$ is continuously differentiable of order α in $t \geq 0$. Show that $S(\hat{t})u$ is Hölder continuous with exponent α on compact subsets of $t \geq 0$.

EXERCISE 8. Let $C(\hat{t})$ be a strongly continuous cosine function. $0 < \alpha \leq 1$. Assume that $C(\hat{t})u$ is continuously differentiable of order α in $t \geq 0$. Show that $C(\hat{t})u$ is Hölder continuous with exponent α on compact subsets of $-\infty < t < \infty$.

EXERCISE 9. (the author, [1969:2]). Let E be an arbitrary Banach space, $C(\hat{t})$ a strongly continuous cosine function in E. Show that Assumption 5.1 holds if and only if $V(\hat{t})E \subseteq D(A)$ and $AV(\hat{t})$ is a strongly continuous function in $t \geq 0$, where

$$V(t) = \int_0^1 \log s(\mathcal{S}(s + t) - \mathcal{S}(s - t))ds . \qquad (9.10)$$

(Hint: use the expression for $(b^2I - A)^{1/2}$ obtained in Exercise 2).

EXERCISE 10. (the author [1969:1]). Let E be an arbitrary Banach space, $C(\hat{t})$ a strongly cosine function satisfying

$$\|C(t)\| \leq C \qquad (-\infty < t < \infty) \qquad (9.11)$$

Using formula (6.18) show that

$$A^{1/2}\mathcal{S}(t)u = \frac{2it}{\pi} \text{ v.p.} \int_0^\infty \frac{1}{t^2 - s^2} C(s)u \, ds \qquad (9.12)$$

in $-\infty < t < \infty$ for every $u \in D(A)$, where v.p. indicates limit as $\varepsilon \to 0+$ of the integral over $|t - s| \geq \varepsilon$. Formula (9.12) is an operator analogue of the scalar formula

$$\sin at = \frac{2t}{\pi} \text{ v.p.} \int_0^\infty \frac{\cos as}{t^2 - s^2} ds \qquad (9.13)$$

valid for $a \geq 0$ (GRADSTEIN-RIDZYK [1963:1, p. 421])

EXERCISE 11. Under the assumptions in Exercise 10, show using formula (6.23) that

$$A^{1/2}u = \lim_{\sigma \to \infty} \frac{2i}{\pi} \int_0^\sigma \frac{1}{s} C'(s)u \, ds \tag{9.14}$$

for every $u \in D(A)$. Formula (9.14) is an operator analogue of the scalar formula

$$\lim_{\sigma \to \infty} \int_0^\sigma \frac{\sin as}{s} ds = \frac{\pi}{2}, \tag{9.15}$$

valid for $a > 0$. (GRADSTEIN-RIDZYK [1963:1, p. 420]).

EXERCISE 12. Using Exercises 5 and 8 show that formula (9.12), as well as its more general version (6.18) hold in $-\infty < t < \infty$ for $u \in E_\gamma$, $\gamma > 0$.

EXERCISE 13. Using Exercises 5 and 8 show that formula (9.14), as well as its more general version (6.23), hold in $-\infty < t < \infty$ for $u \in E_\gamma$ for any $\gamma > 1/2$.

EXERCISE 14. Show that the singular integral operator (8.3) is not bounded in $C_{2\pi}(-\infty,\infty)$.

EXERCISE 15. Prove Theorem 1.3 showing that the Laplace transform of $\mathfrak{S}(\hat{t})$ equals $R(\lambda;\mathfrak{A})$ and applying Theorem I.3.4.

FOOTNOTES TO CHAPTER III

(1) Elements of $E_0 \times E_1$ and similar product spaces will be denoted as "row vectors" or "column vectors" according to convenience.

(2) This estimate can be considerably improved (see Chapter VI, Exercises 3 to 8).

(3) See footnote (2).

(4) See footnote (2).

(5) See footnote (2).

(6) See footnote (2).

(7) See footnote (2).

CHAPTER IV

APPLICATIONS TO PARTIAL DIFFERENTIAL EQUATIONS

§IV.1 Wave equations: the Dirichlet boundary condition.

We consider in the first six sections of this chapter the equation

$$u''(t) = A(\beta)u(t). \tag{1.1}$$

Here

$$Au = \sum_{j=1}^{m} \sum_{k=1}^{m} D^j(a_{jk}(x)D^j u) + \sum_{j=1}^{m} b_j(x)D^j u + c(x)u \tag{1.2}$$

with $x = (x_1, \ldots, x_m)$, $D^j = \partial/\partial x_j$ and the coefficients $a_{jk}(x)$, $b_j(x)$, $c(x)$ are defined in a domain Ω of m-dimensional Euclidean space R^m; $A(\beta)$ denotes the restriction of A obtained by means of a boundary condition β at the boundary Γ of the form

$$u(x) = 0 \qquad (x \in \Gamma), \tag{1.3}$$

or

$$D^{\tilde{\nu}}u(x) = \gamma(x)u(x) \qquad (x \in \Gamma), \tag{1.4}$$

where $D^{\tilde{\nu}}$ denotes the conormal derivative to be defined below (see (4.1)). Since to replace $a_{jk}(x)$ by $\frac{1}{2}(a_{jk}(x) + a_{kj}(x))$ does not change the action of A on smooth functions we shall assume from now on that

$$a_{jk}(x) = a_{kj}(x).$$

We require the a_{jk} to be real-valued; the b_j and c can be complex-valued. If the a_{jk} have first order partial derivatives, we can write A in the more natural form

$$Au = \sum_{j=1}^{m} \sum_{k=1}^{m} a_{jk}(x)D^j D^k u + \sum_{j=1}^{m} \hat{b}_j(x)D^j u + c(x)u , \tag{1.5}$$

where

$$\hat{b}_j(x) = b_j(x) + \sum_{k=1}^{m} D^k a_{jk}(x) .$$

The passage from (1.2) to (1.5) and vice versa is no longer possible if the a_{jk} are not differentiable; in this case (1.2) and (1.5) represent quite different entities (see PUCCI-TALENTI [1974:1]). We shall always assume A is written in the form (1.2), called the divergence or variational form. The coefficients a_{jk}, b_j, c will be required to be merely measurable and bounded; we postulate in addition that A be uniformly elliptic in the sense that

$$\sum\sum a_{jk}\xi_j\xi_k \geq \kappa|\xi|^2 \quad (x \in \Omega, \ \xi \in \mathbb{R}^m) \tag{1.6}$$

for some $\kappa > 0$.

Our first result concerns the Dirichlet boundary condition (1.3). No assumptions whatsoever will be placed on the domain Ω or the boundary Γ. In this high level of generality, it is obvious that Au (as well as the boundary condition $u = 0$) will have to be understood in a suitably generalized sense; for instance, in view of the lack of smoothness of the a_{jk}, it is not clear whether A can be applied to any nonzero function.

The basic space in our treatment is $H = L^2(\Omega)$. An important supporting role will be played by $H_0^1(\Omega)$, the closure of the space $\mathcal{D}(\Omega)$ (of Schwartz test functions) in $H^1(\Omega)$; we recall that $\mathcal{D}(\Omega)$ consists of all infinitely differentiable functions $\varphi(\hat{x})$ with support in Ω, and that the space $H^k(\Omega)$ (k an integer ≥ 1) consists of all functions u having partial derivatives of order $\leq k$ (derivatives understood in the sense of distributions) in $L^2(\Omega)$. The spaces $H^1(\Omega)$ and $H_0^1(\Omega)$ are Hilbert spaces equipped with the scalar product

$$(u,v)_{H^1(\Omega)} = (u,v)_{L^2(\Omega)} + \sum_{j=1}^{m} (D^j u, D^j v)_{L^2(\Omega)} \tag{1.7}$$

(for all necessary facts on the Sobolev spaces $H^k(\Omega)$, $H_0^1(\Omega)$ consult ADAMS [1975:1]).

The first stage of our argument will be the construction of the operator $A_0(\beta)$, where A_0 is the self adjoint part of A,

$$A_0 u = \sum_{j=1}^{m} \sum_{k=1}^{m} D^j(a_{jk}(x) D^j u) + c(x)u . \tag{1.8}$$

With this in mind, we introduce a new scalar product in $H_0^1(\Omega)$ by the formula

$$(u,v)_\alpha = \int_\Omega (\alpha - c)\bar{u}v \, dx + \int_\Omega (\sum\sum a_{jk} D^j \bar{u} D^k v) \, dx , \tag{1.9}$$

where $dx = dx_1 \ldots dx_m$ and

$$\alpha > \nu = \text{ess. sup } c . \tag{1.10}$$

It is obvious that $(u,v)_\alpha = \overline{(v,u)_\alpha}$ and that $(u,v)_\alpha$ is linear in v (and conjugate linear in u). Moreover, we see easily that there exist constants $C > c > 0$ such that

$$c\|u\|_{H_0^1(\Omega)} \leq (u,u)_\alpha \leq C\|u\|_{H_0^1(\Omega)} \tag{1.11}$$

(we use the uniform ellipticity assumption for the first inequality). Accordingly, the norm

$$\|u\|_\alpha = (u,u)_\alpha^{1/2} \tag{1.12}$$

corresponding to the scalar product (1.9) is equivalent to the original norm of H_0^1 defined by (1.7); thus we shall assume from now on $H_0^1(\Omega)$ endowed with (1.12) (of course, the same arguments apply to the space $H^1(\Omega)$, a fact that will be used in §IV.4).

A function $u \in H_0^1(\Omega)$ belongs to $D(A_0(\beta))$ if and only if

$$w \to (u,w)_\alpha \qquad (w \in H_0^1(\Omega)) \tag{1.13}$$

is continuous in the norm of $L^2(\Omega)$: if this is the case, we extend the linear functional (1.13) to all of $L^2(\Omega)$ (since $H_0^1(\Omega)$ is dense in $L^2(\Omega)$ in the topology of $L^2(\Omega)$ this extension is unique). Let $v \in L^2(\Omega)$ be such that

$$(u,w)_\alpha = (v,w) \qquad (w \in H_0^1(\Omega)) . \tag{1.14}$$

Define

$$A_0(\beta)u = \alpha u - v . \tag{1.15}$$

(Motivation is obvious: if the coefficients a_{jk} and the boundary Γ

are smooth and u is a smooth function such that $\alpha u - A_0 u = v$ in Ω and $u = 0$ on Γ, then (1.14) follows for any smooth w such that $w = 0$ on Γ). We check easily that the definition of $A_0(\beta)$ above does not depend on α.

We wish to show that the operator $A_0(\beta)$ just defined is self adjoint. We begin by proving that

$$(\lambda I - A_0(\beta))D(A_0(\beta)) = L^2(\Omega) \tag{1.16}$$

for any $\lambda > \nu$. In fact, let v be an arbitrary element of $L^2(\Omega)$. Define a linear functional by

$$w \to (v,w)_\lambda \, . \tag{1.17}$$

Since (1.17) is continuous in $L^2(\Omega)$ it is as well continuous in $H_0^1(\Omega)$, thus there exists $u \in H_0^1(\Omega)$ such that

$$(u,w)_\lambda = (v,w), \tag{1.18}$$

hence (1.16) follows: note that our construction of u yields the estimate

$$\|u\| \leq C\|v\| \, . \tag{1.19}$$

Rewriting (1.14) in the form

$$(u,w)_\lambda = ((\lambda I - A_0(\beta))u,w) \tag{1.20}$$

and taking $w = u$ we see that $\lambda I - A_0(\beta)$ is one-to-one; combining with (1.19) we deduce that $R(\lambda;A_0(\beta))$ exists for $\lambda > \nu$. It also follows from (1.18) that if $u, v \in D(A_0(\beta))$ then

$$(\lambda u - A_0(\beta)u,v) = (u,v)_\lambda = \overline{(v,u)}_\lambda = (u, \lambda v - A(\beta)v)$$

so that $A_0(\beta)$ is <u>symmetric</u>. We finally prove that $A_0(\beta)$ is densely defined. In order to do this it is sufficient to show that $D(A_0(\beta))$ is dense in $H_0^1(\Omega)$ in the topology of $H_0^1(\Omega)$. If this were not the case, there would exist an element $w \in H_0^1(\Omega)$ with $(u,w)_\lambda = 0$ for all $u \in D(A_0(\beta))$. In view of (1.20) this implies that w is orthogonal to $\lambda I - D(A_0(\beta))$ which, due to (1.16), shows that $w = 0$.

To prove that $A_0(\beta)$ is self adjoint we make use of the following result.

LEMMA 1.1. *Let A be a densely defined symmetric operator in the Hilbert space H. Assume that the resolvent $\rho(A)$ contains a real number λ. Then A is self adjoint.*

Proof: Let u,v be two arbitrary elements of H. Then

$$\begin{aligned}(R(\lambda;A)u,v) &= (R(\lambda;A)u,\ (\lambda I - A)R(\lambda;A)v)\\ &= ((\lambda I - A)R(\lambda;A)u, R(\lambda;A)v) = (u, R(\lambda,A)v)\end{aligned} \tag{1.21}$$

so that $R(\lambda;A)$ is symmetric; thus

$$\lambda I - A^* = (\lambda I - A)^* = (R(\lambda;A)^{-1})^* = (R(\lambda;A)^*)^{-1} = R(\lambda;A)^{-1} = \lambda I - A\,,$$

where the interchange of inverses and adjoints is easily justified (see RIESZ-SZ.-NAGY [1955:1]). This ends the proof.

Having shown that $A_0(\beta)$ is self adjoint and bounded above (by ν) we know that $A_0(\beta)$ generates a strongly continuous cosine function $C(\hat{t})$ given by

$$C(t) = \cosh t A_0(\beta)^{1/2} \tag{1.22}$$

where the expression on the right is defined by means of the functional calculus for self adjoint operators; if $P(d\mu)$ is the spectral resolution of $A_0(\beta)$ then

$$C(t)u = \int_{-\infty}^{\nu+} \cosh t\, \mu^{1/2} P(d\mu)u \tag{1.23}$$

where the upper limit of integration is due to the fact that $A_0(\beta)$ is bounded below by ν $((A_0(\beta)u,u) \leq \nu\|u\|^2$ for $u \in D(A))$.
To handle the full operator $A(\beta)$ we shall construct a phase space for $A_0(\beta)$ and then incorporate the first order terms by means of a perturbation argument to be proved in next section. The proof that $A(\beta)$ generates a strongly continuous cosine function will be given in §IV.3.

§IV.2 The phase space.

Since the operator $A_0(\beta)$ is self adjoint, a square root

$$B = A_0(\beta)^{1/2} \tag{2.1}$$

can be defined by means of the functional calculus for unbounded functions of operators (DUNFORD-SCHWARTZ [1963:1]); if $P(d\mu)$ is the spectral resolution of $A_0(\beta)$, then $u \in D(B)$ if and only if

$$\int_{-\infty}^{\nu+} |\mu|^{1/2} \|P(d\mu)u\|^2 < \infty \tag{2.2}$$

and

$$Bu = \int_{-\infty}^{\nu+} \mu^{1/2} P(d\mu)u . \tag{2.3}$$

There remains the problem of the choice of square root $\mu^{1/2}$ in (2.3); although (as we shall see below) this choice is immaterial, we define $\mu^{1/2} = \mu_+^{1/2}$ for $\mu \geq 0$ ($r_+^{1/2}$ the nonnegative square root of the nonnegative number r) and $\mu^{1/2} = i(-\mu)_+^{1/2}$ for $\mu < 0$. Although B is not easily identifiable in terms of $A_0(\beta)$, $D(B)$ turns out to be a familiar space (Theorem 2.2 below). To facilitate the identification of $D(B)$ we begin by noting that if $\lambda > \nu$ then

$$D(B) = D((\lambda I - A_0(\beta))^{1/2}) . \tag{2.4}$$

To see that (2.4) holds we note that $u \in D((\lambda I - A_0(\beta))^{1/2})$ if and only if

$$\int_{-\infty}^{\nu+} (\lambda - \mu)^{1/2} \|P(d\mu)u\|^2 < \infty . \tag{2.5}$$

The fact that (2.2) and (2.5) are equivalent follows from the boundedness of the function $|\mu|^{1/2} - |\lambda - \mu|^{1/2}$.

LEMMA 2.1. <u>Let Q be a densely defined, invertible operator such that Q^{-1} is everywhere defined and bounded. Then $D(Q^2)$ is dense in $D(Q)$ in the graph norm of $D(Q)$.</u>

<u>Proof</u>: If $u \in D(Q)$ then $u = Q^{-1}v$; select a sequence $\{v_n\}$ in $D(Q)$ with $v_n \to v$ and note that $Q^{-1}v_n \to u$, $Q(Q^{-1}v_n) = v_n \to Qu$.

THEOREM 2.2.

$$D(B) = H_0^1(\Omega) . \tag{2.6}$$

<u>Proof</u>: If $u \in D(A_0(\beta))$, $\lambda > \nu$ we have

$$\begin{aligned} \|(\lambda I - A_0(\beta))^{1/2}u\|^2 &= ((\lambda I - A_0(\beta))^{1/2}u, (\lambda I - A_0(\beta))^{1/2}u) \\ &= (((\lambda I - A_0(\beta))u,u) = (u,u)_\lambda = \|u\|_\lambda^2 . \end{aligned} \tag{2.7}$$

It follows from Lemma 2.1 that $D(A_0(\beta))$ is dense in $D((\lambda I - A_0(\beta))^{1/2})$ in the graph norm of $(\lambda I - A_0(\beta))^{1/2}$; on the other hand, we have shown in §IV.1 that $D(A_0(\beta))$ is dense in $H_0^1(\Omega)$ in the norm of $H_0^1(\Omega)$, thus (2.7) is easily seen to imply that $D((\lambda I - A_0(\beta))^{1/2}) = H_0^1(\Omega)$. Combining with (2.4), Theorem 2.2 follows.

We are now in position to construct a phase space for the equation

$$u''(t) = A_0(\beta)u(t) . \tag{2.8}$$

In fact, Theorem III.5.4 asserts that

$$\mathfrak{E} = D((\lambda I - A_0(\beta))^{1/2}) \times L^2(\Omega) = H_0^1(\Omega) \times L^2(\Omega) \tag{2.9}$$

will be such a state space, the last equality resulting from Theorem 2.2. However, a little more can be said. In fact, it follows from Theorems III.6.4 and 2.2 that

$$\tilde{D}_1 = H_0^1(\Omega) , \tag{2.10}$$

where $\tilde{D}_1$ consists of all $u \in L^2(\Omega)$ such that $C(\hat{t})u = \cosh \hat{t}\, A_0(\beta)^{1/2}u$ is continuously differentiable in $-\infty < t < \infty$. Thus, the state space (2.9) is the same one provided by Theorem III.1.3.

The group $\mathfrak{S}_0(\hat{t})$ in (III.1.10) propagating the solutions of (2.8) is

$$\mathfrak{S}_0(t) = \begin{bmatrix} \cosh t\, A_0(\beta)^{1/2} & A_0(\beta)^{-1/2}\sinh t\, A_0(\beta)^{1/2} \\ A_0(\beta)^{1/2}\sinh t\, A_0(\beta)^{1/2} & \cosh t\, A_0(\beta)^{1/2} \end{bmatrix} ; \tag{2.11}$$

its infinitesimal generator is

$$\mathfrak{A}_0(\beta) = \begin{bmatrix} 0 & I \\ A_0(\beta) & 0 \end{bmatrix} , \tag{2.12}$$

with domain $D(\mathfrak{A}_0(\beta)) = D(A_0(\beta)) \times H_0^1(\Omega)$. To take care of the first order terms we introduce the perturbation operator

$$\mathfrak{P} = \begin{bmatrix} 0 & 0 \\ P & 0 \end{bmatrix} , \tag{2.13}$$

where

$$Pu = \sum_{j=1}^{m} b_j(x)D^j u \ . \tag{2.14}$$

Obviously, $\mathfrak{P}$ is a bounded operator in $\mathfrak{E}$. The necessary properties of the operator $A(\beta) = A_0(\beta) + P$ will be a consequence of the following result, where P does not have the same meaning as in (2.14).

THEOREM 2.3. *Let* A *be the infinitesimal generator of a group* $S(\hat{t})$ *in a Banach space* E *such that*

$$\|S(t)\| \le C_0 e^{\omega|t|} \qquad (-\infty < t < \infty)$$

(that is, let $A \in \mathfrak{C}^1(C_0,\omega)$*), and let* P *be a bounded operator. Then the operator* $A + P$ *(with domain* $D(A + P) = D(A)$*) belongs to* $\mathfrak{C}^1(C_0,\omega + C\|P\|)$.

Proof: By virtue of Theorem I.3.3 $R(\lambda;A)$ exists for $|\lambda| > \omega$ and satisfies

$$\|R(\lambda;A)^n\| \le C_0(|\lambda| - \omega)^{-n} \qquad (|\lambda| > \omega\, , n = 0,1,\ldots)\ . \tag{2.15}$$

In particular, if

$$|\lambda| > \omega + C_0\|P\|\ ,$$

then

$$\|PR(\lambda;A\| \le C_0\|P\|(|\lambda| - \omega)^{-1} < 1\ , \tag{2.16}$$

hence the series

$$Q(\lambda) = R(\lambda;A)\sum_{n=0}^{\infty}(PR(\lambda;A))^n$$

is convergent in the uniform topology of operators. If $u \in D(A)$ we have

$$\begin{aligned} &Q(\lambda)(\lambda I - A - P)u = \\ &= R(\lambda;A)\sum_{n=0}^{\infty}(PR(\lambda;A))^n(\lambda I - A - P)u \\ &= \sum_{n=0}^{\infty}(R(\lambda;A)P)^n u - \sum_{n=1}^{\infty}(R(\lambda;A)P)^n u = u\ . \end{aligned}$$

Similarly, if $u \in E$,

$$(\lambda I - A - P)Q(\lambda)u = (\lambda I - A - P)R(\lambda;A) \sum_{n=0}^{\infty} (PR(\lambda;A))^n u$$

$$= \sum_{n=0}^{\infty} (PR(\lambda;A))^n u - \sum_{n=1}^{\infty} (PR(\lambda;A))^n u = u ,$$

thus

$$Q(\lambda) = R(\lambda;A + P) .$$

We estimate the powers of $R(\lambda;A + P)$:

$$R(\lambda;A + P)^m = \{R(\lambda;A) \sum_{n=0}^{\infty} (PR(\lambda;A))^n\}^m$$

$$= \sum R(\lambda;A)(PR(\lambda;A))^{j_1} \cdots R(\lambda;A)(PR(\lambda;A))^{j_m},$$

where $j_1,\ldots,j_m = 0,1,2,\ldots$. Due to the fact that some of the j_k may be zero, the typical term containing n P's will be of the form

$$R(\lambda;A)^{s_1} PR(\lambda;A)^{s_2} \cdots R(\lambda;A)^{s_n} PR(\lambda;A)^{s_{n+1}} , \tag{2.17}$$

where $s_1 + \cdots + s_{n+1} = n + m$ and $s_k > 0$. Making use of (2.15) we see that (2.17) can be bounded in norm by

$$C_0(|\lambda| - \omega)^{-s_1}\|P\|C_0(|\lambda| - \omega)^{-s_2}\|P\| \cdots C_0(|\lambda| - \omega)^{-s_n}\|P\|C_0(|\lambda| - \omega)^{-s_{n+1}}.$$

Accordingly

$$\|R(\lambda;A + P)^m\| \le \frac{C_0}{(|\lambda| - \omega)^m} \left\{ \sum_{n=0}^{\infty} (C_0\|P\|(|\lambda| - \omega)^{-1})^n \right\}^m$$

$$= \frac{C_0}{(|\lambda| - \omega)^m} \left\{ \frac{1}{1 - C_0\|P\|(|\lambda| - \omega)^{-1}} \right\}^m$$

$$= \frac{C_0}{(\lambda - \omega - C_0\|P\|)^m} \qquad (|\lambda| > \omega + C_0\|P\|) .$$

Applying Theorem I.3.3 (this time backwards) as modified by Remark I.3.5 we conclude the proof.

Putting together the results in this section we obtain.

THEOREM 2.4. _Let_ A _be the operator_ (1.2), β _the Dirichlet_ boundary condition (1.3), and let

$$A(\beta) = A_0(\beta) + P \tag{2.18}$$

with domain $D(A(\beta)) = D(A_0(\beta))$. Then the space $H_0^1(\Omega) \times L^2(\Omega)$ is a phase space for the equation

$$u''(t) = A(\beta)u(t). \tag{2.19}$$

§IV.3 The Cauchy problem.

The results in the previous section do not show that the Cauchy problem for the equation (1.1) is well posed in the sense of §II.1. We shall do so here. The basic product space will be $\mathfrak{D} = L^2(\Omega) \times L^2(\Omega)$. We define a $(\mathfrak{D})$-valued function $\mathfrak{B}_0(\hat{t})$ by

$$\mathfrak{B}_0(t) = \begin{bmatrix} \cosh t\, A_0(\beta)^{1/2} & \sinh t\, A_0(\beta)^{1/2} \\ \sinh t\, A_0(\beta)^{1/2} & \cosh t\, A_0(\beta)^{1/2} \end{bmatrix} =$$

$$= \begin{bmatrix} \cosh t\, B & \sinh t\, B \\ \sinh t\, B & \cosh t\, B \end{bmatrix}, \tag{3.1}$$

where B is the square root of $A_0(\beta)$ constructed in §IV.2. We note that the group $\mathfrak{B}_0(\hat{t})$ does depend on the choice of square root B; in contrast, the group $\mathfrak{S}_0(\hat{t})$ defined by (2.11) is independent of B. To see this we note that $\cosh t\, B$ can be defined directly in terms of $A_0(\beta)$ as follows:

$$\cosh t\, B = c(t;A_0(\beta)) \text{ with } c(t,\mu) = \cosh t\, \mu^{1/2} = 1 + t^2\mu/2! + t^4\mu^2/4! + \cdots$$

The same observation applies to $A_0(\beta)^{-1/2}\sinh t\, A_0(\beta)^{1/2}$ and to $A_0(\beta)^{1/2}\sinh A_0(\beta)^{1/2}$. To see this we simply note that the functions $\mu^{-1/2}\sinh t\, \mu^{1/2} = t + t^3\mu/3! + t^5\mu^2/5! + \cdots$ and $\mu^{1/2}\sinh t\, \mu^{1/2} = $ $= t\mu + t^3\mu^2/3! + t^5\mu^3/5! + \cdots$ depend on μ and not on $\mu^{1/2}$. In contrast, $\sinh t\, \mu^{1/2} = \mu^{1/2} + t^3\mu^{3/2}/3! + t^5\mu^{5/2}/5! + \cdots$ depends on $\mu^{1/2}$, thus (3.1) depends on the value of $\mu^{1/2}$ chosen (equivalently, on the choice of B).

Working as in Theorem III.5.5 we show that the infinitesimal generator $\mathfrak{B}_0$ of $\mathfrak{V}_0$ is

$$\mathfrak{B}_0 = \begin{bmatrix} 0 & B \\ B & 0 \end{bmatrix}. \tag{3.2}$$

To proceed further, we make the assumption that

$$\nu = \text{ess. sup } c(x) < 0; \tag{3.3}$$

(as we shall see later, this can be dispensed with). Since $\lambda I - A_0(\beta)$ has a bounded inverse for $\lambda > \nu$, it follows that $A_0(\beta)$ itself has a bounded inverse $A_0(\beta)^{-1}$, hence B has a bounded inverse B^{-1}. Consider now the operator P in (2.14). We claim that

$$PB^{-1} \tag{3.4}$$

is a bounded operator in $L^2(\Omega)$; since $B^{-1} : L^2(\Omega) \to D(B) = H_0^1(\Omega)$ (see Theorem 2.2) is continuous if $D(B)$ is given the graph norm of B, it is clear that we only have to show that the graph norm of B in $D(B) = H_0^1(\Omega)$ is equivalent to the norm of $H_0^1(\Omega)$. This, in turn, is a consequence of the fact that both norms make $D(B)$ complete and of the open mapping principle (see DUNFORD-SCHWARTZ [1958:1]).

Having proved that (3.4) is in fact bounded, consider the operator

$$\mathfrak{P} = \begin{bmatrix} 0 & PB^{-1} \\ 0 & 0 \end{bmatrix}. \tag{3.5}$$

Applying Theorem 2.3 we deduce that the operator

$$\mathfrak{B} = \mathfrak{B}_0 + \mathfrak{P} \tag{3.6}$$

generates a strongly continuous group $\mathfrak{V}(\hat{t})$ in the space $\mathfrak{D}$. Write

$$\mathfrak{V}(t) = \begin{bmatrix} V_{11}(t) & V_{12}(t) \\ V_{21}(t) & V_{22}(t) \end{bmatrix}. \tag{3.7}$$

The resolvent $R(\lambda;\mathfrak{B}_0 + \mathfrak{P})$ must exist for λ sufficiently large: write

$$R(\lambda;\mathfrak{B}_0 + \mathfrak{P}) = \begin{bmatrix} R_{11}(\lambda) & R_{12}(\lambda) \\ R_{21}(\lambda) & R_{22}(\lambda) \end{bmatrix}. \tag{3.8}$$

Looking at the first column of the matrix equation

$$(\lambda\mathfrak{I} - \mathfrak{B}_0 - \mathfrak{P})R(\lambda;\mathfrak{B}_0 + \mathfrak{P}) = \mathfrak{I}\,, \tag{3.9}$$

$\mathfrak{I}$ the identity operator in $\mathfrak{D}$, we obtain

$$\lambda R_{11}(\lambda) - (B + PB^{-1})R_{21}(\lambda) = I,\ BR_{11}(\lambda) - \lambda R_{21}(\lambda) = 0.$$

Combining these equalities, we see that

$$(\lambda^2 I - B^2 - P)R_{11}(\lambda) = (\lambda^2 I - A_0(\beta) - P)R_{11}(\lambda) = \lambda I. \tag{3.10}$$

Working in the same way with the equation

$$R(\lambda;\mathfrak{B}_0 + \mathfrak{P})(\lambda\mathfrak{I} - \mathfrak{B}_0 - \mathfrak{P})\mathfrak{u} = \mathfrak{u} \tag{3.11}$$

valid for $\mathfrak{u} \in D(\mathfrak{B}_0 + \mathfrak{P})$, we deduce

$$R_{11}(\lambda)(\lambda^2 I - B^2 - P)u = R_{11}(\lambda)(\lambda^2 I - A_0(\beta) - P)u = \lambda u$$

for $u \in D(B^2 + P) = D(B^2) = D(A_0(\beta)) = D(A_0(\beta) + P)$. It follows that

$$R_{11}(\lambda) = \lambda R(\lambda^2;A_0(\beta) + P). \tag{3.12}$$

Finally, observing the element in the upper left corner of the matrix equation

$$R(\lambda;\mathfrak{B}_0 + \mathfrak{P})\mathfrak{u} = \int_0^\infty e^{-\lambda t}\mathfrak{V}(t)\mathfrak{u}\,dt$$

(which results from (I.3.8)) we obtain, making use of (3.12), that

$$\lambda R(\lambda^2;A_0(\beta) + P)u = \int_0^\infty e^{-\lambda t}V_{11}(t)u\,dt\,,$$

which equality, in view of Theorem II.2.3 implies that $S(\hat{t}) = V_{11}(\hat{t})$ is a strongly continuous cosine function with infinitesimal generator

$$A(\beta) = A_0(\beta) + P, \tag{3.13}$$

$D(A(\beta)) = D(A_0(\beta))$.

To get rid of assumption (3.4) it suffices to replace $c(x)$ by

$c(x)-\nu-\delta$ $(\delta > 0)$ in the definition of $A_0(\beta)$; in this case, the operator P is defined by

$$Pu = \sum_{j=1}^{m} b_j(x)D^j u + (\nu+\delta)u \tag{3.14}$$

instead of (2.14).

We have completed the proof of

THEOREM 3.1. *Let* A *be the operator* (1.2), β *the Dirichlet boundary condition* (1.3), *and let*

$$A(\beta) = A_0(\beta) + P \tag{3.15}$$

with domain $D(A(\beta)) = D(A_0(\beta))$. *Then the Cauchy problem for the equation*

$$u''(t) = A(\beta)u(t) \tag{3.16}$$

is well posed in $-\infty < t < \infty$.

§IV.4 Wave equations: other boundary conditions.

We study in the next two sections the operator (1.2) with boundary conditions of the form (1.4),

$$D^{\tilde{\nu}}u(x) = \sum_{j=1}^{m}\sum_{k=1}^{m} a_{jk}(x)\nu_j D^k u(x) = \gamma(x)u(x) \qquad (x \in \Gamma), \tag{4.1}$$

where $\nu = (\nu_1,\ldots,\nu_m)$ is the outer normal vector on the boundary Γ; the expression on the left-hand side of (4.1) is called the *conormal derivative* of u at $x \in \Gamma$. Again, the basic space in our treatment of (1.1) will be $L^2(\Omega)$; the auxiliary space is now $H^1(\Omega)$ instead of $H_0^1(\Omega)$. The operator $A_0(\beta)$ will be again defined by (1.14) and (1.15); however, the scalar product $(u,w)_\alpha$ is different. To guess the correct definition we perform the following formal computation using the divergence theorem, where the boundary Γ and the intervening functions are smooth enough.

$$\int_\Omega A_0(\beta)\bar{u}v\,dx = \int_\Omega \{\sum\sum D^j(a_{jk}D^k\bar{u}) + c\bar{u}\}v\,dx =$$

$$\int_\Omega \{-\sum\sum a_{jk}D^j\bar{u}D^k v + c\bar{u}v\}\,dx + \int_\Gamma (\sum\sum a_{jk}\nu_j D^k\bar{u}v)\,d\sigma =$$

$$= -\int_{\Omega} \{\sum\sum a_{jk} D^j\bar{u} D^k v - c\bar{u}v\}\, dx + \int_{\Gamma} D^{\tilde{\nu}} \bar{u} v \, d\sigma$$

$$= -\int_{\Omega} \{\sum\sum a_{jk} D^j\bar{u} D^k v - c\bar{u}v\}\, dx + \int_{\Gamma} \gamma\bar{u}\, v \, d\sigma \, .$$

Accordingly, we should define

$$(u,v)_{\alpha} = \int_{\Omega} (\alpha - c)\bar{u}v \, dx + \int_{\Omega} (\sum\sum a_{jk} D^j\bar{u} D^k v)\, dx - \int_{\Gamma} \gamma\bar{u}v \, dx. \tag{4.2}$$

However, to give sense to (4.2) for arbitrary $u,v \in H^1(\Omega)$ some assumptions on Ω will be necessary, due to the presence of the boundary integral.

A domain $\Omega \subseteq R^m$ is said to be <u>of class</u> $C^{(k)}$ (k an integer ≥ 0) if and only if, given any point $\bar{x} \in \Gamma$ there exists an open neighborhood V of $\bar{x}$ in R^m and a map $\eta : \bar{V} \to \bar{S}^m$ ($S^m = S^m(0,1) = \{\eta \in R^m ; |\eta| < 1\}$ the open unit sphere in R^m) such that

(a) η is one-to-one and onto $\bar{S}^m$ with $\eta(\bar{x}) = 0$.

(b) The map η (resp. η^{-1}) possesses partial derivatives of order $\leq k$ in V (resp. in S^m) which are continuous in $\bar{V}$ (resp. in $\bar{S}^m$).

(c) $\eta(V \cap \Omega) = S^m_+ = \{\eta \in S^m ; \eta_m > 0\}$.
$\eta(V \cap \Gamma) = S^m_0 = \{\eta \in S^m ; \eta_m = 0\}$.

A little use will be made in the following lines of the Sobolev spaces $W^{1,p}(\Omega)$ consisting of all functions $u \in L^p(\Omega)$ having first partial derivatives in $L^p(\Omega)$; the space $W^{1,p}(\Omega)$ is normed with

$$\|u\|_{W^{1,p}(\Omega)} = \Big(\|u\|^p_{L^p(\Omega)} + \sum_{j=1}^{m} \|D^j u\|^p_{L^p(\Omega)} \Big)^{1/p} .$$

Also, we shall employ the spaces $C^{(1)}(\bar{\Omega})$ consisting of all functions u continuous in $\bar{\Omega}$ having continuous first partial derivatives $D^j u$ in Ω, each derivative admitting a continuous extension to Ω.

THEOREM 4.1 <u>Let</u> Ω <u>be a bounded domain of class</u> $C^{(1)}$, <u>and let</u> $1 \leq p < \infty$. <u>Then</u> (a) <u>if</u>

$$1 \leq p < m, \quad 1 \leq q \leq (m-1)p/(m-p)$$

there exists a constant C *(depending only on* Ω, p,q*) such that*

$$\|u\|_{L^q(\Gamma)} = \left(\int_\Gamma |u|^q \, d\sigma\right)^{1/q} \leq C\|u\|_{W^{1,p}(\Omega)} \tag{4.3}$$

for every $u \in C^{(1)}(\bar{\Omega})$. (b) *if*

$$p \geq m$$

then (4.3) *holds for every* $q \geq 1$.

For the proof of a considerably more general result see ADAMS [1975:1, p. 114]. We note that Theorem 4.1 holds as well for domains which are "piecewise of class $C^{(1)}$" such as, say, parallelepipedons or cylinders whose base is a $(m-1)$-dimensional domain of class $C^{(1)}$; also, the boundedness hypotesis is not essential: for instance, the result holds if Γ (but not Ω) is bounded.

THEOREM 4.2. *Let* Ω *be a domain of class* $C^{(0)}$, $1 \leq p < \infty$. *Then* $\mathscr{D}$ *(or, rather, the set of restrictions of functions of* $\mathscr{D}$ *to* Ω*) is dense in* $W^{1,p}(\Omega)$.

The proof can be seen in ADAMS [1975: 1, p. 54] under less stringent assumptions; recall that the space $\mathscr{D} = \mathscr{D}(R^m)$ is the space of Schwartz test functions in R^m.

Let Ω be a bounded domain of class $C^{(1)}$ (or, more generally, a domain of class $C^{(1)}$ with a bounded boundary Γ). Assuming that $\gamma \in L^\infty(\Gamma)$, the following estimation is justified by Theorem 4.1 (and the comments after it): here u,v are functions in $\mathscr{D}$ and we take $p = q = 1$.

$$\left|\int_\Gamma \gamma \bar{u} v \, d\sigma\right| \leq C'\|\bar{u}v\|_{L^1(\Gamma)} \leq C\|\bar{u}v\|_{W^{1,1}(\Omega)} . \tag{4.4}$$

Now,

$$\|\bar{u}v\|_{W^{1,1}(\Omega)} = \|\bar{u}v\|_{L^1(\Omega)} + \sum \|D^j(\bar{u}v)\|_{L^1(\Omega)}$$

$$\leq \|\bar{u}v\|_{L^1(\Omega)} + \sum \|(D^j\bar{u})v\|_{L^1(\Omega)} + \sum \|\bar{u}D^jv\|_{L^1(\Omega)}$$

$$\leq \|u\|_{L^2(\Omega)} \|v\|_{L^2(\Omega)} + \left(\sum \|D^ju\|_{L^2(\Omega)}\right)\|v\|_{L^2(\Omega)} +$$

$$+ \|u\|_{L^2(\Omega)} \sum \|D^j v\|_{L^2(\Omega)} \leq \|u\|_{L^2(\Omega)} \|v\|_{L^2(\Omega)}$$

$$+ m^{1/2}\Big(\sum \|D^j u\|^2_{L^2(\Omega)}\Big)^{1/2} \|v\|_{L^2(\Omega)} + m^{1/2}\|u\|_{L^2(\Omega)}\Big(\sum \|D^j v\|^2_{L^2(\Omega)}\Big)^{1/2}. \quad (4.5)$$

We go back to (4.2). Assume that the scalar product $(u,v)_\alpha$ is chosen with $\alpha > \nu = \text{ess. sup } c$ as in (1.10). Then, taking the uniform ellipticity condition (1.6) into account we obtain from (4.4) and (4.5) that

$$\left|\int_\Gamma \gamma \bar{u} v \, d\sigma\right| \leq (C_1(\alpha - \nu)^{-1} + C_2(\alpha - \nu)^{-1/2}) \|u\|_\alpha \|v\|_\alpha , \quad (4.6)$$

thus it is obvious that, if α is sufficiently large, the first of the two inequalities

$$c\|u\|_{H^1(\Omega)} \leq (u,u)_\alpha \leq C\|u\|_{H^1(\Omega)} \quad (4.7)$$

will hold for $u \in \mathfrak{D}$; that the second is as well true follows from (4.6) with no particular requirements on α beyond $\alpha > \nu$. The fact that $\mathfrak{D}$ is dense in $H^1(\Omega)$ (Theorem 4.2) and the Schwartz inequality

$$|(u,v)_\alpha| \leq (u,u)_\alpha (v,v)_\alpha \quad (4.8)$$

imply that $(u,v)_\alpha$ can be defined, using an obvious approximation argument for arbitrary $u \in H^1(\Omega)$. Since the norm defined by (4.2) is equivalent to the original norm of $H^1(\Omega)$, we shall assume in what follows that $H^1(\Omega)$ is endowed with $\|\cdot\|_\alpha$.

From this point on, the construction of the operator $A_0(\beta)$ corresponding to the self adjoint part (1.8) of A and to the boundary condition β in (4.1) proceeds exactly in the same way as in the case of the Dirichlet boundary condition: $u \in D(A_0(\beta))$ if and only if the linear functional

$$w \to (u,v)_\alpha \quad (w \in H^1(\Omega)) \quad (4.9)$$

is continuous in the norm of $L^2(\Omega)$; we define

$$A_0(\beta)u = \alpha u - v, \quad (4.10)$$

where v is the unique element of $L^2(\Omega)$ satisfying

$$(u,w)_\alpha = (v,w) \quad (w \in H^1(\Omega)) \tag{4.11}$$

As in §IV.1, motivation for this stems from the fact that if the coefficients a_{jk} and the boundary Γ are smooth and u is a smooth function such that $\alpha u - A_0 u = v$ in Ω and $D^{\tilde{\nu}} u = \gamma u$, then (4.11) follows for any smooth w.

Operating as in §IV.1 we show that

$$(\lambda I - A_0(\beta))D(A_0(\beta)) = L^2(\Omega), \tag{4.12}$$

this time for any $\lambda > \alpha$, α so large that (4.7) holds. We obtain an estimate of the type of (1.19) and prove that $\lambda I - A_0(\beta)$ is one-to-one in the same range of λ, so that $(\lambda I - A_0(\beta))^{-1}$ exists in $\lambda > \alpha$. Finally, $A_0(\beta)$ is symmetric so that, using Lemma 1.1 we show that $A_0(\beta)$ is self adjoint and bounded above by α, where α is a constant depending not only on ν but also on the coefficient γ. Accordingly, the cosine function $\mathfrak{C}_0(t)$ generated by $A_0(\beta)$ is this time given by

$$\mathfrak{C}_0(t)u = \int_{-\infty}^{\alpha+} \cosh t\, \mu^{1/2} P(d\mu)u$$

§IV.5 The phase space.

The arguments in §IV.2 have an obvious counterpart here. The construction of the square root B of $A_0(\beta)$ proceeds in the same way, as does the proof of

THEOREM 5.1

$$D(B) = H^1(\Omega) . \tag{5.1}$$

The phase space for the equation

$$u''(t) = A_0(\beta)u(t) \tag{5.2}$$

is now

$$\mathfrak{E} = D((\lambda I - A_0(\beta))^{1/2}) \times L^2(\Omega) = H^1(\Omega) \times L^2(\Omega) , \tag{5.3}$$

and we have

$$\tilde{D}_1 = H^1(\Omega). \tag{5.4}$$

Again, the phase space (5.3) is the same one provided by Theorem III.1.3. The group $\mathfrak{S}_0(\hat{t})$ propagating the solutions of (5.2) is given by (2.11) with infinitesimal generator (2.12), its domain being identified by $D(\mathfrak{A}_0(\beta)) = D(A_0(\beta)) \times H^1(\Omega)$. To take care of the first order terms we use Theorem 2.3 applied to the bounded perturbation operator (2.13). In this way we obtain:

THEOREM 5.1. Let Ω be a domain of class $C^{(1)}$ with bounded boundary Γ, A the operator (1.2), β the boundary condition (1.4) with γ measurable and bounded on Γ. Let

$$A(\beta) = A_0(\beta) + P \tag{5.5}$$

with domain $D(A(\beta)) = D(A_0(\beta))$. Then the space $H^1(\Omega) \times L^2(\Omega)$ is a phase space for the equation

$$u''(t) = A(\beta)u(t) \ . \tag{5.6}$$

§IV.6 The Cauchy problem.

All the results in Section IV.3 have an immediate counterpart here; we define the semigroup $\mathfrak{B}_0(\hat{t})$ given by (3.1) in the product space $\mathfrak{D} = L^2(\Omega) \times L^2(\Omega)$; again, $\mathfrak{B}_0(t)$ depends on the particular square root of $A_0(\beta)$ chosen. However, we need B to have a bounded inverse. This can be achieved by replacing $c(x)$ by $c(x) - \alpha$ for α sufficiently large in the definition of $A_0(\beta)$; the operator P is then defined by

$$Pu = \sum_{j=1}^{m} b_j(x) D^j u + \alpha u \ . \tag{6.1}$$

THEOREM 6.1. Let Ω be a domain of class $C^{(1)}$ with bounded boundary Γ, A the operator (1.2), β the boundary condition (1.4) with γ measurable and bounded on Γ. Let

$$A(\beta) = A_0(\beta) + P \tag{6.2}$$

with domain $D(A(\beta)) = D(A_0(\beta))$. Then the Cauchy problem for the equation

$$u''(t) = A(\beta)u(t) \tag{6.3}$$

is well posed in $-\infty < t < \infty$.

§IV.7 Higher order equations.

We consider briefly in the rest of the chapter the equation

$$u''(t) = A(\beta)u(t) \tag{7.1}$$

where

$$Au = \sum_{|\alpha|\le p} a_\alpha(x)D^\alpha u \tag{7.2}$$

is an arbitrary partial differential operator of order p (here $\alpha = (\alpha_1,\dots,\alpha_m)$ is a m-ple of nonnegative integers, $|\alpha| = \alpha_1 + \dots + \alpha_m$ and $D^\alpha = (D^1)^{\alpha_1} \dots (D^m)^{\alpha_m}$) whose coefficients $a_\alpha(x)$ are defined in a domain Ω of m-dimensional Euclidean space $\mathbb{R}^m$; $A(\beta)$ denotes the restriction of A obtained by imposition of a boundary condition β at the boundary Γ. Some insight on the equation (7.1) can be obtained examining the constant coefficient case in the whole space; we do this for $m = 1$.

EXAMPLE 7.1. Consider the differential equation

$$u''(t) = Au(t) \tag{7.3}$$

in the space $L^2(\mathbb{R}_x)$. Here A is the differential operator

$$Au = \sum_{j=0}^{p} a_j\left(\frac{d}{dx}\right)^j u \tag{7.4}$$

with $a_0, a_1,\dots,a_p$ complex constants, $a_p \ne 0$; the domain of A consists of all $u(\hat{x}) \in L^2(\mathbb{R})$ such that Au (understood in the sense of distributions) belongs to $L^2(\mathbb{R})$. Through the Fourier-Plancherel transform

$$\tilde{u}(\xi) = \lim_{a\to\infty} \frac{1}{(2\pi)^{1/2}} \int_{-a}^{a} e^{i\xi x}u(x)\,dx \tag{7.5}$$

(see STEIN-WEISS [1971:1]) the equation (7.3) is easily seen to be equivalent to the equation

$$u''(t) = \tilde{A}u(t), \tag{7.6}$$

where $\tilde{A}$ is the multiplication operator

$$\tilde{A}u(\xi) = \Big(\sum_{j=0}^{p} a_j(-i\xi)^j\Big)u(\xi) = p(\xi)u(\xi) \tag{7.7}$$

in $L^2(R_\xi)$. We check that $\tilde{A}$ is a normal operator, thus $\tilde{A} \in \mathfrak{C}^2$ (Exercise II.5) if and only if

$$\omega_0 = \sup\{\operatorname{Re} \lambda^{1/2}; \lambda \in \sigma(\tilde{A})\} < \infty, \tag{7.8}$$

where $\sigma(\tilde{A})$, the spectrum of $\tilde{A}$, is easily identified as

$$\sigma(\tilde{A}) = \{p(\xi); -\infty < \xi < \infty\}.$$

As proved in Exercise II.5, (7.8) is equivalent to the fact that $\sigma(\tilde{A})$ is contained in a region of the form

$$\operatorname{Re} \lambda \le \omega^2 - (\operatorname{Im} \lambda)^2/4\omega^2. \tag{7.9}$$

LEMMA 7.2. $\tilde{A} \in \mathfrak{C}^2$ <u>if and only if</u> (a) p <u>is even</u>, (b) a_p <u>is real with</u>

$$(-1)^{p/2} a_p < 0 , \tag{7.10}$$

(c) a_j <u>is real if</u> j <u>is even</u> $j > p/2$, (d) a_j <u>is imaginary if</u> j <u>is odd</u>, $j > p/2$.

<u>Proof</u>: Assume that (a), (b), (c) and (d) hold. Then we have

$$p(\xi) = (-1)^{p/2} a_0 \xi^p + r(\xi) + q(\xi), \tag{7.11}$$

where $q(\xi)$ is a polynomial of degree $\le p/2$ with complex coefficients, and

$$r(\xi) = \sum_{p/2 < j \le p-1} b_j \xi^j \tag{7.12}$$

with real coefficients b_j. Obviously, for $|\xi|$ sufficiently large we shall have

$$|\operatorname{Im} p(\xi)|^2 = |\operatorname{Im} q(\xi)|^2 \le |q(\xi)|^2 \le C|\xi|^p \le C'(-\operatorname{Re} p(\xi)) \tag{7.13}$$

for large $|\xi|$, which implies that $\sigma(A)$ is contained in a region of the form (7.9), thus $\tilde{A} \in \mathfrak{C}^2$.

We prove the converse. Assume that (a) is violated, that is, that p is odd. Since

$$p(\xi) = i(-1)^{(p-1)/2} a_0 \xi^p (1 + O(|\xi|^{-1})) \qquad (|\xi| \to \infty), \tag{7.14}$$

and $a_0 \neq 0$, it is obvious that $\mathfrak{p}(\xi)$ tends to infinity along the imaginary axis in opposite directions $\xi \to \pm \infty$, hence $\sigma(\tilde{A})$ cannot be contained in a region of the form (7.9). To show the necessity of (7.10) when p is even we use the corresponding version of (7.14):

$$\mathfrak{p}(\xi) = (-1)^{p/2} a_0 \xi^p (1 + 0(|\xi|^{-1})) \qquad (|\xi| \to \infty); \tag{7.15}$$

if $(-1)^{p/2} a_0$ is not real and negative, $\mathfrak{p}(\xi)$ will tend to infinity in the direction of the positive real axis.

Finally, we show that (c) and (d) are as well necessary. Note first that both conditions are necessary and sufficient in order that the polynomial $r(\xi)$ in (7.11) have real coefficients. Assume that $r(\xi)$ has at least one complex coefficient, and let b_k be the complex coefficient of highest order; we can write then

$$\operatorname{Im} \mathfrak{p}(\xi) = (\operatorname{Im} b_k) \xi^k (1 + 0(|\xi|^{-1})) \qquad (|\xi| \to \infty) \tag{7.16}$$

where $p/2 < k \leq p - 1$. Since $2k > p$, an inequality of the type of (7.13) will not hold in this case for large $|\xi|$. This ends the proof of Lemma 7.1.

We note the curious consequences of Lemma 7.1: although the operator

$$A = -\left(\frac{d}{dx}\right)^8 + \left(\frac{d}{dx}\right)^6$$

belongs to $\mathfrak{C}^2$, the operator

$$A = -\left(\frac{d}{dx}\right)^8 + \left(\frac{d}{dx}\right)^5$$

does not, in spite of the fact that $(d/dx)^5$ is a "tamer" perturbation of $(d/dx)^8$ than $(d/dx)^6$.

In the following section we shall attempt a theory of the equation (7.1), but only in the case where β is the higher order version of the Dirichlet boundary condition. Lemma 7.1 indicates that the coefficients of A of order $> p/2$ will have to be suitably restricted.

§IV.8 Higher order equations (continuation)

We study here the equation (7.1) with an operator A of the form

$$Au = \sum_{|\alpha|\leq k} \sum_{|\alpha|\leq k} (-1)^{|\alpha|-1} D^{\alpha}(a_{\alpha\beta}(x) D^{\beta}u) + \sum_{|\alpha|\leq k} b_{\alpha}(x) D^{\alpha}u \,. \tag{8.1}$$

The coefficients $a_{\alpha\beta}$, b_{α} are real and defined in a bounded domain Ω of m-dimensional Euclidean space R^m. We shall assume that the coefficients $a_{\alpha\beta}$ of the principal part of the operator A,

$$\sum_{|\alpha|=k} \sum_{|\beta|=k} (-1)^{|\alpha|-1} D^{\alpha}(a_{\alpha\beta} D^{\beta}) \,, \tag{8.2}$$

are continuous in $\bar{\Omega}$; the rest of the $a_{\alpha\beta}$, as well as the b_{α} are simply measurable and bounded in Ω. The operator $A(\beta)$ denotes the restriction of A obtained by imposition at the boundary Γ of the Dirichlet boundary condition

$$u = D^{\nu}u = \ldots = (D^{\nu})^{k-1}u = 0 \qquad (x \in \Gamma) \tag{8.3}$$

(although (8.3) will be satisfied only in a generalized sense to be clarified later). We assume that

$$a_{\alpha\beta} = a_{\beta\alpha} \,,$$

and that A is uniformly elliptic, which in this case means that

$$\sum_{|\alpha|=k} \sum_{|\beta|=k} a_{\alpha\beta}(x) \xi^{\alpha}\xi^{\beta} \geq \kappa |\xi|^{2k} \quad (x \in \bar{\Omega},\ \xi \in R^m) \tag{8.4}$$

for some $\kappa > 0$.

The following result (Gårding's inequality) will be basic. To state it we introduce the Sobolev spaces $W^{k,p}(\Omega)$ $(1 \leq p < \infty)$ consisting of all functions u defined in Ω and having partial derivatives of order $\leq k$ (understood in the sense of distributions) in $L^p(\Omega)$; the norm of $W^{k,p}(\Omega)$ is

$$\|u\|_{W^{k,p}(\Omega)} = \Big(\sum_{|\alpha|\leq k} \|D^{\alpha}u\|^p_{L^p(\Omega)} \Big)^{1/p} \,. \tag{8.5}$$

For $p = 2$ (the only case of interest to us) we shall write $W^{k,2}(\Omega) = H^k(\Omega)$. The space $H^k_0(\Omega)$ is the closure of $\mathcal{D}(\Omega)$ in $H^k(\Omega)$. (in the norm of $H^k(\Omega)$). The statement that $u \in H^k_0(\Omega)$ is the weak version of the boundary conditions (8.3).

THEOREM 8.1 *Let* L *be a differential operator*

$$\sum_{|\alpha|\le k}\ \sum_{|\beta|\le k} (-1)^{|\alpha|-1} D^{\alpha}(a_{\alpha\beta} D^{\beta} u)$$

<u>in a bounded domain</u> $\Omega \subseteq \mathbb{R}^m$. <u>Assume that all the coefficients</u> $a_{\alpha\beta}$ <u>are measurable and bounded and that</u> $a_{\alpha\beta}$ <u>is continuous in</u> $\overline{\Omega}$ <u>when</u> $|\alpha| = |\beta| = k$. <u>Then there exist constants</u> c, α <u>such that</u>

$$\sum_{|\alpha|\le k}\ \sum_{|\beta|\le k} \int_{\Omega} a_{\alpha\beta}(x) D^{\alpha}\bar{u} D^{\beta} u \, dx \ge$$

$$\ge c\|u\|_{H_0^k(\Omega)} - \alpha\|u\|_{L^2(\Omega)} \qquad (u \in H_0^k(\Omega)) . \tag{8.6}$$

For a proof see FRIEDMAN [1969:1, p.32].

We proceed to the construction of a phase space for the equation

$$u''(t) = A_0(\beta)u(t), \tag{8.7}$$

where A_0 is the self adjoint part of A,

$$A_0 = \sum_{|\alpha|\le k}\ \sum_{|\beta|\le k} (-1)^{|\alpha|-1} D^{\alpha}(a_{\alpha\beta} D^{\beta} u). \tag{8.8}$$

The definition of $A_0(\beta)$ follows that for the second order case. We renorm the space $H_0^k(\Omega)$ by means of the scalar product

$$(u,v)_{\alpha} = \alpha \int_{\Omega} \bar{u} v \, dx + \int_{\Omega} (\sum \sum a_{\alpha\beta} D^{\alpha}\bar{u} D^{\alpha} v) \, dx , \tag{8.9}$$

where α is the constant in (8.6). We have

$$c\|u\|_{H_0^k(\Omega)} \le (u,u)_{\alpha} \le C\|u\|_{H_0^k(\Omega)} . \tag{8.10}$$

The second inequality follows from the boundedness of the coefficients of A_0; the first is a consequence of Theorem 8.1. An element $u \in H_0^k(\Omega)$ belongs to $D(A_0(\beta))$ if and only if the linear functional $w \to (u,w)_{\alpha}$ is continuous in the norm of $L^2(\Omega)$, $A_0(\beta)u$ being the only element of $L^2(\Omega)$ that satisfies

$$(u,w)_{\alpha} = ((\alpha I - A_0(\beta))u, w) \qquad (w \in H_0^k(\Omega)) \tag{8.11}$$

We show in the same way as in the case $k = 2$ that $A_0(\beta)$ is self adjoint and that $A_0(\beta)$ is bounded above (by α), so that $A_0(\beta)$ generates the cosine function

$$C(t) = \cosh t\, A_0(\beta)^{1/2}, \tag{8.12}$$

and a square root $B = A_0(\beta)^{1/2}$ can be defined as in §IV.2: we have

$$D(B) = H_0^k(\Omega) = D((\lambda I - A_0(\beta))^{1/2}), \tag{8.13}$$

the last inequality holding for $\lambda > \alpha$. Theorem III.5.4, combined with (8.13) implies that

$$\mathfrak{E} = H_0^k(\Omega) \times L^2(\Omega) \tag{8.14}$$

is a state space for (8.7). To show that $\mathfrak{E}$ is as well a state space for the full equation

$$u''(t) = A(\beta)u(t), \tag{8.15}$$

we incorporate the lower order terms in (8.1) through perturbation (Theorem 2.3) defining

$$P = \sum_{|\alpha| \le k} b_\alpha(x)D^\alpha u \tag{8.16}$$

and

$$\mathfrak{P} = \begin{bmatrix} 0 & 0 \\ P & 0 \end{bmatrix} \tag{8.17}$$

We obtain in this way:

THEOREM 8.2. <u>Let</u> A <u>be the operator</u> (8.1), β <u>the Dirichlet boundary condition</u> (8.3), <u>and let</u>

$$A(\beta) = A_0(\beta) + P \tag{8.18}$$

<u>with domain</u> $D(A(\beta)) = D(A_0(\beta))$. <u>Then the space</u> $H_0^k(\Omega) \times L^2(\Omega)$ <u>is a phase space for the equation</u> (8.15).

The treatment of the Cauchy problem for (8.15) follows word by word that for second order equations in §IV.3; we only state the final result.

THEOREM 8.3. <u>Let</u> A <u>be the operator</u> (8.1), β <u>the Dirichlet boundary condition</u> (8.3), <u>and let</u>

$$A(\beta) = A_0(\beta) + P \tag{8.19}$$

with domain $D(A(\beta)) = D(A_0(\beta))$. Then the Cauchy problem for the equation (8.15) is well posed in $-\infty < t < \infty$.

§IV.9 Miscellaneous comments.

Wave equations like (1.1) (and its higher order counterpart (7.1)) were treated by semigroup methods in GOLDSTEIN [1969:1].

EXERCISE 1. A linear operator B with domain $D(B) = E$ in a Banach space E is said to be compact (in older terminology, completely continuous) if $\{Au_n\}$ contains a convergent subsequence for every bounded sequence $\{u_n\}$. Show that the spectrum $\sigma(A)$ of a compact operator consists of a (finite or countable) sequence $\{\lambda_1, \lambda_2, \ldots\}$ of complex numbers having an accumulation point at zero when the sequence is infinite. (See KATO [1976:1, Sec. 3.7]).

EXERCISE 2. Let A be as in Exercise 1, and let $\lambda \in \sigma(A)$, $\lambda \neq 0$. Prove that λ is an eigenvalue of A, that is, there exists $u \in E$, $u \neq 0$ such that

$$(\lambda I - A)u = 0 \tag{9.1}$$

Show that the space $F_g(\lambda)$ of all generalized eigenvectors with eigenvalue λ (i.e. the set of all vectors u such that

$$(\lambda I - A)^m u = 0 \tag{9.2}$$

for some integer $m \geq 1$) is finite dimensional. Show that there exists an integer $m(\lambda)$ such that if (9.2) holds for any integer $m \geq m(\lambda)$ then

$$(\lambda I - A)^{m(\lambda)} u = 0 \tag{9.3}$$

EXERCISE 3. Show that there exists a compact operator A in separable Hilbert space such that (a) $\sigma(A) = \{0\}$ (b) 0 is not an eigenvalue of A (Hint: try the Volterra operator

$$Vu(x) = \int_0^x u(\xi)d\xi \tag{9.4}$$

in $L^2(0,1)$).

EXERCISE 4. Let A be an operator in a Banach space E such that $R(\lambda;A)$ is compact for some $\lambda \in \rho(A)$. Then $R(\mu;A)$ is compact for all $\mu \in \rho(A)$ (Hint: use the second resolvent equation $R(\mu;A) = R(\lambda;A) + (\lambda-\mu)R(\mu;A)R(\lambda;A)$ and the fact that the sum of two compact operators and the product of a compact operator and a bounded operator are compact; see KATO [1976:1]).

EXERCISE 5. Let A be as in Exercise 4. Show that $\sigma(A)$ is empty or consists of a sequence $\lambda_1, \lambda_2, \ldots$ of complex numbers such that $|\lambda_k| \to \infty$ if the sequence is infinite. Show that if $\lambda \in \sigma(A)$ then λ is an eigenvalue of A and the space $E_g(\lambda)$ of generalized eigenvectors of A enjoys the properties described in Exercise 2.

EXERCISE 6. Show that there exists an operator as in Exercise 4 with $\sigma(A) = \emptyset$ (Hint: try the inverse of the Volterra operator (9.4)).

EXERCISE 7. Let Ω be a bounded domain in m-dimensional Euclidean space $\mathbb{R}^m$, B a linear bounded operator from $L^2(\Omega)$ into $H^1(\Omega)$. Show that B, thought of as an operator from $L^2(\Omega)$ into $L^2(\Omega)$, is compact (See MIHAILOV [1976:1]).

EXERCISE 8. Using Exercise 7 show that the second order operators in (3.15) and (6.2) and the higher order operators in (8.19) enjoy the spectral properties in Exercise 5 (Hint: show that $R(\lambda;A_0(\beta))$ is compact using Exercise 7 and then apply Exercise 5).

CHAPTER V

UNIFORMLY BOUNDED GROUPS AND COSINE FUNCTIONS IN HILBERT SPACE

§ V.1 *The Hahn-Banach theorem: Banach limits.*

Let E be an arbitrary real linear space. A functional $p : E \to \mathbb{R}$ is called *sublinear* if

$$p(u + v) \leq p(u) + p(v), \; p(\alpha u) = \alpha p(u) \quad (\alpha \geq 0) \tag{1.1}$$

for $u, v \in E$ arbitrary.

THEOREM 1.1. (Hahn-Banach). *Let* F *be a subspace of* E, $\varphi : F \to \mathbb{R}$ *a linear functional. Assume that*

$$\varphi(u) \leq p(u) \quad (u \in F). \tag{1.2}$$

Then there exists a linear functional $\Phi : E \to \mathbb{R}$ *such that*

$$\Phi(u) = \varphi(u) \quad (u \in E), \tag{1.3}$$

$$\Phi(u) \leq p(u) \quad (u \in E). \tag{1.4}$$

For a proof see BANACH [1932: 1, p. 28].

With the help of Theorem 1.1 we can construct an intriguing extension of the notion of limit. Let $B = B[0, \infty)$ be the space of all bounded complex functions defined in $t \geq 0$. A *Banach limit* or *generalized limit* in B is a functional $\underset{s \to \infty}{\mathrm{LIM}} : B \to \mathbb{C}$ enjoying the following properties, where $f(\hat{t}), g(\hat{t}) \in B$ and α, β are complex numbers.

(a) $\underset{s \to \infty}{\mathrm{LIM}} (\alpha f(s) + \beta g(s)) = \alpha \underset{s \to \infty}{\mathrm{LIM}} f(s) + \beta \underset{s \to \infty}{\mathrm{LIM}} g(s).$

(b) $\underset{s \to \infty}{\mathrm{LIM}} f(s + t) = \underset{s \to \infty}{\mathrm{LIM}} f(s) \quad (t \geq 0).$

(c) $\underset{s \to \infty}{\mathrm{LIM}} f(s) \geq 0$ if $f(s) \geq 0$.

(d) $\liminf_{s\to\infty} f(s) \le \operatorname{LIM}_{s\to\infty} f(s) \le \limsup_{s\to\infty} f(s)$ if f is real.

(e) $|\operatorname{LIM}_{s\to\infty} f(s)| \le \limsup_{s\to\infty} |f(s)|$.

(f) $\operatorname{LIM}_{s\to\infty} f(s) = \lim_{s\to\infty} f(s)$ if the latter exists.

THEOREM 1.2. <u>Banach limits in</u> B <u>exist</u>.

<u>Proof</u>: Obviously, it is enough to construct $\operatorname{LIM}_{s\to\infty} f(s)$ for f real valued: once this done we simply set

$$\operatorname{LIM}_{s\to\infty} f(s) = \operatorname{LIM}_{s\to\infty} \operatorname{Re} f(s) + i \operatorname{LIM}_{s\to\infty} \operatorname{Im} f(s)$$

for $f \in B$ arbitrary. Let $B_{\mathbb{R}}$ be the subspace of B consisting of real valued functions. Define

$$p(f) = \inf \limsup_{s\to\infty} \frac{1}{n} \sum_{k=1}^{n} f(s + \xi_k) \tag{1.5}$$

for $f \in B_{\mathbb{R}}$, the infimum taken over all possible finite sequences $\{\xi_k\}$ of nonnegative numbers. We check instantly that p satisfies (1.1). Using Theorem 1.1 for $F = \{0\}$ and $\varphi = 0$ we deduce the existence of a linear functional $\Phi : B_{\mathbb{R}} \to \mathbb{R}$ such that

$$\Phi(f) \le p(f) \qquad (f \in B_{\mathbb{R}}). \tag{1.6}$$

Set $\operatorname{LIM}_{s\to\infty} = \Phi$. Obviously, (a) holds. Replacing f by $-f$ in (1.6) we obtain

$$\Phi(f) \ge -p(-f). \tag{1.7}$$

This yields (c). Since $p(f) \le \limsup f(t)$ and $-p(-f) \ge \liminf f(t)$, (d) follows from (1.6) and (1.7); obviously, (d) implies (f). To check (b) we take $\xi_1 = h$, $\xi_2 = 2h, \ldots, \xi_n = nh$ in (1.5) so that

$$p(f(\hat{t} + h) - f(\hat{t})) \le \frac{1}{n} \limsup_{s\to\infty} (f(s + nh) - f(t)).$$

Since n is arbitrary, $p(\hat{f}(t + h) - f(\hat{t})) \le 0$. We deduce in the same way that $p(f(\hat{t}) - f(\hat{t} + h)) \le 0$, thus it follows from (1.6) and (1.7) that $\Phi(f(\hat{t} + h)) = \Phi(f(\hat{t}))$. Finally, we show (e) as follows. Let θ be such that $e^{i\theta} \operatorname{LIM} f(s) = |\operatorname{LIM} f(s)|$. Then

$$|\text{LIM}\, f(s)| = \text{LIM}\, (e^{i\theta} f(s)) =$$

$$= \text{LIM}\, \text{Re}(e^{i\theta} f(s)) \leq \limsup_{s\to\infty} \text{Re}(e^{i\theta} f(s)) \leq \limsup_{s\to\infty} |f(s)| .$$

This concludes the proof of Theorem 1.2.

A Banach limit $\text{LIM}_{n\to\infty}$ in the space ℓ^∞ of complex bounded sequences $\zeta = (\zeta_0, \zeta_1, \ldots)$ is a functional in ℓ^∞ enjoying the properties corresponding to (a) - (f) in the definition of Banach limits in B:

(a') $\text{LIM}_{n\to\infty} (\alpha\zeta_n + \beta\tau_n) = \alpha \text{LIM}_{n\to\infty} \zeta_n + \beta \text{LIM}_{n\to\infty} \tau_n$.

(b') $\text{LIM}_{n\to\infty} \zeta_{n+m} = \text{LIM}_{n\to\infty} \zeta_n \quad (m \geq 0)$.

(c') $\text{LIM}_{n\to\infty} \zeta_n \geq 0$ if $\zeta_n \geq 0$.

(d') $\liminf_{n\to\infty} \zeta_n \leq \text{LIM}_{n\to\infty} \zeta_n \leq \limsup_{n\to\infty} \zeta_n$ if $\{\zeta_n\}$ is real.

(e') $|\text{LIM}_{n\to\infty} \zeta_n| \leq \limsup_{n\to\infty} |\zeta_n|$.

(f') $\text{LIM}_{n\to\infty} \zeta_n = \lim_{n\to\infty} \zeta_n$ if the latter exists.

COROLLARY 1.3. Banach limits in ℓ^∞ exist.

Proof: Define

$$\text{LIM}_{n\to\infty} \zeta_n = \text{LIM}_{s\to\infty} f(s),$$

where $\text{LIM}_{s\to\infty}$ is one of the Banach limits constructed in Theorem 1.2 and $f(s) = \zeta_n$ in $n \leq s < n + 1$.

§V.2 Uniformly bounded groups in Hilbert space.

Throughout the rest of this chapter (except in Section V.3) we shall assume that $E = H$ is a complex Hilbert space.

Let B be a self adjoint operator in H. Then it follows easily from the functional calculus for self adjoint operators that $U(\hat{t})$, where

$$U(t) = \exp(itB) \qquad (-\infty < t < \infty), \tag{2.1}$$

is a strongly continuous group in H. Moreover, since $U(t)^* = \exp(itB)^* = \exp(-itB) = U(-t) = U(t)^{-1}$, each $U(t)$ is unitary; in particular

$$\|U(t)u\| = \|u\| \ . \quad (-\infty < t < \infty, \ u \in E) \ . \tag{2.2}$$

(see DUNFORD-SCHWARTZ [1963:1, Ch. XII] for the necessary details on the functional calculus). It was first proved by Stone that the converse is as well true (See Exercise I.11)

THEOREM 2.1. <u>Let</u> $U(\hat{t})$ <u>be a strongly continuous group. Assume that each</u> $U(t)$ <u>is a unitary operator. Then</u> (2.1) <u>holds for</u> $B = -iA$, B <u>the infinitesimal generator of</u> $U(\hat{t})$.

<u>Proof</u>: Let A be the infinitesimal generator of $U(\hat{t})$. We show that $-iA$ is self adjoint. To this end we use formula (I.3.8) for $n = 1$ and $\lambda > 1$ real. We have

$$(R(\lambda;A)u,v) = \int_0^\infty e^{-\lambda t}(U(t)u,v)\,dt = \int_0^\infty e^{-\lambda t}(u,U(t)^*v)\,dt$$
$$= \int_0^\infty e^{-\lambda t}(u,U(-t)v)\,dt = -(u,R(-\lambda;A)v) \ , \tag{2.3}$$

where we have used in the last equality the fact that $U(-\hat{t})$ is a strongly continuous semigroup with infinitesimal generator $-A$. Accordingly, $R(\lambda;A)^* = -R(-\lambda;A)$. Taking inverses, this implies that $(\lambda I - A)^* = \lambda I + A$ so that $A^* = -A$, thus $B = -iA$ is self adjoint. If $V(t) = \exp(itB)$ (computed by the functional calculus) we show easily that

$$\int_0^\infty e^{-\lambda t}V(t)u\,dt = R(\lambda;iB)u = R(\lambda;A)u \qquad (\lambda > 1)$$

for each $u \in H$. Since $U(\hat{t})$ satisfies the same equality it follows that $U(\hat{t})$ and $V(t)$ coincide by uniqueness of Laplace transforms.

The following result shows that uniform boundedness of the group $S(\hat{t})$ implies that $S(t)$ is unitary after a "change in geometry" in the space H. Surprisingly enough, there are no conditions whatsoever on the t-dependence of $S(\hat{t})$.

THEOREM 2.2 <u>Let</u> $S(\hat{t})$ <u>be a group in</u> H $(S(0) = I,\ S(s + t) = S(s)S(t)$

for $-\infty < s,\ t < \infty$). Assume that

$$\|S(t)\| \le C \qquad (-\infty < t < \infty). \tag{2.4}$$

Then there exists a bounded, self adjoint operator Q satisfying

$$C^{-1}\|u\|^2 \le (Qu,u) \le C\|u\|^2 \qquad (u \in E)\ , \tag{2.5}$$

and such that

$$U(t) = QS(t)Q^{-1} \tag{2.6}$$

is unitary for every t.

Proof: We define a sesquilinear functional $\mathfrak{h}$ (that is, a map $\mathfrak{h} : H \times H \to \mathbb{C}$ linear in the second variable, conjugate linear in the first) by

$$\mathfrak{h}(u,v) = \underset{t\to\infty}{\mathrm{LIM}}\ (S(t)u, S(t)v), \tag{2.7}$$

where LIM is one of the Banach limits constructed in Theorem 1.2 (note that, since $|(S(t)u,S(t)v)| \le C^2\|u\|\ \|v\|$ the function $(S(\hat{t})u,S(\hat{t})v)$ belongs to $B[0,\infty)$). By (e) we have

$$|\mathfrak{h}(u,v)| \le C^2\|u\|\ \|v\| \qquad (u,v \in H), \tag{2.8}$$

thus there exists a bounded operator P in H such that

$$\mathfrak{h}(u,v) = (Pu,v)\ . \tag{2.9}$$

Moreover, since $\overline{\mathfrak{h}(v,u)} = \mathfrak{h}(v,u)$, the operator P is self adjoint, and it follows from (2.4) (which implies in particular that $\|S(t)u\| \ge C^{-1}\|u\|$) and from property (d) of Banach limits that $C^{-2}\|u\|^2 \le (Pu,u) \le C^2\|u\|^2$ so that P is positive as well and hence possesses a positive, self adjoint square root Q satisfying (2.5). Moreover, we have

$$(PS(t)u,v) = \underset{s\to\infty}{\mathrm{LIM}}\ (S(s+t)u,\ S(s)v) = \underset{s\to\infty}{\mathrm{LIM}}\ (S(s)u,\ S(s-t)v) =$$

$$\underset{s\to\infty}{\mathrm{LIM}}\ (S(s)u,S(s)S(-t)v) = (Pu,S(-t)v)$$

for $u,v \in H$ by virtue of (b). Hence

$$S(t)^* P = PS(-t) \qquad (-\infty < t < \infty) . \tag{2.10}$$

Multiplying by Q^{-1} left and right, this equality becomes

$$Q^{-1}S(t)^*Q = QS(-t)Q^{-1}, \tag{2.11}$$

which is nothing but $(QS(t)Q^{-1})^* = (QS(t)Q^{-1})^{-1}$. This ends the proof of Theorem 2.2.

The fact that $QS(t)Q^{-1}$ is unitary deserves some comment. Define

$$(u,v)_Q = (Qu,Qv) = (Pu,v) = (u,Pv). \tag{2.12}$$

Then $(u,v)_Q$ is a scalar product in H which produces a norm $\|u\|_Q$ equivalent to the original norm $(C^{-1}\|u\| \leq \|u\|_Q \leq C\|u\|)$. Now,

$$(S(t)u,v)_Q = (QS(t)u,Qv) = (S(t)u,Pv) = (u,S(t)^*Pv)$$

$$= (u,PS(-t)u) = (Qu,QS(-t)v) = (u,S(-t)v)_Q \, .$$

Hence $S(t)^{*Q} = S(-t)$, where $*Q$ denotes adjoint with respect to the scalar product (2.12). Accordingly, Theorem 2.2 can be interpreted as stating that each $S(t)$ <u>will in fact be unitary after replacing the original scalar product by</u> (2.12); since both norms are equivalent the topology of the space induced by any of the two norms is the same.

In the absence of hypoteses on the t-dependence of $S(\hat{t})$, no further conclusions on $U(\hat{t})$ are possible. On the other hand, we have

COROLLARY 2.3. *Let* $S(\hat{t})$ *satisfy the assumptions of Theorem* 2.2. *Assume that* $S(\hat{t})u$ *is continuous (or, more generally, strongly measurable) for each* $u \in E$. *Then there exists a self adjoint operator* B *and a bounded self adjoint operator* Q *satisfying* (2.5) *and such that*

$$S(t) = Q^{-1}\exp(itB)Q \qquad (-\infty < t < \infty). \tag{2.13}$$

Conversely, every $S(\hat{t})$ *of the form* (2.13) *is a strongly continuous group satisfying* (2.4).

Proof: By virtue of Theorem I.1.1 strong continuity and strong measurability of $S(\hat{t})u$ for each $u \in E$ are equivalent. Applying

Theorem 2.1 to the strongly continuous group $QS(\hat{t})Q^{-1}$ of unitary operators (2.13) results instantly.

We close this section with a discrete version of Theorem 2.2.

THEOREM 2.4. *Let* $\{S_n; -\infty < n < \infty\}$ *be a sequence of bounded operators satisfying* $S_0 = I$, $S_{m+n} = S_m S_n$ (*so that* $S_n = S_1^n$). *Assume that*

$$\|S_n\| \le C \qquad (-\infty < n < \infty). \tag{2.14}$$

Then there exists a bounded, self adjoint operator Q *such that* (2.5) *holds and*

$$U_n = QS_nQ^{-1} \tag{2.15}$$

is unitary for every n. *Equivalently, there exists a unitary operator* U *such that*

$$S_n = Q^{-1}U^nQ\,. \tag{2.16}$$

The proof imitates that of Theorem 2.2. The operator Q is defined as the positive, self adjoint square root of the operator P defined by

$$(Pu,v) = \operatorname*{LIM}_{n\to\infty} (S_nu, S_nv), \tag{2.17}$$

where LIM is one of the Banach limits of sequences constructed in Corollary 1.3. We prove arguing as in Theorem 2.2 that $U_n = QS_nQ^{-1}$ is unitary for all n; since $U_n = U_1^n$, we have $QS_nQ^{-1} = U_1^n$ whence (2.16) follows for $U = U_1$.

We note the following restatement of Corollary 2.3 in the language of differential equations in Banach spaces .

THEOREM 2.5. *Let* A *be a closed, densely defined operator in the Hilbert space* H *such that the Cauchy problem for*

$$u'(t) = Au(t) \tag{2.18}$$

is well posed in $-\infty < t < \infty$. *Assume, moreover, that for every generalized solution of* (2.18) *we have*

$$\|u(t)\| \leq C \qquad (-\infty < t < \infty) \tag{2.19}$$

(where C may depend on $u(\hat{t})$). Then there exists a self adjoint operator B and a bounded self adjoint operator Q satisfying inequalities (2.5) and such that $D(A) = \{u; Qu \in D(B)\}$ and

$$A = iQ^{-1}BQ. \tag{2.20}$$

The converse is as well true.

Theorem 2.4 has an obvious formulation in relation to the difference equation

$$S_{n+1} = AS_n \qquad (-\infty < n < \infty), \tag{2.21}$$

where A is a bounded invertible operator in H (so that (2.21) can be solved forward and backwards). We omit the details.

REMARK 2.6. It is useful to know whether the conclusions of Corollary 2.3 hold under weaker conditions on the t-dependence of $S(\hat{t})$. If the space H is separable it is enough to require that $S(\hat{t})$ be weakly measurable (that is, that $(S(t)u,v)$ be measurable as a complex valued function for every $u,v \in H$) since this implies that $S(\hat{t})u$ is strongly measurable for every $u \in E$ (see HILLE-PHILLIPS [1957:1, p. 73]). In a general Hilbert space weak and strong measurability do not coincide. However, it is enough to assume that $S(\hat{t})$ is weakly continuous (i.e. that $(S(\hat{t})u,v)$ is continuous for each $u,v \in H$) since weak continuity of $S(\hat{t})$ implies strong continuity. To see this we may use (2.6) (which does not require any assumption on t-dependence); since $S(\hat{t})$ will be strongly (weakly) continuous if and only if $U(\hat{t})$ is, we may assume that $S(\hat{t})$ itself is unitary. Then

$$\begin{aligned} \|S(t)u - S(s)u\|^2 &= ((S(t) - S(s))u,\ (S(t) - S(s))u) \\ &= 2(u,u) - ((S(s-t) + S(t-s))u,u). \end{aligned} \tag{2.22}$$

§V.3 Almost periodic functions.

Let $\ell > 0$. A set $e \subseteq (-\infty,\infty)$ is said to be ℓ-dense in $(-\infty,\infty)$ if and only if every (open) interval I of length $\geq \ell$ contains some point of e.

A function $f(\hat{t})$ defined in $-\infty < t < \infty$ with values in a Banach space E is called almost periodic (in the sense of H. Bohr) if it is continuous and for every $\varepsilon > 0$ there exists $\ell = \ell(\varepsilon;f)$ and a ℓ-dense set $e = e(\varepsilon;f)$ such that

$$\|f(t + \tau) - f(\tau)\| \leq \varepsilon \qquad (-\infty < t < \infty, \tau \in e). \tag{3.1}$$

We shall ordinarily denote by $e(\varepsilon;f)$ the set of all τ which satisfy (3.1) for all t; every element of $e(\varepsilon;f)$ will be denominated a ε-translation number of f.

Obviously, a periodic function is almost periodic (for every $\varepsilon > 0$ we may take $e = \{np, -\infty < n < \infty\}$ where p is the period of $f(\hat{t})$).

We prove below a number of elementary facts on vector valued almost periodic functions which suffice to illuminate somewhat the definition. On the other hand, the proofs of the fundamental properties (Theorems 3.7, 3.8 and 3.9) will be omitted.

LEMMA 3.1. *Let* $f(\hat{t})$ *be almost periodic. Then* (a) $f(\hat{t})$ *is uniformly continuous in* $-\infty < t < \infty$. (b) *The range of* $f(\hat{t})$,

$$\Re f(\hat{t}) = \{f(t); -\infty < t < \infty\} \tag{3.2}$$

is precompact in E.

Proof: To show (a) we select $\varepsilon > 0$, a number $\ell = \ell(\varepsilon/3;f)$ and a ℓ-dense set $e = e(\varepsilon/3;f)$ such that $\|f(t + \tau) - f(t)\| \leq \varepsilon/3$ for $-\infty < t < \infty$ and $\tau \in e$. Using uniform continuity of f on compact intervals we select next δ, $0 < \delta < 1$ such that $\|f(t) - f(t')\| \leq \varepsilon$ for $0 \leq t, t' \leq \ell + 1$ and $|t - t'| \leq \delta$.

Let t,t' be arbitrary points in the real line such that $|t - t'| \leq \delta$. Choose $\tau \in e$ such that $0 < t + \tau, t' + \tau < \ell + 1$. Then

$$\|f(t) - f(t')\| \leq \|f(t) - f(t + \tau)\|$$

$$+ \|f(t + \tau) - f(t' + \tau)\| + \|f(t' + \tau) - f(t')\| \leq \varepsilon.$$

This completes the proof of (a). To show (b), we select again $\varepsilon > 0$ and obtain from the definition of almost periodic function a number $\ell = \ell(\varepsilon/2,f)$ and a ℓ-dense set $e = e(\varepsilon/2;f)$ such that $\|f(t + \tau) - f(t)\| \leq \varepsilon/2$ for $\tau \in e$ and all t. Since the image of the interval $0 \leq t \leq \ell$ by $f(\hat{t})$ is compact in E we can select a finite sequence $t_1, \ldots, t_n$ in $0 \leq t \leq \ell$ such that

$$f(t) \in \mathcal{S}(f(t_1);\varepsilon/2) \cup \ldots \cup \mathcal{S}(f(t_n);\varepsilon/2),$$

where $\mathcal{S}(u,\rho)$ denotes the sphere $\{v; \|v-u\| \leq \rho\}$. Let t be an arbitrary real number, $\tau = \tau(t)$ a translation number in e such that $0 < t + \tau < \ell$, t_j an element of the sequence $t_1, \ldots, t_n$ such that $f(t+\tau) \in \mathcal{S}(f(t_j);\varepsilon/2)$. Then

$$\|f(t) - f(t_j)\| \leq \|f(t) - f(t+\tau)\| + \|f(t+\tau) - f(t_j)\| \leq \varepsilon$$

so that

$$\mathcal{R}f(\hat{t}) \subseteq \mathcal{S}(f(t_1);\varepsilon) \cup \ldots \cup \mathcal{S}(f(t_n);\varepsilon) .$$

Since ε is arbitrary, $\mathcal{R}f(\hat{t})$ is totally bounded; equivalently, $\mathcal{R}f(\hat{t})$ is precompact (see DUNFORD-SCHWARTZ [1958:1, p. 22]).

As a consequence of (b) we obtain that <u>every almost periodic function is bounded</u>.

LEMMA 3.2. <u>Let</u> $\{f_n(\hat{t})\}$ <u>be a sequence of almost periodic functions such that</u> $f_n(t) \to f(t)$ <u>uniformly in</u> $-\infty < t < \infty$. <u>Then</u> $f(\hat{t})$ <u>is almost periodic</u>.

<u>Proof</u>: Choose $\varepsilon > 0$. Let n be so large that $\|f_n(t) - f(t)\| \leq \varepsilon/3$ for $-\infty < t < \infty$. Once n has been selected, let $\ell = \ell(\varepsilon/3; f_n)$ and $e = e(\varepsilon/3; f_n)$ a ℓ-dense set such that $\|f_n(t+\tau) - f_n(t)\| \leq \varepsilon/3$ for $-\infty < t < \infty$ and $\tau \in e$. Then

$$\|f(t+\tau) - f(t)\| \leq \|f(t+\tau) - f_n(t+\tau)\| +$$
$$+ \|f_n(t+\tau) - f_n(t)\| + \|f_n(t) - f(t)\| \leq \varepsilon$$

for all t and $\tau \in e$, which shows that $f(\hat{t})$ is almost periodic as claimed.

The following "topological" characterization of almost periodic functions will simplify some of the proofs that follow. A continuous E-valued function $f(\hat{t})$ defined in $-\infty < t < \infty$ is <u>translation precompact</u> if for every sequence of real numbers $\{h_n\}$ there exists a subsequence

$h_{n(k)}$ such that $\{f(\hat{t} + h_{n(k)})\}$ is convergent, uniformly in $-\infty < t < \infty$.

THEOREM 3.3. <u>A function</u> $f(\hat{t})$ <u>is almost periodic if and only if it is translation precompact</u>.

<u>Proof</u>: Assume $f(\hat{t})$ is almost periodic. Let $\mathfrak{D}$ be a countable dense set in $(-\infty,\infty)$ (say, the rationals). If s_0 is an element of $\mathfrak{D}$ we may take advantage of the precompactness of the range $\mathfrak{R}f(t)$ (Lemma 3.1) to select a subsequence $\{h_{n(k)}\}$ of $\{h_n\}$ such that $f(s_0 + h_{n(k)})$ is convergent. Using then the method of diagonal sequences we can better this with a subsequence, denoted $\{k_n\}$, such that $f(s + k_n)$ converges for <u>all</u> $s \in \mathfrak{D}$ (at this stage we cannot assert that there is convergence for $s \notin \mathfrak{D}$, or even that convergence in $\mathfrak{D}$ is uniform).

Select $\varepsilon > 0$, and let $\ell = \ell(\varepsilon/5;f) > 0$, $e = e(\varepsilon/5;f)$ a ℓ-dense set such that $\|f(t + \tau) - f(t)\| \leq \varepsilon/5$ for all t and all $\tau \in e$: choose then $\delta = \delta(\varepsilon/5)$ such that $\|f(t) - f(t')\| \leq \varepsilon/5$ if $|t - t'| \leq \delta$ (that this is possible follows from Lemma 3.1 (a)). Finally, cover the interval $[0,\ell]$ with a finite number r of subintervals of length $\leq \delta$ and pick rationals $t_1,\ldots,t_r$ in each of these intervals; once this is done, determine n_0 such that $\|f(t_j + k_n) - f(t_j + k_m)\| \leq \varepsilon/5$ for $m,n \geq n_0$, $j = 1,2,\ldots,r$.

Let t be an arbitrary real number. Pick a translation number $\tau = \tau(t) \in e$ in the interval $(-t, -t + \ell)$ (so that $0 < t + \tau < \ell$) and a point t_j such that $|t + \tau - t_j| < \delta$. We have

$$\|f(t + k_n) - f(t + k_m)\| \leq \|f(t + k_n) - f(t + \tau + k_n)\|$$

$$+ \|f(t + \tau + k_n) - f(t_j + k_n)\| + \|f(t_j + k_n) - f(t_j + k_m)\|$$

$$+ \|f(t_j + k_m) - f(t + \tau + k_m)\|$$

$$+ \|f(t + \tau + k_m) - f(t + k_m)\| \leq \varepsilon \qquad (-\infty < t < \infty)$$

if $n,m \geq n_0$. This shows that $f(\hat{t})$ is translation precompact as claimed.

We prove the converse. Assume that $f(\hat{t})$ is translation precompact. If $f(\hat{t})$ is not almost periodic then there exists $\varepsilon > 0$ such that for no $\ell > 0$ there exists a ℓ-dense set e of ε-translation numbers τ. Let h_1 be arbitrary and let (a_1,b_1) be an interval of lenght $> 2|h_1|$ not containing any ε-translation number of $f(\hat{t})$. Let $h_2 = (a_1 + b_1)/2$ be the midpoint of (a_1, b_1). Obviously, $h_2 - h_1$ belongs to (a_1, b_1), thus $h_2 - h_1$ cannot be an ε-translation number of $f(\hat{t})$. We select next an interval (a_2, b_2) of length $> 2(|h_1| + |h_2|)$ not containing any ε-translation number of $f(\hat{t})$ and define $h_3 = (a_2 + b_2)/2$. Then $h_3 - h_1$ and $h_3 - h_2$ belong to (a_2, b_2), hence cannot be ε-translation numbers of $f(\hat{t})$. Arguing in the same way we construct a sequence $\{h_n\}$ of real numbers such that, for each n, $h_n - h_1, h_n - h_2, \ldots, h_n - h_{n-1}$ are not ε-translation numbers of $f(\hat{t})$. Then, if $n > m$ we have

$$\sup \|f(t + h_n) - f(t + h_m)\| = \sup \|f(t + h_n - h_m) - f(t)\| \geq \varepsilon,$$

which shows that the sequence $\{f(\hat{t} + h_n)\}$ cannot contain any subsequence uniformly convergent in $-\infty < t < \infty$. This ends the proof of Theorem 3.3.

COROLLARY 3.4. <u>Let</u> $f(t), g(t)$ <u>be almost periodic. Then</u> $\alpha f(\hat{t}) + \beta g(\hat{t})$ <u>is almost periodic for arbitrary complex</u> α, β.

<u>Proof</u>: Let $\{h_n\}$ be an arbitrary numerical sequence. Using Theorem 3.3 select a subsequence $\{h_{n(k)}\}$ such that both $f(\hat{t} + h_{n(k)})$, $g(\hat{t} + h_{n(k)})$ are uniformly convergent in $-\infty < t < \infty$. Then $\alpha f(t + h_{n(k)}) + \beta g(t + h_{n(k)})$ is uniformly convergent in $-\infty < t < \infty$, thus $\alpha f(\hat{t}) + \beta g(\hat{t})$ is almost periodic by Theorem 3.3.

COROLLARY 3.5. <u>Let</u> $f(\hat{t})$ <u>be almost periodic, and let</u> $\varphi(\hat{t})$ <u>be a complex valued almost periodic function. Then</u> $\varphi(\hat{t})f(\hat{t})$ <u>is almost periodic</u>.

The proof is very similar to that of Corollary 3.4 and is therefore omitted.

COROLLARY 3.6. <u>Let</u>

$$f(t) = \sum e^{i\lambda_n t} u_n, \tag{3.3}$$

<u>where</u> $\{u_n\}$ <u>is a sequence of elements of</u> E, $\{\lambda_n\}$ <u>is a sequence of real numbers and the series</u> (3.3) <u>converges uniformly in</u> $-\infty < t < \infty$. <u>Then</u> $f(\hat{t})$ <u>is almost periodic</u>.

Proof: Partial sums of (3.3) are almost periodic by Corollary 3.4; the limit of the partial sums is almost periodic by Lemma 3.2.

THEOREM 3.7. Let $f(\hat{t})$ be almost periodic. Then the mean value

$$M(f) = \lim_{T\to\infty} \frac{1}{2T}\int_{-T}^{T} f(s)\, ds \tag{3.4}$$

exists. Moreover,

$$M(f) = \lim_{T\to\infty} \frac{1}{2T}\int_{-T}^{T} f(s+t)\, ds \tag{3.5}$$

for every t, $-\infty < t < \infty$.

For a proof see CORDUNEANU [1968:1, p. 145] or AMERIO-PROUSE [1971:1]; we shall only use the result for complex-valued functions. The same holds true of the following two theorems, where $f(\hat{t})$ is again assumed to be almost periodic.

THEOREM 3.8. The mean value

$$M(e^{-i\lambda\hat{t}} f(\hat{t})) \qquad (-\infty < \lambda < \infty) \tag{3.6}$$

is non null only for a finite or countable set of λ's.

See CORDUNEANU [1968:1, p. 149] or AMERIO-PROUSE [1971:1]. Note that $\exp(-i\lambda\hat{t})f(\hat{t})$ is almost periodic by Corollary 3.5.

THEOREM 3.9. Assume that $M(\exp(-i\lambda\hat{t})f(\hat{t})) = 0$ for all λ, $-\infty < \lambda < \infty$. Then $f(\hat{t}) = 0$.

The proof is in CORDUNEANU [1968:1, p. 151].

§V.4 Almost periodic groups in Hilbert space.

Let B be a self adjoint operator in the Hilbert space H. If B has pure point spectrum (i.e. if all elements of $\sigma(B)$ are eigenvalues of B) then there exists a family $\{P_\lambda; -\infty < \lambda < \infty\}$ of mutually orthogonal (self adjoint) projections such that

$$\sum_{-\infty<\lambda<\infty} P_\lambda u = u \qquad (u \in H), \tag{4.1}$$

and

$$Bu = \sum_{-\infty<\lambda<\infty} \lambda P_\lambda u \qquad (u \in D(B)), \tag{4.2}$$

both series convergent in the norm of H with $\|u\|^2 = \sum \|P_\lambda u\|^2$ and $\|Bu\|^2 = \sum \lambda^2 \|P_\lambda u\|^2$. Of course this implies that $P_\lambda u = 0$ except for λ in a finite or countable set $\sigma(u)$, in general depending on u. The domain of B consists precisely of those u for which the series on the right hand side of (4.2) converges in the norm of H or, equivalently, $\sum \lambda^2 \|P_\lambda u\|^2 < \infty$.

In case the space H is separable the family of projections $\{P_\lambda\}$ is actually <u>countable</u>: to see this, let $\{u_n\}$ be a dense sequence in H; consider the (clearly countable) set $\sigma = \sigma(u_1) \cup \sigma(u_2) \cup \ldots$ Then, if $\lambda \notin \sigma$ and $u \in E$ we have $P_\lambda u = \lim P_\lambda v_n = 0$, where v_n is a sequence in $\{u_n\}$ converging to u.

It has already been observed in §V.2 that

$$U(t) = \exp(itB) \qquad (-\infty < t < \infty) \tag{4.3}$$

is a strongly continuous group with infinitesimal generator B (with each $U(t)$ unitary). Under the present assumptions we have

$$U(t)u = \sum_{-\infty<\lambda<\infty} e^{i\lambda t} P_\lambda u \qquad (u \in E) \tag{4.4}$$

with

$$\|U(t)u\|^2 = \sum_{-\infty<\lambda<\infty} \|P_\lambda u\|^2 \qquad (u \in E). \tag{4.5}$$

Condition (4.5) obviously implies that convergence of the series (4.4) is uniform in $(-\infty, \infty)$ thus it follows from Lemma 3.2 and Corollary 3.4 that $U(\hat{t})u$ is almost periodic for every $u \in E$. Obviously the same conclusion holds for a strongly continuous group $S(t)$ which admits the representation

$$S(t) = Q^{-1}\exp(itB)Q\ , \tag{4.6}$$

where B is a self adjoint operator having pure point spectrum and Q a bounded invertible operator, or, equivalently, for a strongly continuous group whose infinitesimal generator is of the form

$$A = iQ^{-1}BQ\ , \tag{4.7}$$

with B and Q as before.

We show below that the representation (4.6) with B having pure point spectrum is a necessary condition in order that each $S(t)u$ be almost periodic.

THEOREM 4.1. <u>Let</u> $S(\hat{t})$ <u>be a group in</u> H. <u>Assume that, for each</u> $u,v \in H$ <u>the function</u> $(S(t)u,v)$ <u>is almost periodic. Then</u> $S(\hat{t})$ <u>admits the representation</u> (4.6) <u>with</u> Q <u>a bounded self adjoint operator satisfying inequalities of the type of</u> (2.5) <u>and</u> B <u>a self adjoint operator of the form</u> (4.2); <u>if</u> H <u>is separable the family</u> $\{P_\lambda\}$ <u>is countable</u>.

<u>Proof</u>: It follows from the fact that almost periodic functions are bounded in $-\infty < t < \infty$ and from the uniform boundedness principle (HILLE-PHILLIPS [1957:1, p.26]) that

$$\|S(t)\| \leq C \qquad (-\infty < t < \infty) . \tag{4.8}$$

For each real λ define a bounded operator M_λ in H by the formula

$$(M_\lambda u,v) = \lim_{T\to\infty} \frac{1}{2T}\int_{-T}^{T} e^{-i\lambda s}(S(s)u,v)\, ds . \tag{4.9}$$

By virtue of Theorem 3.7 the limit exists for all λ (and $(M_\lambda u,v) = 0$ for all λ except for a finite or countable set). For $-\infty < \lambda,\ t < \infty$ and $u,v \in H$ we have

$$(M_\lambda S(t)u,v) = \lim_{T\to\infty} \frac{1}{2T}\int_{-T}^{T} e^{-i\lambda s}(S(s)S(t)u,v)\, dt$$

$$= e^{i\lambda t} \lim_{T\to\infty} \frac{1}{2T}\int_{-T}^{T} e^{-i\lambda(s+t)}(S(s+t)u,v)\, dt = e^{i\lambda t}(M_\lambda u,v) \tag{4.10}$$

using Theorem 3.7: accordingly,

$$M_\lambda S(t) = e^{i\lambda t} M_\lambda . \tag{4.11}$$

Now,

$$\lim_{T\to\infty} \frac{1}{2T}\int_{-T}^{T} e^{-i\mu t}e^{i\lambda t}\, dt = \begin{cases} 1 & \text{if } \lambda = \mu \\ 0 & \text{if } \lambda \neq \mu \end{cases}, \tag{4.12}$$

so that if we multiply (4.11) by $\exp(i\mu t)/2T$ and integrate in $-T \leq t \leq T$ we obtain

$$M_\lambda M_\mu = \begin{cases} M_\lambda & \text{if} \quad \lambda = \mu \\ 0 & \text{if} \quad \lambda \neq \mu \end{cases} . \tag{4.13}$$

Accordingly $\{M_\lambda ; -\infty < \lambda < \infty\}$ is a family of mutually orthogonal (not necessarily self adjoint) projections in H.

Let P be the self adjoint operator defined in the proof of Theorem 2.2. Keeping (2.10) in mind we deduce that

$$(PM_\lambda u,v) = (M_\lambda u,Pv) = \lim_{T\to\infty} \frac{1}{2T}\int_{-T}^{T} e^{-i\lambda s}(S(s)u,Pv)\,ds$$

$$= \lim_{T\to\infty} \frac{1}{2T}\int_{-T}^{T} e^{-i\lambda s}(u,S(s)^*Pv)\,ds = \lim_{T\to\infty} \frac{1}{2T}\int_{-T}^{T} e^{-i\lambda s}(Pu,S(-s)v)\,ds$$

$$= \lim_{T\to\infty} \frac{1}{2T}\int_{-T}^{T} e^{-i\lambda s}\overline{(S(-s)v,Pu)}\,ds = \overline{(M_\lambda v,Pu)} = (Pu,M_\lambda v). \tag{4.14}$$

Accordingly,

$$PM_\lambda = M_\lambda^* P , \tag{4.15}$$

so that, if Q is the positive self adjoint square root of P, $P_\lambda = QM_\lambda Q^{-1}$ is self adjoint for all λ. We define then a self adjoint operator B by (4.2) and a group V(t) by

$$V(t) = Q^{-1}\exp(itB)Q \qquad (-\infty < t < \infty). \tag{4.16}$$

Noting that

$$V(t)u = \sum_{-\infty<\lambda<\infty} e^{i\lambda t}Q^{-1}P_\lambda Qu = \sum_{-\infty<\lambda<\infty} e^{i\lambda t}M_\lambda u \tag{4.17}$$

we deduce, using uniform convergence of (4.17) that

$$\lim_{T\to\infty} \frac{1}{2T}\int_{-T}^{T} e^{-i\lambda s}(V(s)u,v)\,ds = (M_\lambda u,v) = \lim_{T\to\infty} \frac{1}{2T}\int_{-T}^{T} e^{-i\lambda s}(S(s)u,v)\,ds$$

for all real λ. Application of Theorem 3.8 then yields the equality $(V(s)u,v) = (S(s)u,v)$ for all $u,v \in H$ hence $S(s) = V(s)$, establishing in full the claims of Theorem 4.1.

We restate the result in the style of Theorem 2.5:

THEOREM 4.2. Let A be a closed, densely defined operator in the Hilbert space H such that the Cauchy problem for

$$u'(t) = Au(t) \tag{4.18}$$

is well posed in $-\infty < t < \infty$. Assume, moreover, that for every generalized solution of (4.18) and every $v \in H$ the function

$$(u(t),v) \tag{4.19}$$

is almost periodic. Then there exists a self adjoint operator B with pure point spectrum and a bounded self adjoint operator Q satisfying inequalities of the type of (2.5) and such that

$$D(A) = \{u; Qu \in D(B)\} ,$$

$$A = iQ^{-1}BQ = i\sum_{-\infty<\lambda<\infty} \lambda Q^{-1}P_\lambda Q ,$$

(the sum understood elementwise) where $\{P_\lambda\}$ is the family of self adjoint projections in (3.2).

§V.5 Banach integrals.

Consider the space $B_1 = B_1(-\infty,\infty)$ of bounded complex valued functions periodic with period 1 defined in $-\infty < t < \infty$. We define below anextension of the notion of integral similar to the extension of limit introduced in §1. A Banach integral or generalized integral in B_1 is a functional $\int(\cdot)ds : B_1 \to \mathbb{C}$ such that, for $f,g \in B_1$ and α,β complex we have

(a) $\int (\alpha f(s) + \beta g(s))\, ds = \alpha \int f(s)\, ds + \beta \int g(s)\, ds$.

(b) $\int f(s+t)\, ds = \int f(s)\, ds \quad (-\infty < t < \infty)$.

(c) $\int f(s)\, ds \geq 0$ if $f \geq 0$.

(d) $\int 1\, ds = 1$.

(e) $\left|\int f(s)\, ds\right| \leq \int |f(s)|\, ds$.

(f) $\int f(s)\, ds = \int_0^1 f(s)\, ds$ if f is locally Riemann integrable.

THEOREM 5.1. Banach integrals in B_1 exist.

Proof: As in the case of Banach limits it is obviously sufficient to define $\int (\cdot)\,ds$ for real valued functions. Let $B_{1,\mathbb{R}}$ be the corresponding subspace of B_1. For $f \in B_{1,\mathbb{R}}$ define

$$p(f) = \inf \sup_{-\infty < s < \infty} \frac{1}{n} \sum_{k=1}^{n} f(s + \xi_k), \tag{5.1}$$

the infimum taken with respect to all finite sets $\xi_1, \ldots, \xi_n$ of real numbers. Again we use Theorem 1.1 for $F = \{0\}$ and the functional $\varphi = 0$ obtaining a linear functional $\Phi : B_{1,\mathbb{R}} \to \mathbb{R}$ such that

$$\Phi(f) \le p(f) \qquad (f \in B_{1,\mathbb{R}}), \tag{5.2}$$

so that

$$\Phi(f) \ge -p(-f) \qquad (f \in B_{1,\mathbb{R}}). \tag{5.3}$$

We define $\int (\cdot)\,ds = \Phi$. Obviously (a) holds and (c) follows from (5.3). To show (d) it suffices to observe that $p(1) = 1$, $p(-1) = -1$ and use (5.2) and (5.3). To prove (b) we select t arbitrary and take $\xi_1 = t$, $\xi_2 = 2t, \ldots, \xi_n = nt$. Then

$$p(f(\hat{s} + t) - f(\hat{s})) \le \sup_{-\infty < s < \infty} \frac{1}{n}(f(s + nt) - f(s)),$$

so that $p(f(\hat{s} + t) - f(\hat{s})) \le 0$; we prove in the same way that $p(f(\hat{s}) - f(\hat{s} + t)) \le 0$ thus (b) follows from (5.2) and (5.3). Let now f be a locally Riemann integrable function in $B_{1,\mathbb{R}}$. Taking $\xi_1 = 1/n$, $\xi_2 = 2/n, \ldots, \xi_n = 1$ and letting $n \to \infty$ we deduce that

$$p(f) \le \int_0^1 f(s)\,ds, \quad p(-f) \le -\int_0^1 f(s)\,ds$$

so that (f) follows. Finally, let $f \in B_1$ be arbitrary. Choose θ such that $e^{i\theta}\int f(s)\,ds$ is positive. Then, by (c),

$$\left|\int f(s)\,ds\right| = \int e^{i\theta} f(s)\,ds = \int \mathrm{Re}(e^{i\theta} f(s))\,ds \le \int |f(s)|\,ds .$$

This shows (e) and ends the proof of Theorem 5.1.

We shall need in the sequel a version of the Banach integral for the space $B = B(-\infty, \infty)$ of all bounded functions in $(-\infty, \infty)$ satisfying properties similar to $\int (\cdot)\,ds$ plus an interval additivity property and a rudimentary change-of-variable formula.

THEOREM 5.2. <u>There exists a family of functionals</u>

$$\int_a^b (\cdot)ds : B \to C$$

<u>defined for</u> $-\infty < a \leq b < \infty$ <u>satisfying the following properties</u>:

(i) $\int_a^b (\alpha f(s) + \beta g(s))\, ds = \alpha \int_a^b f(s)\, ds + \beta \int_a^b g(s)\, ds$.

(ii) $\int_{a+t}^{b+t} f(s)\, ds = \int_a^b f(s + t)\, ds \quad (-\infty < t < \infty)$.

(iii) $\int_a^b f(s)\, ds \geq 0$ if $f(s) \geq 0$.

(iv) $\int_a^b 1\, ds = b - a$.

(v) $\left|\int_a^b f(s)\, ds\right| \leq \int_a^b |f(s)|\, ds$.

(vi) $\int_a^b f(s)\, ds + \int_b^c f(s)\, ds = \int_a^c f(s)\, ds$ if $a < b < c$.

(vii) $\int_{a/2}^{b/2} f(2s)\, ds = \frac{1}{2}\int_a^b f(s)\, ds$.

<u>Proof</u>: Let $c > 0$. If

$$-c < 2a,\ 2b < c \tag{5.4}$$

define

$$f(s;a,b,c) = 0 \quad (-\tfrac{1}{2} \leq s < \tfrac{a}{c}\,,\ \tfrac{b}{c} \leq s < \tfrac{1}{2})$$

$$f(s;a,b,c) = cf(cs) \quad (\tfrac{a}{c} \leq s < \tfrac{b}{c})$$

and extend $f(\hat{s};a,b,c)$ to $-\infty < s < \infty$ in such a way that the resulting function is 1-periodic (that is, belongs to B_1). In case one of the inequalities (5.4) fails to hold we define $f(s;a,b,c) = 0$ for all s. The functionals in question are defined by means of any of the Banach integrals $\int$ in Theorem 5.1 by means of the formula

$$\int_a^b f(s)\, ds = \operatorname*{LIM}_{n\to\infty} \int f(s;a,b,2^n)\, ds\,,$$

where $\operatorname*{LIM}_{n\to\infty}$ is any of the Banach limits constructed in Corollary 1.3. The checking of properties (i) to (vii) is fairly routine and is thus left to the reader.

The integrals in Theorem 5.2 will also be called Banach integrals.

§V.6 Uniformly bounded cosine functions in Hilbert space.

Let B be a self adjoint operator in H such that $B \geq 0$ (that is, $(Bu,u) \geq 0$ for $u \in D(B)$) and let

$$\mathcal{C}(t) = \cos(tB^{1/2}) = \xi(t;B) , \tag{6.1}$$

where $\xi(t,\lambda) = \cos(t\lambda^{1/2}) = \sum (-1)^n t^{2n}\lambda^n/(2n)!$ (note that the choice of square root of λ or of that of B is irrelevant; in fact, the use of $B^{1/2}$ in (6.1) is purely symbolic). Then $\mathcal{C}(\hat{t})$ is a strongly continuous cosine function such that each $\mathcal{C}(t)$ is self adjoint; moreover,

$$\|\mathcal{C}(t)\| \leq 1 \qquad (-\infty < t < \infty). \tag{6.2}$$

We have

$$\mathcal{S}(t) = B^{-1/2}\sin(tB^{1/2}) = \eta(t;B) \tag{6.3}$$

where $\eta(t,\lambda) = \lambda^{-1/2}\sin(t\lambda^{1/2}) = \sum (-1)^n t^{2n+1}\lambda^n/(2n+1)!$ (again, the use of $B^{1/2}$ is purely symbolic). Since

$$\sup_{\lambda \geq 0} |\lambda^{-1/2}\sin t\lambda^{1/2}| = t \qquad (t \geq 0) ,$$

we can only assert that

$$\|\mathcal{S}(t)\| \leq |t| \qquad (-\infty < t < \infty) . \tag{6.4}$$

(which inequality follows anyway from direct integration of (6.2)). However, if $\varepsilon > 0$,

$$|\lambda^{-1/2}\sin t\lambda^{1/2}| \leq C \qquad (\lambda \geq \varepsilon),$$

so that, if $\sigma(B)$ is contained in $\lambda \geq \varepsilon$,

$$\|\mathcal{S}(t)\| \leq C \qquad (-\infty < t < \infty). \tag{6.5}$$

The converse of these observations holds as well.

THEOREM 6.1. <u>Let</u> $C(\hat{t})$ <u>be a strongly continuous cosine function.</u> <u>Assume that each</u> $C(t)$ <u>is a self adjoint operator. Then</u> (6.1) <u>holds with</u> $B = -A$, A <u>the infinitesimal generator of</u> $C(\hat{t})$. <u>If</u>

$$\|C(\hat{t})\| \leq C \qquad (-\infty < t < \infty). \tag{6.6}$$

<u>then</u> $B \geq 0$ <u>and the stronger estimate</u> (6.2) <u>holds. Inequality</u> (6.5) <u>for</u> $\mathcal{S}(\hat{t})$ <u>holds if and only if</u> $B \geq \varepsilon I$ <u>with</u> $\varepsilon > 0$ (<u>that is if</u> $(Bu,u) \geq \varepsilon\|u\|^2$, $u \in D(B)$ <u>with</u> $\varepsilon > 0$).

<u>Proof</u>: It imitates that of Theorem 2.1. Let A be the infinitesimal generator of $C(\hat{t})$. Making use of formula (II.2.11) for $n = 1$ and λ real and large enough we obtain

$$(\lambda R(\lambda^2;A)u,v) = \int_0^\infty e^{-\lambda t}(C(t)u,v)dt$$

$$= \int_0^\infty e^{-\lambda t}(u,C(t)v)dt = (u,\lambda R(\lambda^2;A)v) ,$$

so that $R(\lambda^2;A)$ is self adjoint, thus it follows that A itself is self adjoint. Moreover, $R(\lambda^2;A)$ exists for λ sufficiently large so that $\sigma(A)$ is contained in an interval $(-\infty,a]$, $a < \infty$. Accordingly, $\mathcal{D}(t) = \xi(t;-A)$ is a strongly continuous operator valued function satisfying $\|\mathcal{D}(t)\| \leq \exp(\omega|t|)$ with $\omega = a^{1/2}$. The fact that $\mathcal{D}(t) = C(t)$ is proved by uniqueness of Laplace transforms as in Theorem 2.1.

Assume that $\sigma(A)$ contains some $a > 0$. Then $\|C(t)\|$ is the supremum of $\cos t\lambda^{1/2}$ in $\lambda \geq -a$ and (6.6) has no chance to hold. The fact that (6.5) will be true when $\sigma(A) \subseteq (-\infty,-\varepsilon]$, $\varepsilon < 0$ has been already established in the remarks preceding the statement of Theorem 6.1. Conversely, assume that (6.5) holds. Integrating by parts the first formula (II.2.11) we obtain

$$R(\lambda^2;A)u = \int_0^\infty e^{-\lambda t}\mathcal{S}(t)u\, dt$$

for $\lambda > 0$ and $u \in E$. Taking norms, it results that

$$\|R(\lambda^2;A)\| = 0(\lambda^{-1}) \quad \text{as} \quad \lambda \to 0+,$$

which implies that $0 \in \rho(A)$ (DUNFORD-SCHWARTZ [1958:1, p. 567]), hence $\sigma(A) \subseteq (-\infty,-\varepsilon]$ for some $\varepsilon > 0$ as claimed. This ends the proof of Theorem 6.1.

The next result is an exact counterpart of Theorem 2.2 for cosine functions. However, the method of proof is somewhat different.

THEOREM 6.2. Let $C(\hat{t})$ be a cosine function in H ($C(0) = I$, $C(s + t) + C(s - t) = 2C(s)C(t)$ for $-\infty < s, t < \infty$). Assume that

$$\|C(t)\| \leq C \qquad (-\infty < t < \infty) . \tag{6.7}$$

Then there exists a bounded, self adjoint operator Q such that

$$3^{-1/2}(2C + 1)^{-1}\|u\|^2 \leq (Qu,u) \leq C\|u\|^2 \qquad (u \in E), \tag{6.8}$$

and

$$\mathcal{D}(t) = QC(t)Q^{-1} \tag{6.9}$$

is self adjoint for all t.

The proof uses some techniques already met in connection with the theory of almost periodic functions, although nothing is almost periodic here. We begin by selecting one of the Banach integrals constructed in Theorem 5.2 and one of the Banach limits in Theorem 1.2, and observing that

$$\operatorname*{LIM}_{T\to\infty} \frac{1}{T}\int_0^T f(s)\, ds \tag{6.10}$$

exists for any $f \in B$; in fact, due to property (v) of Banach integrals we have

$$\left|\int_0^T f(s)\, ds\right| \leq \int_0^T |f(s)|\, ds \leq CT$$

where C is a bound for $|f(s)|$, so that the function of T inside the Banach limit in (6.10) is bounded in $T \geq 0$.

LEMMA 6.3. Let $f \in B$. Then

$$\operatorname*{LIM}_{T\to\infty} \frac{1}{T}\int_0^T f(s + t)\, ds = \operatorname*{LIM}_{T\to\infty} \frac{1}{T}\int_0^T f(s)\, ds \tag{6.11}$$

for any t in $(-\infty,\infty)$.

Proof: Assume for the moment that $t \geq 0$. By virtue of (ii) and

(vi) we have

$$\frac{1}{T}\int_0^T f(s+t)\,ds - \frac{1}{T}\int_0^T f(s)\,ds = \frac{1}{T}\int_t^{T+t} f(s)\,ds - \frac{1}{T}\int_0^T f(s)\,ds$$
$$= \frac{1}{T}\int_T^{T+t} f(s)\,ds - \frac{1}{T}\int_0^t f(s)\,ds ,$$

which expression is seen to tend to zero as $T \to \infty$ using (v); hence its Banach limit must be zero as well, proving (6.11).

LEMMA 6.4. *Let* $T > 0$, $\varepsilon \leq 1/(2C+1)$, $u \in H$,

$$e = e(\varepsilon, T, u) = \{t; 0 \leq t \leq T,\ \|C(t)u\| < \varepsilon\|u\|\} . \tag{6.12}$$

Then

$$\int_0^T \chi_e(s)\,ds \leq \frac{2T}{3} , \tag{6.13}$$

χ_e *the characteristic function of* e.

Proof: Set $t = s = \sigma/2$ in the (second) cosine functional equation (II.1.9). The result is

$$2C(\sigma/2)^2 = C(\sigma) + I .$$

Hence, if $\sigma \in e$ we have

$$2C\|C(\sigma/2)u\| \geq 2\|C(\sigma/2)^2 u\| \geq (1-\varepsilon)\|u\| .$$

Accordingly, if $\varepsilon \leq 1/(2C+1)$ we deduce that

$$\|C(\sigma/2)u\| \geq \varepsilon\|u\| ,$$

so that $\sigma/2 \notin e$. It follows that if $\sigma \in e$ then $2\sigma \notin e$, which shows that the functions $\chi_e(\hat{s})$ and $\chi_e(2\hat{s})$ have disjoint support. Hence

$$\int_0^T \chi_e(s)\,ds + \int_0^T \chi_e(2s)\,ds \leq T$$

by (i) and (v). Taking the change-of-variable property in consideration we obtain (6.13), thus ending the proof of Lemma 6.4.

Proof of Theorem 6.2. The operator P is this time defined by

$$(Pu,v) = \operatorname*{LIM}_{T\to\infty} \frac{1}{T}\int_0^T (C(s)u,C(s)v)\,ds\ . \tag{6.14}$$

By virtue of Lemma 6.4 with $\varepsilon = 1/(2C+1)$ we have

$$\int_0^T \|C(s)u\|^2\,ds \geq \frac{\|u\|^2}{(2C+1)^2}\int_0^T (1-\chi_e(s))\,ds \geq \frac{T}{3(2C+1)^2}\|u\|^2 ,$$

thus it follows from the definition of P that

$$(Pu,u) \geq \frac{1}{3(2C+1)^2}\|u\|^2 \qquad (u \in E)\ . \tag{6.15}$$

On the other hand, it is obvious that

$$(Pu,u) \leq C^2\|u\|^2 \qquad (u \in E)\ . \tag{6.16}$$

Accordingly, if Q is the positive, self adjoint square root of P, inequalities (6.8) hold.

Let now t be a real number, u,v elements of H. Using the cosine functional equations and Theorem 5.2 we deduce that

$$\begin{aligned}
(PC(t)u,v) &= \operatorname*{LIM}_{T\to\infty} \frac{1}{T}\int_0^T (C(s)C(t)u,C(s)v)\,ds \\
&= \frac{1}{2}\operatorname*{LIM}_{T\to\infty} \frac{1}{T}\int_0^T (C(s+t)u,C(s)v)\,ds \\
&\quad + \frac{1}{2}\operatorname*{LIM}_{T\to\infty} \frac{1}{T}\int_0^T (C(s-t)u,C(s)v)\,ds \\
&= \frac{1}{2}\operatorname*{LIM}_{T\to\infty} \frac{1}{T}\int_0^T (C(s)u,C(s-t)v)\,ds \\
&\quad + \frac{1}{2}\operatorname*{LIM}_{T\to\infty} \frac{1}{T}\int_0^T (C(s)u,C(s+t)v)\,ds \\
&= \operatorname*{LIM}_{T\to\infty} \frac{1}{T}\int_0^T (C(s)u,C(s)C(t)v)\,ds = (Pu,C(t)v)
\end{aligned} \tag{6.17}$$

for $u,v \in H$. Accordingly,

$$PC(t) = C(t)^*P \qquad (-\infty < t < \infty)\ . \tag{6.18}$$

Pre- and post-multiplying by Q^{-1} we obtain

$$QC(t)Q^{-1} = Q^{-1}C(t)^*Q = (QC(t)Q^{-1})^*. \tag{6.19}$$

This completes the proof of Theorem 6.2.

The comments following Theorem 2.2 apply here as well: replacing the original scalar product by the (topologically equivalent) scalar product (2.12) renders each $C(t)$ _self adjoint_. We omit the details.

COROLLARY 6.5. _Let_ $C(\hat{t})$ _satisfy the hypoteses of Theorem_ 6.2. _Assume in addition that_ $C(\hat{t})u$ _is continuous (or, more generally strongly measurable) for each_ $u \in E$. _Then there exists a self adjoint operator_ B _with_ $B \geq 0$ _and a bounded self adjoint operator_ Q _satisfying inequalities of the form_ (6.8) _and such that_

$$C(t) = Q^{-1}\cos(tB^{1/2})Q \quad (-\infty < t < \infty). \tag{6.20}$$

Conversely, every $C(\hat{t})$ _of the form_ (6.20) _is a strongly continuous cosine function satisfying_ (6.7). _Inequality_ (6.5) _holds if and only if_ $B \geq \varepsilon I$ _for some_ $\varepsilon > 0$.

The following discrete version of Theorem 6.2 corresponds to Theorem 2.4.

THEOREM 6.6. _Let_ $\{C_n; -\infty < n < \infty\}$ _be a sequence of bounded operators in_ H _satisfying the "discrete cosine equations"_ $C_0 = I$, $2C_mC_n = C_{m+n} + C_{m-n}$ _for all_ m,n. _Assume that_

$$\|C_n\| \leq C \quad (-\infty < n < \infty). \tag{6.21}$$

Then there exists a bounded, self adjoint operator Q _satisfying_

$$2^{-1}(2C+1)^{-1}\|u\|^2 \leq (Qu,u) \leq C\|u\|^2 \tag{6.22}$$

and such that

$$\mathcal{D}_n = QC_nQ^{-1} \tag{6.23}$$

is self adjoint for all n. _Equivalently, there exists a unitary operator_ U _such that_

$$C_n = \frac{1}{2}Q^{-1}(U^n + U^{-n})Q. \tag{6.24}$$

The proof is rather similar to that for the continuous version. The operator P is now defined by

$$(Pu,v) = \underset{n\to\infty}{\mathrm{LIM}} \frac{1}{n+1} \sum_{k=0}^{n} (\mathcal{C}_n u, \mathcal{C}_n v) ,$$

where LIM is one of the Banach limits of sequences constructed in Corollary 1.3. Proceeding as in the proof of Lemma 6.4 we can show that if $\|\mathcal{C}_m u\| < \varepsilon\|u\|$ for an arbitrary integer m and $\varepsilon \leq (2C+1)^{-1}$ then $\|\mathcal{C}_{2m} u\| \geq \varepsilon\|u\|$; hence the number of integers between 0 and n for which $\|\mathcal{C}_m u\| \leq \varepsilon\|u\|$ is at least equal to $[(n-5)/4]$, where $[s]$ indicates the largest integer $\leq s$. Taking $\varepsilon = 1/(2C+1)$ we obtain

$$(Pu,u) \geq \frac{1}{4(2C+1)^2}\|u\|^2 \qquad (u \in E), \tag{6.25}$$

and it is obvious that

$$(Pu,u) \leq C^2\|u\|^2 \qquad (u \in E) , \tag{6.26}$$

thus Q, the positive self adjoint square root of P satisfies (6.22). A computation entirely similar to (6.17) shows that

$$P\mathcal{C}_n = \mathcal{C}_n^* P , \tag{6.27}$$

hence each $\mathcal{D}_n$ in (6.23) is self adjoint.

Consider now the sequence of operators $\{\mathcal{D}_n; -\infty < n < \infty\}$. Obviously $\{\mathcal{D}_n\}$ satisfies as well the discrete cosine equations so that, taking $m = n$ we obtain

$$\mathcal{D}_{2n} = 2\mathcal{D}_n^2 - I .$$

Making use of the spectral mapping theorem for bounded operators (DUNFORD-SCHWARTZ [1958:1, p. 569]) we deduce that if $\lambda \in \sigma(\mathcal{D}_1)$ and $\{\lambda_n\}$ is defined inductively by $\lambda_1 = \lambda$, $\lambda_n = 2\lambda_{n-1}^2 - 1$ then $\lambda_n \in \sigma(2\mathcal{D}_n - 1)$. But $|\lambda_n| \to \infty$ if $|\lambda| > 1$; since $\sigma(\mathcal{D}_n)$ must be bounded independently of n, we deduce that

$$\sigma(\mathcal{D}_1) \subseteq [-1,1] .$$

We can thus define $B = \arccos \mathcal{D}_1$, where $\arccos\lambda$ is the function inverse to $\cos\lambda$ in the interval $[-\pi/2, \pi/2]$. Let

$$\mathcal{E}_n = \cos(nB) . \tag{6.28}$$

Then it follows from the functional calculus for self adjoint operators that $\{\mathcal{E}_n\}$ satisfies as well the discrete cosine functional equation; in particular,

$$\mathcal{E}_{n+1} = 2\mathcal{E}_n\mathcal{E}_1 - \mathcal{E}_{n-1} ,$$

which shows inductively that $\mathcal{E}_n = \mathcal{D}_n$ for all n since $\mathcal{E}_1 = \mathcal{D}_1$. We only have to combine (6.28) (in the form $\mathcal{E}_n = \frac{1}{2}\{\exp(inB) + \exp(-inB)\}$) with (6.23) to obtain (6.24), where $U = \exp(iB)$.

Theorem 6.1 is obviously equivalent to the following result for second order abstract differential equations.

THEOREM 6.7. *Let* A *be a closed, densely defined operator in the Hilbert space* H *such that the Cauchy problem for*

$$u''(t) = Au(t) \tag{6.29}$$

is well posed in $-\infty < t < \infty$ *(equivalently, in* $t \geq 0$). *Assume, moreover, that for every generalized solution of* (6.29) *with* $u'(0) = 0$ *we have*

$$\|u(t)\| \leq C \qquad (t \geq 0) \tag{6.30}$$

(*where* C *may depend on* $u(\hat{t})$). *Then there exists a self adjoint operator* B *with* $B \geq 0$ *and a bounded self adjoint operator* Q *satisfying inequalities of the form* (6.8) *and such that* $D(A) = \{u;Qu \in D(B)\}$ *and*

$$A = -Q^{-1}BQ . \tag{6.31}$$

Moreover, all solutions of (6.29) *are bounded in* $t \geq 0$ (*equivalently, in* $-\infty < t < \infty$) *if and only if* $B \geq \varepsilon I$ *for some* $\varepsilon > 0$. *The converse is as well true.*

REMARK 6.8. As in the group case it is of interest to know whether the conclusions of Corollary 6.5 hold under less stringent conditions on the t-dependence of $C(\hat{t})$. The answers are essentially the same. Weak measurability of $C(\hat{t})$ (that is, measurability of the function $(C(t)u,v)$ for all $u,v \in H$) suffices when H is separable: in the general case, it is enough to assume weak continuity, since it implies strong continuity. In view of the representation (6.9) with $\mathfrak{D}(\hat{t})$ self adjoint, it is obviously sufficient to prove this when $C(\hat{t})$ is self adjoint, in which case we have

$$2\|C(t)u - C(s)u\|^2 = 2((C(t) - C(s))u, (C(t) - C(s))u) =$$

$$= ((C(2t) + I)u, u) - 2((C(s + t) + C(s - t))u, u) + ((C(2s) + I)u, u)$$

thus $C(\hat{t})$ is strongly continuous if it is weakly continuous.

§V.7 Almost periodic cosine functions in Hilbert space.

Let B be a self adjoint operator in H. Assume that $B \geq 0$ and that it has pure point spectrum; then there exists a family $\{P_\lambda; \lambda \geq 0\}$ of mutually orthogonal projections such that

$$\sum_{\lambda \geq 0} P_\lambda u = u \qquad (u \in H) \tag{7.1}$$

and

$$Bu = \sum_{\lambda \geq 0} \lambda P_\lambda u \qquad (u \in D(B)) \tag{7.2}$$

for $u \in D(B)$ (see the comments following (4.1) and (4.2)). It results from the initial remarks in §V.6 that

$$C(t) = \cos (tB^{1/2}) \tag{7.3}$$

is a strongly continuous, uniformly bounded cosine function with infinitesimal generator $A = -B$. We have

$$C(t)u = \sum_{\lambda \geq 0} \cos t\lambda^{1/2} P_\lambda u \qquad (u \in E), \tag{7.4}$$

the series convergent in the norm of H uniformly in $-\infty < t < \infty$; this follows from the equality $\sum \|P_\lambda u\|^2 = \|u\|^2$ which also implies that only a countable number of the $P_\lambda u$ (in general depending on u) may be non null (as pointed out in §V.4, if the space H is separable all but a countable number of the projections P_λ must of needs vanish). Uniform convergence of (7.4), Lemma 3.2 and Corollary 3.4 imply that $C(\hat{t})u$ is an almost periodic function for every $u \in H$. The corresponding property for $\mathfrak{S}(\hat{t})u$ may not hold since $\|\mathfrak{S}(\hat{t})\|$ may not be bounded (see (6.4) and following comments). However, the condition on $\sigma(B)$ that guarantees boundedness of $\|\mathcal{S}(\hat{t})\|$ guarantees almost periodicity as well; in fact, if $\sigma(B)$ is contained in $\lambda \geq \varepsilon$, we have

$$\mathcal{S}(t)u = \sum_{\lambda \geq \varepsilon} \lambda^{-1/2} \sin t\lambda^{1/2} P_\lambda u ,$$

which is as well almost periodic. Under these conditions, an almost periodicity property holds as well in phase spaces; we look only at the "principal value square root" phase space $\mathfrak{Z}$ constructed in Theorem III.5.4 for $\omega = 0$. Observe first that the square root $(-A)^{1/2}$ constructed in §III.3 coincides with the positive, self adjoint square root of B. Since $(-A)^{1/2}$ commutes with each $C(t)$ we have

$$(-A)^{1/2}C(t)u = C(t)(-A)^{1/2}u = \sum_{\lambda \geq \varepsilon} \cos t\lambda^{1/2} P_\lambda (-A)^{1/2}u . \quad (7.5)$$

This, combined with (7.4) shows that $C(\hat{t})$ is almost periodic in the space $E_0 = D((-A)^{1/2})$ (endowed with the graph norm of $(-A)^{1/2}$). Since $\mathcal{S}(\hat{t})$ is almost periodic in E, it follows that $\mathfrak{S}(\hat{t})u$ is almost periodic in $\mathfrak{Z}$ for all $u \in \mathfrak{Z}$, where $\mathfrak{S}(\hat{t})$ is the group (III.1.10) corresponding to the phase space $\mathfrak{Z}$. Noting that

$$(-A)^{1/2}\mathcal{S}(t)u = \sum_{\lambda \geq \varepsilon} \sin t\lambda^{1/2} P_\lambda u \quad (7.6)$$

we see that $(-A)^{1/2}\mathcal{S}(t)u$ is almost periodic in E for every $u \in E$. Accordingly, the group $\mathfrak{U}_0(\hat{t})$ in the group decomposition (III.6.24) corresponding to the principal value square root $(-A)^{1/2}$ is such that $\mathfrak{U}_0(\hat{t})$ is almost periodic for any $u \in H$ and the group $\mathfrak{U}_0(\hat{t})$ in the product space $\mathfrak{D} = E \times E$ used in the reduction to a first order system in Theorem III.5.5 enjoys the same privileges.

The following result is a sort of converse of the preceding observations.

THEOREM 7.1. Let $C(\hat{t})$ be a cosine function in H. Assume that for each $u,v \in H$ the function $(C(\hat{t})u,v)$ is almost periodic. Then $C(\hat{t})$ admits the representation (6.20) with Q a bounded self adjoint operator satisfying inequalities of the type of (6.8) and B a self adjoint operator of the form (7.2): if H is separable, the family $\{P_\lambda\}$ is countable.

Proof: It follows from the fact that almost periodic functions are bounded in $-\infty < t < \infty$ and from the uniform boundedness principle that

$$\|C(t)\| \leq C \qquad (-\infty < t < \infty). \tag{7.7}$$

Noting that the assumptions imply that $C(\hat{t})$ is weakly continuous and taking heed of Remark 6.8 we notice that $C(\hat{t})$ fits into the assumptions of Corollary 6.5, thus we might adopt its conclusion as a starting point. However, as in the group case in Theorem 4.1 we shall prefer a direct construction featuring a fairly explicit formula for the spectral family of B. For each $\lambda \geq 0$ define a bounded operator N_λ in H by the formula

$$\begin{aligned}(N_\lambda u,v) &= \lim_{T\to\infty} \frac{1}{T}\int_0^T \cos\lambda t(C(s)u,v)\,dt \\ &= \lim_{T\to\infty} \frac{1}{2T}\int_{-T}^T e^{i\lambda t}(C(t)u,v)\,dt\,.\end{aligned} \tag{7.8}$$

We see applying Theorem 3.7 that the limit in (7.8) exists for all λ (incidentally, $(N_\lambda u,v) = 0$ for all but a finite or countable number of λ's.) For any $\lambda \geq 0$, $-\infty < t < \infty$, $u,v \in H$ we have

$$\begin{aligned}(N_\lambda C(t)u,v) &= \lim_{T\to\infty} \frac{1}{2T}\int_{-T}^T e^{i\lambda s}(C(s)C(t)u,v)\,ds \\ &= \frac{1}{2}e^{-i\lambda t}\lim_{T\to\infty}\frac{1}{2T}\int_{-T}^T e^{i\lambda(s+t)}(C(s+t)u,v)\,ds \\ &+ \frac{1}{2}e^{i\lambda t}\lim_{T\to\infty}\frac{1}{2T}\int_{-T}^T e^{i\lambda(s-t)}(C(s-t)u,v)\,ds \\ &= \cos\lambda t(N_\lambda u,v)\,,\end{aligned}$$

or

$$C(t)N_\lambda = N_\lambda C(t) = (\cos\lambda t)N_\lambda \tag{7.9}$$

by virtue of (3.5); the fact that N_λ and $C(t)$ commute follows from the fact that $C(s)$ and $C(t)$ commute for arbitrary s,t. Multiply (7.9) by $\exp(i\mu t)/2T$ and integrate in $-T \leq t \leq T$. Since

$$\lim_{T\to\infty}\frac{1}{2T}\int_{-T}^T e^{i\mu t}\cos\lambda t\,dt = \begin{cases} 1 & \text{if } \lambda = \mu = 0 \\ \frac{1}{2} & \text{if } \lambda = \mu > 0\,, \\ 0 & \text{if } \lambda \neq \mu \end{cases} \tag{7.10}$$

we have

$$N_\lambda N_\mu = \begin{cases} N_\lambda & \text{if} \quad \lambda = \mu = 0 \\ \frac{1}{2} N_\lambda & \text{if} \quad \lambda = \mu > 0 \\ 0 & \text{if} \quad \lambda \neq \mu \end{cases} . \tag{7.11}$$

Accordingly the operators

$$M_\lambda = \begin{cases} N_\lambda & \text{if} \quad \lambda = 0 \\ 2^{1/2} N_\lambda & \text{if} \quad \lambda > 0 \end{cases} \tag{7.12}$$

are a family of mutually orthogonal (not necessarily self adjoint) projections in H.

Let P be the self adjoint operator defined in the proof of Theorem 3.1:

$$(Pu,v) = \operatorname*{LIM}_{T\to\infty} \frac{1}{T}\int_0^T (C(s)u, C(s)v)\, ds ,$$

and let Q be the positive self adjoint square root of P. Making use of (7.8) and (7.9) we obtain

$$\begin{aligned}(PN_\lambda u,v) &= \lim_{T\to\infty} \frac{1}{T}\int_0^T \cos \lambda s(N_\lambda u, C(s)v)\, ds \\ &= (N_\lambda u, N_\lambda v) = \lim_{T\to\infty} \frac{1}{T}\int_0^T \cos \lambda s(C(s)u, N_\lambda v)\, ds \\ &= (Pu, N_\lambda v)\end{aligned}$$

which is $PN_\lambda = N_\lambda^* P$; accordingly,

$$PM_\lambda = M_\lambda^* P \qquad (\lambda \geq 0) , \tag{7.13}$$

so that $P_\lambda = QM_\lambda Q^{-1}$ is self adjoint for all λ. We define now a self adjoint operator B from the family $\{P_\lambda ; \lambda \geq 0\}$ of self adjoint projections by the formula

$$Bu = \sum_{\lambda \geq 0} \lambda P_\lambda u , \tag{7.14}$$

(the domain of B consists of all u which make the series (7.14) convergent in the norm of E) and set

$$\mathcal{E}(t) = Q^{-1}\cos(tB^{1/2})Q\ . \qquad (7.15)$$

Since

$$\mathcal{E}(t)u = \sum_{\lambda\geq 0} \cos t\lambda^{1/2} Q^{-1}P_\lambda Qu = \sum_{\lambda\geq 0} \cos t\lambda M_\lambda u \qquad (7.16)$$

an obvious computation based on the uniform convergence of (7.16) and on (7.10) shows that if $u,v \in H$ we have

$$\lim_{T\to\infty} \frac{1}{2T}\int_{-T}^{T} e^{i\lambda s}(\mathcal{E}(s)u,v)\,ds = (N_\lambda u,v)$$

$$= \lim_{T\to\infty} \frac{1}{2T}\int_{-T}^{T} e^{i\lambda s}(C(s)u,v)\,ds \qquad (7.17)$$

so that both almost periodic functions $(\mathcal{E}(\hat{s})u,v)$ and $(C(\hat{s})u,v)$ have the same Fourier series. By Theorem 3.8 we must have $(\mathcal{E}(s)u,v) = (C(s)u,v)$, thus $\mathcal{E}(s) = C(s)$ by arbitrariness of u,v. This concludes the proof of Theorem 7.1.

In the language of abstract differential equations, this result can be formulated as follows.

THEOREM 7.2. *Let* A *be a closed, densely defined operator in the Hilbert space* H *such that the Cauchy problem for*

$$u''(t) = Au(t) \qquad (7.18)$$

is well posed in $-\infty < t < \infty$ *(equivalently, in* $t \geq 0$*). Assume, moreover, that for every generalized solution of* (7.18) *with* $u'(0) = 0$ *and every* $v \in H$ *the function*

$$(u(t),v) \qquad (7.19)$$

is almost periodic. Then there exists a self adjoint operator B *with pure point spectrum and such that* $B \geq 0$ *and a self adjoint operator* Q *satisfying inequalities of the form* (6.8) *and such that* $D(A) = \{u; Qu \in D(B)\}$ *and*

$$A = -Q^{-1}BQ = -\sum_{\lambda\geq 0} \lambda Q^{-1}P_\lambda Q$$

(the sum understood elementwise) where $\{P_\lambda\}$ *is the family of self adjoint projections in* (7.2).

§V.8 Miscellaneous comments.

Banach limits and Banach integrals were introduced by BANACH [1932:1] as applications of the Hahn-Banach theorem. As anything based on Theorem 1.1 (whose proof uses the axiom of choice or, equivalently, the Hausdorff-Zorn lemma or the principle of transfinite induction), "construction" of Banach limits or integrals is totally nonconstructive and vaguely incredible even to mathematicians without a philosophical turn of mind. Two of the most intriguing applications of Banach limits to analysis are Theorems 2.2 and 2.4, both due to SZ.-NAGY [1947:1]. These results are statements on representations of groups G (the real line and the integers respectively) and have been generalized by DIXMIER [1950:1] to uniformly bounded representations of certain groups G into the algebra of linear bounded operators in a Hilbert space; for related material see GREENLEAF [1969:1]. The corresponding generalization of Theorem 6.2 is due to KUREPA [1972:1]. We note the results of LORCH [1941:1] on spectral analysis of operators satisfying the assumptions of Theorem 2.4 in reflexive Banach spaces; the main result is a spectral resolution modelled on that for unitary operators in Hilbert space.

Theorem 2.2 can be stated in the following equivalent form:

THEOREM 8.1. *Let* $S(\hat{t})$ *be a group in a Hilbert space* H *satisfying* (2.4). *Then there exists a Hilbert norm* $\|\cdot\|'$ *in* H *of the form* (2.12), *equivalent to the original one and such that*

$$\|S(t)\|' \leq 1 \qquad (-\infty < t < \infty) \tag{8.1}$$

In fact, Theorem 8.1 is an obvious consequence of Theorem 2.2: on the other hand, if (8.1) holds then we must have

$$\|S(t)u\|' = \|u\|' \quad (-\infty < t < \infty) \tag{8.2}$$

In fact, if $\|S(t)u\|' < \|u\|'$ then $\|u\|' = \|S(-t)S(t)u\|' < \|u\|'$, a contradiction. Since each $S(t)$ is invertible, (8.2) implies that each $S(t)$ is unitary, hence Theorem 2.2 follows. The same argument shows that Theorem 2.4 is equivalent to

THEOREM 8.2. *Let* S *be a bounded operator in the Hilbert space* H *satisfying* (2.14). *Then there exists a Hilbert norm* $\|\cdot\|'$ *in* H *of the form* (2.12), *equivalent to the original one and such that*

$$\|S^n\|' \leq 1 \qquad (-\infty < n < \infty) \tag{8.3}$$

It is natural to ask whether Theorem 8.2 has a counterpart for semigroups, i.e. whether a semigroup satisfying (2.4) in $t \geq 0$ will satisfy (8.1) after a renorming of the space along the lines of Theorem 8.1. The answer is in the negative; see PACKEL [1969:1]. The answer for the discrete version was known earlier to be also in the negative. (FOGUEL [1964:1]).

We note that the problem of renorming a Banach space with an equivalent Banach norm that improves the constant in (2.4) to 1 has a simple affirmative answer for semigroups, groups or cosine functions alike (see Exercise 5) as well as for their discrete counterparts.

The theory of almost periodic groups in Hilbert space is due to the author [1970:4] although the fact that $U(\hat{t})u$, u given by (4.3) is almost periodic was known before (see AMERIO-PROUSE [1971:1] for references and for numerous results on almost periodic solutions of abstract differential equations). Likewise, the results presented here on uniformly bounded cosine functions (Theorem 6.2 and 6.5) as well as the results on almost periodic cosine functions (Theorem 7.1) are due to the author [1970:3], although the remarks on almost periodicity of functions of the form (7.4) go back essentially to MUCKENHOUPT [1928-1929:1]. The discrete version (Theorem 6.6) is also due to the author [1970:3].

Remark 2.6 is in SZ.-NAGY [1938:1]; the analogue for cosine functions (Remark 6.6 is in the author [1974:3]).

The theory of periodic and almost periodic semigroups, groups and cosine functions in Banach space was iniated in the late sixties. DA PRATO [1968:1] and BART [1977:1] treat periodic strongly continuous semigroups and groups while CIORĂNESCU [1982:1] treats periodic distribution semigroups. Periodic cosine functions were considered by GIUSTI [1967:1]. Finally, almost periodic semigroups and cosine functions were considered respectively by BART-GOLDBERG [1978:1] and by PISKAREV [1982:1]. Some of their results will be found below (Exercises 6 to 15).

EXERCISE 1. Let E be an arbitrary vector space. Show that E has a basis, that is, a linearly independent subset $\{u_\alpha\}$ such that, for each $u \in E$ we have

$$u = \sum \lambda_\alpha u_\alpha,$$

the scalars $\{\lambda_\alpha\}$ all zero except for a finite number. See JACOBSON [1953:1, p. 239]

EXERCISE 2. Using Exercise 1 show that there exists a function $f : \mathbb{R} \to \mathbb{R}$ such that

$$f(s + t) = f(s) + f(t) \tag{8.4}$$

but

$$f(t) \not\equiv ct \tag{8.5}$$

for any c (Hint: $\mathbb{R}$ is a vector space over $\mathbb{Q}$, the field of rational numbers. Any nonzero linear functional $f : \mathbb{R} \to \mathbb{Q}$ will do. To construct one, let $\{t_\alpha\}$ be a base of $\mathbb{R}$ as in Exercise 1 and define

$$f(\sum r_\alpha t_\alpha) = r_{\bar{\alpha}} \qquad (r_\alpha \in \mathbb{Q})$$

where $\bar{\alpha}$ is fixed .

EXERCISE 3. Using Exercise 2 show that there exists a unitary group $U(\hat{t})$ (in one-dimensional Banach space) that does not admit the representation (2.1).

EXERCISE 4. Give an example of a self-adjoint operator B such that (a) $\sigma(B) = (-\infty, \infty)$. (b) Every $\lambda \in \sigma(B)$ is an eigenvalue of B (Hint: consider the Hilbert space H of all functions $u(\hat{\lambda})$ defined in $-\infty < \lambda < \infty$, vanishing for all but for a finite or countable number of λ's and such that

$$\|u\|^2 = \sum_{-\infty < \lambda < \infty} |u(\lambda)|^2 < \infty . \tag{8.6}$$

Define

$$Au(\lambda) = \lambda u(\lambda) \qquad (-\infty < \lambda < \infty) , \tag{8.7}$$

the domain of A consisting of all $u(\hat{\lambda}) \in H$ such that (8.7) belongs to H).

EXERCISE 5. (a) Let $S(\hat{t})$ be a semigroup in a Banach space E such that

$$\|S(t)\| \leq Ce^{\omega t} \qquad (t \geq 0) .$$

Show that there exists an equivalent norm $\|\cdot\|'$ such that

$$\|S(t)\|' \leq e^{\omega t} \qquad (t \geq 0) .$$

(b) Let $S(\hat{t})$ be a group in a Banach space E such that

$$\|S(t)\| \leq Ce^{\omega|t|} \qquad (-\infty < t < \infty) .$$

Show that there exists an equivalent norm $\|\cdot\|'$ in E such that

$$\|S(t)\|' \leq e^{\omega|t|} \qquad (-\infty < t < \infty) .$$

(c) State and prove a version of (b) for cosine functions both with $C\exp(\omega|t|)$ and with $\cosh \omega t$ on the right-hand side.

(Hint: in (a) use

$$\|u\|' = \sup_{t\geq 0} e^{-\omega t}\|S(t)u\| ,$$

and in (b),

$$\|u\|' = \sup_{-\infty<t<\infty} e^{-\omega|t|}\|S(t)u\| ,$$

with a similar norm in (c)).

EXERCISE 6. Let $S(\hat{t})$ be a strongly continuous semigroup in a Banach space E such that $S(\hat{t})u$ is almost periodic (that is, a strongly almost periodic semigroup). Then (a) Each $S(t)$ is one-to-one (in fact, $\|S(t)u\| = \|u\|$ $(t \geq 0)$ for $u \in E$, thus each $S(t)$ is an <u>isometry</u>) (b) Each $S(t)$ is onto (in view of (a) it is enough to show that $S(t)E$ is dense in E) (c) The group $S(t) = S(t)$ for $t \geq 0$, $S(t) = S(-t)^{-1}$ for $t < 0$ is strongly almost periodic (The definition of almost periodic function in $t \geq 0$ is the same as in $-\infty < t < \infty$; see BART-GOLDBERG [1978:1]).

EXERCISE 7. Let $S(\hat{t})$ be a strongly almost periodic group in a Banach space E. For $-\infty < \lambda < \infty$ define a bounded operator M_λ in E by

$$M_\lambda u = \lim_{T\to\infty} \frac{1}{2T}\int_{-T}^{T} e^{-i\lambda t}S(t)u\,dt . \tag{8.8}$$

(a) Show that $\{M_\lambda;\ -\infty < \lambda < \infty\}$ is a family of mutually orthogonal projections, that is

$$M_\lambda M_\mu = \begin{cases} M_\lambda & \text{if} \quad \lambda = \mu \\ 0 & \text{if} \quad \lambda \neq \mu \end{cases} . \tag{8.9}$$

(b) Show that

$$M_\lambda S(t) = S(t)M_\lambda = e^{i\lambda t}S(t) , \tag{8.10}$$

and that

$$E_\lambda = M_\lambda E \subseteq D(A), \tag{8.11}$$

with

$$AM_\lambda = i\lambda M_\lambda, \tag{8.12}$$

where A is the infinitesimal generator of $S(\hat{t})$. (c) Show that the subspace generated by the eigenspaces $E_\lambda(-\infty < \lambda < \infty)$ is dense in E; more precisely, for each $u \in E$ there exists a sequence of integers $\{r_n\}$ and a double sequence of complex numbers $\{a_{nk}\}$ such that

$$u = \lim_{n\to\infty} \sum_{k=1}^{r_n} a_{nk} M_{\lambda_k} u , \tag{8.13}$$

where $\{\lambda_k\}$ is the (finite or countable) sequence of λ's for which $M_\lambda u \neq 0$ (Hint: use the approximation theorem for the E-valued almost periodic function $S(\hat{t})u$; see AMERIO-PROUSE [1971:1, p. 24]). (d) Show that if A has an isolated eigenvalue λ_0, then $R(\lambda;A)$ has a simple pole at λ_0.

EXERCISE 8. State and prove a suitable converse of Exercise 7; that is, give sufficient conditions on an infinitesimal generator A to generate a strongly almost periodic group.

EXERCISE 9. Let $S(\hat{t})$ be a strongly periodic semigroup (i.e. a strongly continuous semigroup such that $S(\hat{t})u$ is periodic for each u, the period depending in general on u) (a) Show that $S(\hat{t})$ can be extended to a strongly continuous group $S(\hat{t})$ periodic in $-\infty < t < \infty$. (See BART [1977:1]). (b) Let p be the least positive period of $S(\hat{t})$. Then if A is the infinitesimal generator of $S(\hat{t})$, $\sigma(A)$ is contained in the set $\{ikp;\ k = \ldots, -1, 0, 1, \ldots\}$ with each ikp an eigenvalue where $R(\lambda;A)$ has a pole of order one. (c) Show that (c) in Exercise 7 can be strenghtened to

$$u = \sum_{k=-\infty}^{\infty} M_{kp} u \tag{8.14}$$

if $u \in D(A)$. If $u \in E$, (8.14) holds in Cèsaro mean,

$$u = \lim_{n\to\infty} \sum_{k=-n}^{n} \left(1 - \frac{|k|}{n+1}\right) M_{kp} u , \tag{8.15}$$

or in Abel mean,

$$u = \lim_{\lambda \to 1-} 2(1-\lambda) \sum_{k=-\infty}^{\infty} \lambda^{|k|} M_{kp} u . \tag{8.16}$$

EXERCISE 10. State and prove a suitable converse of Exercise 9; that is, give sufficient conditions on an infinitesimal generator A to generate a (strongly) periodic group.

EXERCISE 11. Using Exercise III.3 show that (8.14) holds if $u \in D((I-A)^{\alpha})$, $\alpha > 0$; see §III.2 for definitions and properties of fractional powers (Hint: as shown in Exercise III.7, if $u \in D((I-A)^{\alpha})$ then $S(\hat{t})u$ is Hölder continuous with exponent α; use then the Dini-Lipschitz theorem (ZYGMUND [1959:1, p. 63]), generalized to vector valued functions, for the Fourier series of $S(\hat{t})u$).

The following exercises are analogues of Exercises 7, 8, 9, 10 and 11 for cosine functions.

EXERCISE 12. Let $C(\hat{t})$ be a strongly almost periodic cosine function in a Banach space E. For $0 \leq \lambda < \infty$ define a bounded operator $N_{\lambda}u$,

$$N_{\lambda} = \lim_{T\to\infty} \frac{1}{T} \int_0^T \cos \lambda t (C(s)u, v)dt =$$

$$= \lim_{T\to\infty} \frac{1}{2T} \int_{-T}^{T} e^{i\lambda t} (C(t)u, v)dt .$$

(a) Show that $\{M_{\lambda}; \lambda \geq 0\}$ is a family of mutually orthogonal projections, where

$$M_{\lambda} = N_{\lambda} \text{ for } \lambda > 0, \ M_0 = 2^{1/2} N_0 .$$

(b) Show that

$$M_{\lambda} C(t) = C(t) M_{\lambda} = \cos \lambda t C(t) \tag{8.17}$$

and that

$$E_\lambda = M_\lambda E \subseteq D(A), \tag{8.18}$$

with

$$AM_\lambda = -\lambda^2 M_\lambda , \tag{8.19}$$

where A is the infinitesimal generator of $S(\hat{t})$. (c) Show that (c) of Exercise 7 holds for the eigenspaces E_λ; in particular, the representation (8.13) is valid for each $u \in E$.

EXERCISE 13. State and prove a suitable converse of Exercise 12.

EXERCISE 14. Let $C(\hat{t})$ be a strongly periodic cosine function (defined in the same way as for semigroups in Exercise 9).
(a) Show that $C(\hat{t})$ is periodic in $-\infty < t < \infty$. (b) Let p be the least positive period of $C(\hat{t})$. Then, if A is the infinitesimal generator of $C(\hat{t})$, $\sigma(A)$ is contained in the set $\{-(kp)^2;\ k = 0,1,\dots\}$ with each $-(kp)^2$ an eigenvalue of A where $R(\lambda;A)$ has a pole of order one. (c) Show that (c) in Exercise 12 can be strengthened to

$$u = \sum_{k=0}^{\infty} M_{-(kp)^2} u \tag{8.20}$$

if $u \in D(A)$ (or, more generally, if $u \in D((I-A)^\alpha)$ for some $a > 0$; cf. Exercise III.5). For $u \in E$, (8.20) holds in Cèsaro mean,

$$u = \lim_{n\to\infty} \sum_{k=0}^{n} \left(1 - \frac{|k|}{n+1}\right) M_{-(kp)^2} u , \tag{8.21}$$

or in Abel mean

$$u = \lim_{\lambda\to 1-} (1-\lambda) \sum_{k=0}^{\infty} \lambda^k M_{-(kp)^2} u . \tag{8.22}$$

EXERCISE 15. State and prove a suitable converse of Exercise 14.

CHAPTER VI

THE PARABOLIC SINGULAR PERTURBATION PROBLEM

§VI.1 Vibrations of a membrane in a viscous medium.

Consider a uniform membrane fixed to the boundary Γ of a two dimensional domain Ω and immersed in a viscous medium. The small oscilations of the membrane are described by the equation

$$\rho v_{tt} + \gamma v_t = \sigma \Delta v \qquad (x \in \Omega), \tag{1.1}$$

where $v = v(x,t) = v(x_1,x_2,t)$ is the vertical movement of the membrane and the constants ρ,γ,σ are, respectively, the mass density per unit area of the membrane, the coefficient of viscosity of the medium and the tension of the membrane. The displacement v satisfies the boundary condition

$$v(x,t) = 0 \qquad (x \in \Gamma), \tag{1.2}$$

and the initial conditions

$$v(x,0) = u_0(x), v_t(x,0) = u_1(x) \qquad (x \in \Omega). \tag{1.3}$$

If we define $v(x,t) = u(x, (\sigma/\gamma)t)$ then u satisfies the equation

$$\frac{\sigma\rho}{\gamma^2} u_{tt} + u_t = \Delta u \qquad (x \in \Omega) \tag{1.4}$$

and the boundary condition (1.2); the initial conditions are

$$u(x,0) = u_0(x), \ u_t(x,0) = \frac{\gamma}{\sigma} u_1(x) \qquad (x \in \Omega). \tag{1.5}$$

Setting $(\sigma\rho)^{1/2}/\gamma = \varepsilon$ we can write (1.5) in the form

$$\varepsilon^2 u_{tt} + u_t = \Delta u \qquad (u \in \Omega), \tag{1.6}$$

where ε will be small if the medium is highly viscous $(\gamma >> 1)$. We can write the second initial condition in the form

$$u_t(x,0) = \varepsilon^{-1}(\frac{\rho}{\sigma})^{1/2} u_1(x).$$

This suggests the problem of studying the behavior of the solution of (1.6) as $\varepsilon \to 0$, allowing for dependence of the initial conditions on ε. This will be done in the rest of the chapter for an abstract model encompassing equations like (1.6).

§VI.2 Singular perturbation. Explicit solution of the perturbed equation.

Throughout this chapter A will be the infinitesimal generator of a strongly continuous cosine function $C(\hat{t})$ in the complex Banach space E (that is, $A \in \mathfrak{C}^2$; see Exercise 2). We consider the Cauchy problems

$$\varepsilon^2 u''(t;\varepsilon) + u'(t;\varepsilon) = Au(t;\varepsilon) + f(t;\varepsilon) \qquad (t \geq 0),$$
$$u(0;\varepsilon) = u_0(\varepsilon),\ u'(0,\varepsilon) = u_1(\varepsilon)\,, \tag{2.1}$$

and

$$u'(t) = Au(t) + f(t) \qquad (t \geq 0), \qquad u(0) = u_0. \tag{2.2}$$

Roughly speaking, the results in the following sections establish that $u(t;\varepsilon) \to u(t)$ as $\varepsilon \to 0$ if $u_0(\varepsilon) \to u$, $f(t;\varepsilon) \to f(t)$ and if $u_1(\varepsilon)$ is suitably restricted. Results of the same type hold for derivatives.

As a first step, we compute the explicit solution of (2.1) using simple changes of dependent and independent variable. Writing $u(t;\varepsilon) = e^{-t/2\varepsilon^2}v(t/\varepsilon;\varepsilon)$ (or, equivalently, $v(t;\varepsilon) = e^{t/2\varepsilon}u(\varepsilon t;\varepsilon)$) we easily show that $v(\hat{t};\varepsilon)$ is a solution of the initial value problem

$$v''(t;\varepsilon) = \left(A + \frac{1}{4\varepsilon^2}I\right)v(t;\varepsilon) + e^{t/2\varepsilon}f(\varepsilon t;\varepsilon),$$
$$v(0,\varepsilon) = u_0(\varepsilon), \quad v'(0,\varepsilon) = \frac{1}{2\varepsilon}\,u_0(\varepsilon) + \varepsilon u_1(\varepsilon). \tag{2.3}$$

Conversely, every solution $v(\hat{t};\varepsilon)$ of (2.3) gives rise to a solution $u(\hat{t};\varepsilon)$ of the initial value problem (2.1). The (generalized) solution of the initial value problem (2.3) is given by

$$v(t;\varepsilon) = C(t;\varepsilon)v(0;\varepsilon) + \mathcal{S}(t;\varepsilon)v'(0;\varepsilon)$$
$$+ \int_0^t \mathcal{S}(t - s;\, \varepsilon)e^{s/2\varepsilon}\, f(\varepsilon s;\varepsilon)\, ds\,, \tag{2.4}$$

where $C(\hat{t};\varepsilon)$ is the cosine function generated by $A + (2\varepsilon)^{-2}I$ and $\mathcal{S}(t;\varepsilon)u = \int_0^t C(s;\varepsilon)u\,ds$. We know from Lemma III.4.1 that the Cauchy problem for

$$u''(t) = \left(A + \frac{1}{4\varepsilon^2}I\right) u(t) \tag{2.5}$$

is well posed. However, the series representation (III.4.7) is not convenient here and we provide a different representation below which, incidentally, proves afresh that $A + (2\varepsilon)^{-2}I \in \mathfrak{C}^2$.

LEMMA 2.1. *Let* $A \in \mathfrak{C}^2$, *that is, let the Cauchy problem for*

$$u''(t) = Au(t) \tag{2.6}$$

be well posed, and let b *be an arbitrary complex number. Then the Cauchy problem for*

$$u''(t) = (A + b^2 I)u(t) \tag{2.7}$$

is well posed; the propagators $C_b(\hat{t})$, $\mathcal{S}_b(\hat{t})$ of (2.6) [1] *are given by*

$$C_b(t)u = C(t)u + bt\int_0^t \frac{I_1(b(t^2 - s^2)^{1/2})}{(t^2 - s^2)^{1/2}} C(s)u\,ds, \tag{2.8}$$

and

$$\mathcal{S}_b(t)u = \int_0^t I_0(b(t^2 - s^2)^{1/2})C(s)u\,ds, \tag{2.9}$$

where I_0, I_1 are the Bessel functions defined by

$$I_\nu(x) = \sum_{n=0}^{\infty} \frac{1}{n!\Gamma(\nu + n + 1)} \left(\frac{x}{2}\right)^{\nu+2n}. \tag{2.10}$$

The proof can be carried out in (at least) three ways, all of which we sketch briefly below. In the first, one takes advantage of the differential equation

$$I_\nu''(x) + \frac{1}{x} I_\nu'(x) + \left(1 - \frac{\nu^2}{x^2}\right) I_\nu(x) = 0 \tag{2.11}$$

satisfied by $I_\nu(x)$ and of a little integration by parts to show that if $u, v \in D(A)$ (so that $C(\hat{t})u$, $\mathcal{S}(\hat{t})u$ are genuine solutions of (2.6)) then $u(\hat{t}) = C_b(\hat{t})u + \mathcal{S}_b(\hat{t})u$ is a genuine solution of (2.7): uniqueness of solutions of (2.7) is proved transforming them back into solutions of (2.6) by means of formulas (2.8) and (2.9) with b replaced by ib.

For details on this line of approach (although the context there is somewhat different) see SOVA [1970:4]. The second uses Theorem II.2.3; we show that the Laplace transform of $C_b(\hat{t})u$ given by (2.7) equals $\lambda R(\lambda^2; A + b^2 I)u = \lambda R(\lambda^2 - b^2; A)u$. The third is based on Lemma III.4.1, explicit construction of each of the terms C_n in (III.4.7) combined with the series representation (2.10) for $I_0(x)$, $I_1(x)$. We omit the details.

Applying the translation formulas (2.8), (2.9) to the equation (2.5) we obtain

$$C(t;\varepsilon)u = C(t)u + \frac{t}{2\varepsilon}\int_0^t \frac{I_1((t^2 - s^2)^{1/2}/2\varepsilon)}{(t^2 - s^2)^{1/2}} C(s)u\, ds \tag{2.12}$$

and

$$S(t;\varepsilon)u = \int_0^t I_0((t^2 - s^2)^{1/2}/2\varepsilon)C(s)u\, ds\,, \tag{2.13}$$

both valid for arbitrary u and t. Using these formulas in (2.4) and making an obvious change of variable in the iterated integral resulting from the integral in (2.4) we obtain the final formula for $u(\hat{t};\varepsilon)$, where we have used (II.4.3) for the solution of the nonhomogeneous equation:

$$\begin{aligned} u(t;\varepsilon) &= e^{-t/2\varepsilon^2} C(t/\varepsilon)u_0(\varepsilon) \\ &+ \frac{te^{-t/2\varepsilon^2}}{\varepsilon^2}\int_0^{t/\varepsilon} \frac{I_1(((t/\varepsilon)^2 - s)^{1/2}/2\varepsilon)}{((t/\varepsilon)^2 - s^2)^{1/2}} C(s)(\tfrac{1}{2}\, u_0(\varepsilon))\, ds \\ &+ \frac{e^{-t/2\varepsilon^2}}{\varepsilon}\int_0^{t/\varepsilon} I_0(((t/\varepsilon)^2 - s^2)^{1/2}/2\varepsilon)C(s)(\tfrac{1}{2}\, u_0(\varepsilon) + \varepsilon^2 u_1(\varepsilon))\, ds \\ &+ e^{-t/2\varepsilon^2}\int_0^{t/\varepsilon}\left(\int_0^{t/\varepsilon - s} I_0(((t/\varepsilon - s)^2 - \sigma^2)^{1/2}/2\varepsilon)C(\sigma)\, d\sigma\right) e^{s/2\varepsilon} f(\varepsilon s;\varepsilon)\, ds \\ &= e^{-t/2\varepsilon^2} C(t/\varepsilon)u_0(\varepsilon) + \int_0^{t/\varepsilon} \varphi(t,s;\varepsilon)C(s)(\tfrac{1}{2}\, u_0(\varepsilon))\, ds \\ &+ \int_0^{t/\varepsilon} \psi(t,s;\varepsilon)C(s)(\tfrac{1}{2}\, u_0(\varepsilon) + \varepsilon^2 u_1(\varepsilon))\, ds \\ &+ \int_0^t\left(\int_0^{(t-s)/\varepsilon} \psi(t - s,\sigma;\varepsilon)C(\sigma)\, d\sigma\right) f(s;\varepsilon)\, ds \\ &= e^{-t/2\varepsilon^2} C(t/\varepsilon)u_0(\varepsilon) + \mathfrak{R}(t;\varepsilon)(\tfrac{1}{2}\, u_0(\varepsilon)) + \mathfrak{S}(t;\varepsilon)(\tfrac{1}{2}\, u_0(\varepsilon) + \varepsilon^2 u_1(\varepsilon)) \\ &+ \int_0^t \mathfrak{S}(t - s;\varepsilon)\, f(s,\varepsilon)\, ds\,, \end{aligned} \tag{2.14}$$

(2)

where the definitions of the scalar functions $\varphi(t,s;\varepsilon)$, $\psi(t,s;\varepsilon)$ and the operator valued functions $\mathfrak{R}(t;\varepsilon)$, $\mathfrak{S}(t;\varepsilon)$ are easily read off the formula (3).

The following result is an immediate consequence of formula (2.14) and the correspondence between solutions of (2.1) and those of (2.3):

LEMMA 2.2. Every generalized solution $u(t;\varepsilon)$ of (1.1) is continuous in $t \geq 0$. If $u_0(\varepsilon) = 0$ then $u(t;\varepsilon)$ is continuously differentiable.

We examine now the initial value problem (2.2), precisely, its relation with the second order initial value problem for (2.6).

THEOREM 2.3. Let $A \in \mathfrak{C}^2$, that is, let the Cauchy problem for (2.6) be well posed. Then the initial value problem for

$$u'(t) = Au(t) \tag{2.15}$$

is well posed in $t \geq 0$, the propagator of (2.15) given by the abstract Weierstrass formula

$$S(t)u = \frac{1}{(\pi t)^{1/2}} \int_0^\infty e^{-s^2/4t} C(s)u \, ds \qquad (t > 0). \tag{2.16}$$

The propagator $S(\hat{t})$ can be extended to a (E)-valued analytic function in $\operatorname{Re} \zeta > 0$; more precisely, $A \in \mathfrak{G}(\varphi)$ for $0 < \varphi < \pi/2$. Finally, if $A \in \tilde{\mathfrak{C}}^2(C_0;\omega)$ then $A \in \mathfrak{C}^1_+(C_0;\omega^2)$.

The proof of Theorem 2.3 (which was proposed as Exercise 11 in Chapter II) goes as follows. Recall that, by definition, $A \in \tilde{\mathfrak{C}}^2(C_0;\omega)$ if and only if

$$\|C(t)\| \leq C_0 \cosh \omega t \qquad (-\infty < t < \infty). \tag{2.17}$$

Thus, if $S(\hat{t})$ is the (E)-valued function defined by (2.16) we have

$$\|S(t)\| \leq C_0 \frac{1}{(\pi t)^{1/2}} \int_0^\infty e^{-s^2/4t} \cosh \omega s \, ds = C_0 e^{\omega^2 t} \quad (t > 0). \tag{2.18}$$

We check easily that $S(\hat{t})$ is continuous in the norm of (E) for $t > 0$ due to the fast telescoping of the integrand as $s \to \infty$. Defining

$$S(0) = I, \tag{2.19}$$

we check next that $S(\hat{t})$ is strongly continuous in $t \geq 0$; trivially, only continuity at $t = 0$ must be proved. To do this we note first that, if $\rho > 0$,

$$\frac{1}{(\pi t)^{1/2}}\int_\rho^\infty e^{-s^2/4t}\cosh \omega s\, ds$$

$$= \frac{1}{(\pi t)^{1/2}}\int_{\rho/2t^{1/2}}^\infty e^{-\sigma^2}\cosh 2\omega t^{1/2}\sigma\, d\sigma\ , \qquad (2.20)$$

which expression tends to zero as $t \to 0+$. For $t > 0$ we have

$$\|S(t)u - u\| \le \frac{1}{(\pi t)^{1/2}}\int_0^\infty e^{-s^2/4t}\|C(s)u - u\|\, ds\ . \qquad (2.21)$$

The argument below is standard in approximation theory. Once $\varepsilon > 0$ is given, we divide the interval of integration in (2.21) at $s = \rho$. In $0 \le s \le \rho$ we use (2.18), taking ρ so small that we shall have $\|C(s)u - u\| < \varepsilon/2$ there; in $s \ge \rho$ we estimate by $2C_0$ times (2.20) taking $0 \le t \le \delta$, δ so small that (2.20) does not surpass $\varepsilon/4C_0$. This shows that $\|S(t)u - u\| \le \varepsilon$ for $0 \le t \le \delta$, so that $S(\hat{t})$, its definition completed by (2.19), is strongly continuous in $t \ge 0$.

Let $\lambda > \omega^2$. Using the equality in (2.18) we obtain

$$\int_0^\infty e^{-\lambda t}\left(\frac{1}{(\pi t)^{1/2}}\int_0^\infty e^{-s^2/4t}\cosh \omega s\, ds\right) dt$$

$$= \int_0^\infty e^{-\lambda t}e^{\omega^2 t}dt = \frac{1}{\lambda - \omega^2}\ . \qquad (2.22)$$

By Tonelli's theorem, convergence of the integral (2.22) implies that the order of integration can be inverted. This justifies the following computation:

$$\int_0^\infty e^{-\lambda t}S(t)u\, dt = \int_0^\infty e^{-\lambda t}\left(\frac{1}{(\pi t)^{1/2}}\int_0^\infty e^{-s^2/4t}C(s)u\, ds\right) dt$$

$$= \int_0^\infty\left(\int_0^\infty e^{-\lambda t}\frac{e^{-s^2/4t}}{(\pi t)^{1/2}}\, dt\right) C(s)u\, ds = \frac{1}{\lambda^{1/2}}\int_0^\infty e^{-\lambda^{1/2}s}C(s)u\, ds$$

$$= R(\lambda;A)u \quad (\lambda > \omega^2) \qquad (2.23)$$

by the first equality (2.11). We apply then Theorem 2.3 to deduce that $S(\hat{t})$ is a strongly continuous semigroup with infinitesimal generator A. That $S(\hat{t})$ can be extended to a (E)-valued function analytic in $\operatorname{Re}\zeta > 0$, as well as the estimates corresponding to the classes $\mathcal{G}(\varphi)$ can be proved replacing t by ζ in (2.23). If $\operatorname{Re}\zeta > 0$ we have

$$\|S(\zeta)\| \le \frac{1}{(\pi|\zeta|)^{1/2}} \int_0^\infty e^{-\mathrm{Re}(s^2/4\zeta)} \|C(s)\|\, ds$$

$$\le C_0 \frac{1}{(\pi|\zeta|)^{1/2}} \int_0^\infty e^{-s^2 \mathrm{Re}\, \zeta/4|\zeta|^2} \cos \omega s\, ds = C_0(|\zeta|/\mathrm{Re}\, \zeta), \tag{2.24}$$

which implies the desired estimates.

§VI.3 The homogeneous equation: convergence of $u(\hat{t};\varepsilon)$.

We investigate here the convergence of $u(\hat{t};\varepsilon)$ to $u(\hat{t})$ as $\varepsilon \to 0$ in the homogeneous case $f(\hat{t}) = f(\hat{t};\varepsilon) = 0$. As a first step we obtain uniform bounds on $u(\hat{t};\varepsilon)$. Using formula (2.14) and noting that $\varphi, \psi \ge 0$ we obtain from (2.17) that

$$\begin{aligned} \|u(t;\varepsilon)\| &\le C_0 e^{-t/2\varepsilon^2} \cosh(\omega t/\varepsilon) \|u_0(\varepsilon)\| \\ &+ C_0 \int_0^{t/\varepsilon} \varphi(t,s;\varepsilon) \cosh \omega s (\tfrac{1}{2} \|u_0(\varepsilon)\|)\, ds \\ &+ C_0 \int_0^{t/\varepsilon} \psi(t,s;\varepsilon) \cosh \omega s (\tfrac{1}{2} \|u_0(\varepsilon)\| + \varepsilon^2 \|u_1(\varepsilon)\|)\, ds, \end{aligned} \tag{3.1}$$

where C_0 is the constant in (2.17). Since $\cosh \omega\hat{t}$ is a cosine function (in the sense of the preceding section) with infinitesimal generator ω^2, the function on the right hand side of (3.1) equals $C_0\xi(t;\varepsilon)$, where ξ is the solution of the scalar initial value problem

$$\varepsilon^2 \xi''(t;\varepsilon) + \xi'(t;\varepsilon) = \omega^2 \xi(t;\varepsilon),$$

$$\xi(0,\varepsilon) = \xi_0(\varepsilon),\ \xi'(0,\varepsilon) = \xi_1(\varepsilon), \tag{3.2}$$

with $\xi_0(\varepsilon) = \|u_0(\varepsilon)\|$, $\xi_1(\varepsilon) = \|u_1(\varepsilon)\|$. We compute this function explicitly. Let $\lambda^{\pm}(\varepsilon) = (-1 \pm (1 + 4\omega^2\varepsilon^2)^{1/2})/2\varepsilon^2$ be the two (real, different) roots of the characteristic equation $\varepsilon^2\lambda^2 + \lambda - \omega^2 = 0$. Then

$$\xi(t,\varepsilon) = \frac{\lambda^+(\varepsilon)e^{\lambda^-(\varepsilon)t} - \lambda^-(\varepsilon)e^{\lambda^+(\varepsilon)t}}{\lambda^+(\varepsilon) - \lambda^-(\varepsilon)} \xi_0(\varepsilon) + \frac{e^{\lambda^+(\varepsilon)t} - e^{\lambda^-(\varepsilon)t}}{\lambda^+(\varepsilon) - \lambda^-(\varepsilon)} \xi_1(\varepsilon)$$

$$= \Phi_\omega(t;\varepsilon)\xi_0(\varepsilon) + \Psi_\omega(t;\varepsilon)(\varepsilon^2\xi_1(\varepsilon)). \tag{3.3}$$

Since $(1 + 4\omega^2\varepsilon^2)^{1/2} < 1 + 2\omega^2\varepsilon^2$ we have $0 \leq \lambda^+(\omega) < \omega^2$. On the other hand,, $\lambda^-(\varepsilon) < 0$, thus we may replace $\lambda^+(\varepsilon)e^{\lambda^-(\varepsilon)t}$ by $\lambda^+(\varepsilon)e^{\lambda^+(\varepsilon)t}$ in the first term of (3.3) for estimation purposes. In the second, we simply delete $-e^{\lambda^-(\varepsilon)t}$ noting that $\lambda^+(\varepsilon) - \lambda^-(\varepsilon) = (1 + 4\omega^2\varepsilon^2)^{1/2}/\varepsilon^2 > 1/\varepsilon^2$. We obtain in this way the following result:

LEMMA 3.1. _For every_ $\varepsilon > 0$ _we have_

$$\begin{aligned} \|u(t;\varepsilon)\| &\leq C_0(\Phi_\omega(t;\varepsilon)\xi_0(\varepsilon) + \Psi_\omega(t;\varepsilon)(\varepsilon^2\xi_1(\varepsilon))) \\ &\leq C_0 e^{\omega^2 t}(\|u_0(\varepsilon)\| + \varepsilon^2\|u_1(\varepsilon)\|) \qquad (t > 0) \end{aligned} \tag{3.4}$$

where C_0, ω _are the constants in_ (2.17).

To study convergence of $u(t;\varepsilon)$ we shall use the asymptotic series for the Bessel functions,

$$I_\nu(x) = \frac{e^x}{(2\pi x)^{1/2}}\left(\sum_{j=0}^{m} \frac{(-1)^j[\nu,j]}{(2x)^j} + 0\left(\frac{1}{x^{m+1}}\right)\right) \tag{3.5}$$

where

$$[\nu,j] = \frac{\Gamma(\nu + j + 1/2)}{j!\Gamma(\nu - j + 1/2)} \tag{3.6}$$

(see WATSON [1948:1, pp. 203 and 198]). The asymptotic series (3.5) will be used with

$$x = \frac{1}{2\varepsilon}\left(\left(\frac{t}{\varepsilon}\right)^2 - s^2\right)^{1/2}$$

both to calculate limits (as in (3.10)-(3.11), (3.25)-(3.26), (4.14)-(4.15)) and for estimation purposes (as in (3.14)-(3.15), (3.27)-(3.28), (4.16)). The second application deserves some comment. Since the functions to be estimated are regular at $x = 0$, (3.5) will not provide good bounds near zero. However, this is not very significant, since these functions converge to limits that do have singularities at $x = 0$. To improve the estimations, we shall replace remainders of the form $0(x^{-\alpha})$ by

$$0((x + a)^{-\alpha})$$

which are regular at the origin.

It is plain that the observations above can be applied as well to the asymptotic series obtained from the series (3.5) for functions of the form

$x^{-\beta}I_\nu(x)$, where β and ν are arbitrary real numbers. Likewise, we shall differentiate (3.5) term by term to provide asymptotic series for derivatives of arbitrary order. This can be easily justified.

We proceed to the estimation of $u(t;\varepsilon)-u(t)$. To this end we divide the interval of integration in (2.14) in $0 \leq s \leq s(\varepsilon)$, where the asymptotic development (2.5) can be used, and the "small" interval $s(\varepsilon) \leq t \leq t/\varepsilon$ where rougher bounds will suffice. The divisory point between the "inner" and "outer" integrals is defined by

$$\frac{1}{2\varepsilon}\left(\left(\frac{t}{\varepsilon}\right)^2 - s(\varepsilon)^2\right)^{1/2} = \frac{\eta t}{\varepsilon^2}, \tag{3.7}$$

where $\eta < 1/2$ will be specified later. The length of the second interval is

$$\frac{t}{\varepsilon} - s(\varepsilon) = \frac{4\eta^2 t^2}{\varepsilon(t + \varepsilon s(\varepsilon))} \leq \frac{4\eta^2 t}{\varepsilon} . \tag{3.8}$$

We use the asymptotic development (3.5) to order $m = 0$ in $0 \leq s \leq s(\varepsilon)$:

$$I_1(x) = \frac{e^x}{(2\pi x)^{1/2}}\left(1 + 0\left(\frac{1}{x}\right)\right) . \tag{3.9}$$

After dividing by x and performing a few cancellations we obtain

$$\varphi(t,s;\varepsilon) = \chi(t,s;\varepsilon)\ \exp\left\{-\frac{t}{2\varepsilon^2}\left(1-\left(1-\left(\frac{\varepsilon s}{t}\right)^2\right)^{1/2}\right)\right\}, \tag{3.10}$$

where

$$\chi(t,s;\varepsilon) = (\pi t)^{-1/2}\left(1-\left(\frac{\varepsilon s}{t}\right)^2\right)^{-3/4}\left(1 + 0\left(\frac{\varepsilon^2}{\eta t}\right)\right). \tag{3.11}$$

To obtain (3.11) from (3.9) we have used the inequality

$$x = \frac{1}{2\varepsilon}\left(\left(\frac{t}{\varepsilon}\right)^2 - s^2\right)^{1/2} \geq \frac{1}{2\varepsilon}\left(\left(\frac{t}{\varepsilon}\right)^2 - s(\varepsilon)^2\right)^{1/2} = \frac{\eta t}{\varepsilon^2}. \tag{3.12}$$

Obviously, (3.10) - (3.11) is uniform in the interval $0 \leq s \leq s(\varepsilon)$. To obtain an estimate for φ we use formula (3.5) for the function $x^{-1}I_1(x)$, $m = 0$. Since $x^{-1}I_1(x)$ is regular for $x = 0$, there exists a constant C such that

$$\frac{I_1(x)}{x} \leq C\,\frac{e^x}{(2 + x)^{3/2}} \quad (x \geq 0) . \tag{3.13}$$

where C does not depend on x; the choice of 2 in the denominator of (3.13) makes the function $(2+x)^{-3/2}e^x$ increasing, since $(a+x)^{-\alpha}e^x$ is increasing for $a > \alpha$. Essentially the same manipulations leading to (3.10)-(3.11) reveal that the estimate

$$\varphi(t,s;\varepsilon) \leq C\rho(t,s;\varepsilon)\exp\left\{-\frac{t}{2\varepsilon^2}\left(1-\left(1-\left(\frac{\varepsilon s}{t}\right)^2\right)^{1/2}\right)\right\} \qquad (3.14)$$

holds in $0 \leq s \leq t$, where

$$\rho(t,s;\varepsilon) = t^{-1/2}\left(\frac{4\varepsilon^2}{t} + \left(1-\left(\frac{\varepsilon s}{t}\right)^2\right)^{1/2}\right)^{-3/2} \qquad (3.15)$$

and the constant C does not depend on ε, t. Inequality (3.14) leads to the estimates below.

LEMMA 3.2. *Let* $t, \varepsilon > 0$. *Then the estimate*

$$\varphi(t,s;\varepsilon) \leq C\eta^{-3/2}\,t^{-1/2}\,e^{-s^2/4t} \qquad (0 \leq s \leq s(\varepsilon)) \qquad (3.16)$$

holds, where C *is a constant that does not depend on* s,t,ε.

Proof. We make use of (3.14)-(3.15) observing that $(1-(\varepsilon s/t)^2)^{1/2} \leq 1-(\varepsilon s/t)^2/2$ in the exponent and that $(1-(\varepsilon s/t)^2)^{1/2} \geq (1-(\varepsilon s(\varepsilon)/t)^2)^{1/2} = 2\eta$ in the denominator of (2.14); the term $4\varepsilon^2/t$ inside the parenthesis is positive and can be dropped.

LEMMA 3.3. *Let* $t, \varepsilon > 0$. *Then*

$$\varphi(t,s;\varepsilon) \leq C\eta^{-3/2}\,t^{-1/2}\,e^{-t/2\varepsilon^2}\,e^{\eta t/\varepsilon^2} \qquad (s(\varepsilon) \leq s \leq t/\varepsilon)\,, \quad (3.17)$$

where the constant C *does not depend of* s,t,ε.

Proof. We use again (3.14)-(3.15) keeping in mind that the right hand side of the inequality (3.13) is an increasing function of x. Accordingly, we can estimate the right hand side of (3.14) by the value obtained inserting the highest possible value of $(1-(\varepsilon s/t^2)^{1/2}$ (which is $(1-(\varepsilon s(\varepsilon)/t)^2)^{1/2} = 2\eta$). Once this is done we discard the summand $4\varepsilon^2/t$ in the outer parenthesis of (3.15). The result is (3.17).

As an immediate consequence of (3.17) and of the estimation (3.8) for the length of the interval $s(\varepsilon) \leq s \leq t$ we obtain

$$\int_{s(\varepsilon)}^{t/\varepsilon} \varphi(t,s,\varepsilon)\|C(s)u\|\,ds \leq C\eta^{1/2}\|u\| \frac{t^{1/2}}{\varepsilon} \exp\left\{-\frac{t}{\varepsilon^2}\left(\frac{1}{2} - \eta - \omega\varepsilon\right)\right\}; \quad (3.18)$$

there and in other inequalities C denotes a constant independent of $s,t,\varepsilon,$ not necessarily the same in different inequalities.

Let $t(\varepsilon) > 0$ for each $\varepsilon > 0$. We say that a family of functions $g(\hat{t};\varepsilon)$ converges uniformly in $t \geq t(\varepsilon)$ to a function $g(\hat{t})$ if and only if

$$\lim_{\varepsilon\to 0}\ \sup_{t\geq t(\varepsilon)} \|g(t;\varepsilon) - g(t)\| = 0. \quad (3.19)$$

If the supremum is taken in $t(\varepsilon) \leq t \leq a$ for $a > 0$ arbitrary we say that $g(\hat{t};\varepsilon)$ converges to $g(\hat{t})$ uniformly on compacts of $t \geq t(\varepsilon)$.

We prove below that for every $u \in E$, $\mathfrak{R}(t;\varepsilon)u \to S(t)u$ uniformly on compacts of $t \geq t(\varepsilon)$ uniformly with respect to u on bounded subsets of E, as long as

$$t(\varepsilon)/\varepsilon^2 \to \infty \qquad (\varepsilon \to 0). \quad (3.20)$$

In fact, assume this is false. Then there exists a bounded sequence $\{u_n\} \subset E$, a sequence $\{\varepsilon_n\}$ with $\varepsilon_n \to 0$ and a bounded sequence $\{t_n\}$ such that $t_n/\varepsilon_n^2 \to \infty$ and $\|\mathfrak{R}(t_n;\varepsilon_n)u_n - S(t_n)u_n\| \geq \delta > 0$. For each n we choose η_n such that

$$\frac{1}{2} - \eta_n - \omega\varepsilon_n = \frac{\varepsilon_n}{t_n^{1/2}} \quad (3.21)$$

(note that $\eta_n < 1/2$: moreover, since both ε_n and $\varepsilon_n t_n^{-1/2}$ tend to zero, $\eta_n \to 1/2$ as $n \to \infty$). We divide the interval of integration in (2.14) according to the equality (3.7) with $\eta = \eta_n$. We have

$$\begin{aligned}
&\|\mathfrak{R}(t_n;\varepsilon_n)u_n - S(t_n)u_n\| \\
&\leq \int_0^{s(\varepsilon_n)} |\varphi(t_n,s;\varepsilon_n) - (\pi t_n)^{-1/2}e^{-s^2/4t_n}|\ \|C(s)u_n\|\,ds \\
&+ \int_{s(\varepsilon_n)}^{t_n/\varepsilon_n} \varphi(t_n,s;\varepsilon_n)\ \|C(s)u_n\|\,ds \qquad (3.22)\\
&+ (\pi t_n)^{-1/2}\int_{s(\varepsilon_n)}^{\infty} e^{-s^2/4t_n}\ \|C(s)u\|\,ds.
\end{aligned}$$

The first integral tends to zero as $n \to \infty$ due to the dominated convergence theorem: in fact, the asymptotic relation (3.10) shows that $\varphi(t_n, s; \varepsilon_n) - (\pi t_n)^{-1/2} e^{-s^2/4t_n} \to 0$ as $n \to \infty$ for s fixed (note that $1 - (\varepsilon s/t)^2 \geq 2\eta_n \to 1$, hence $\varepsilon s/t \to 0$) and we have the uniform estimate (3.16). The second integral tends to zero by (3.18). As for the third it is easily seen to telescope making the change of variable $t_n^{-1/2} s = \sigma$ and recalling that the t_n are bounded. In fact,

$$
\begin{aligned}
t_n^{-1/2} \int_{s(\varepsilon_n)}^{\infty} & e^{-s^2/4t_n} \|C(s)u\| \, ds \\
& \leq C_0 \|u\| \, t_n^{-1/2} \int_{s(\varepsilon_n)}^{\infty} e^{-s^2/4t_n + \omega s} \, ds \\
& C_0 \|u\| \int_{t_n^{-1/2} s(\varepsilon_n)}^{\infty} e^{-\sigma^2/4 + \omega t_n^{1/2} \sigma} \, d\sigma
\end{aligned}
\tag{3.23}
$$

Now, it follows from (3.8) and (3.20) that

$$
s(\varepsilon_n) = \frac{t_n}{\varepsilon_n} (1 - 4\eta_n^2)^{1/2} \geq \frac{t_n}{\varepsilon_n} (1 - 2\eta_n)^{1/2} \geq 2 \, \frac{t_n^{3/4}}{\varepsilon_n^{1/2}} \, . \tag{3.24}
$$

Thus $t_n^{-1/2} s(\varepsilon_n) \geq 2 \, t_n^{1/4} \varepsilon_n^{-1/2} \to \infty$ as $n \to \infty$ so that (3.23) tends to zero (we note that if $\omega = 0$ the integral (3.23) tends to zero under the sole assumption that $t_n/\varepsilon_n^2 \to \infty$, where the t_n may be unbounded; this fact bears on a result below). We have then obtained a contradiction and justified our claim about $\mathfrak{K}$.

We prove next the corresponding statement for $\mathfrak{S}(t;\varepsilon)$. The estimates are obtained in a similar fashion, thus we only state the final results. Formula (3.10)-(3.11) has the following counterpart:

$$
\psi(t,s;\varepsilon) = \chi(t,s;\varepsilon) \exp\left\{- \frac{t}{2\varepsilon^2}\left(1 - \left(1 - \left(\frac{\varepsilon s}{t}\right)^2\right)^{1/2}\right\} \tag{3.25}
$$

with

$$
\chi(t,s;\varepsilon) = (\pi t)^{-1/2}\left(1 - \left(\frac{\varepsilon s}{t}\right)\right)^{-1/4}\left(1 + O\left(\frac{\varepsilon^2}{\eta t}\right)\right) . \tag{3.26}
$$

The estimate is uniform in $0 \leq s \leq s(\varepsilon)$.

The inequality

$$
\psi(t,s;\varepsilon) \leq C\rho(t,s;\varepsilon) \exp\left\{- \frac{t}{2\varepsilon^2} \left(1 - \left(1 - \left(\frac{\varepsilon s}{t}\right)^2\right)^{1/2}\right)\right\} \tag{3.27}
$$

holds in $0 \leq s \leq t$, where

$$\rho(t,s;\varepsilon) = t^{-1/2}\left(\frac{2\varepsilon^2}{t} + \left(1 + \left(\frac{\varepsilon s}{t}\right)^2\right)^{1/2}\right)^{-1/2} . \tag{3.28}$$

and the constant C does not depend on ε, t. To obtain (3.25)-(3.26) we use the asymptotic formula (3.5) for $m = 1$; the same formula with $m = 0$ yields the inequality

$$I_0(x) \leq C \frac{e^x}{(1 + x)^{1/2}} \quad (x \geq 0) . \tag{3.29}$$

Using the inequality (3.26)-(3.27) we easily obtain the following counterparts of Lemma 3.2 and Lemma 3.3:

LEMMA 3.4. <u>Let</u> $t, \varepsilon > 0$. <u>Then the estimate</u>

$$\psi(t,s;\varepsilon) \leq C\eta^{-1/2}t^{-1/2}e^{-s^2/4t} \qquad (0 \leq s \leq s(\varepsilon)) \tag{3.30}$$

<u>holds, where the constant</u> C <u>does not depend on</u> s,t,ε.

LEMMA 3.5. <u>Let</u> $t, \varepsilon > 0$. <u>Then</u>

$$\psi(t,s;\varepsilon) \leq C\eta^{-1/2}t^{-1/2}e^{-t/2\varepsilon^2}e^{\eta t/\varepsilon^2} \qquad (s(\varepsilon) \leq s \leq t), \tag{3.31}$$

<u>where the constant</u> C <u>does not depend on</u> s,t,ε.

Using (3.31) and (3.8) we obtain

$$\int_{s(\varepsilon)}^{t/\varepsilon} \psi(t,s;\varepsilon)\|C(s)u\|\, ds \leq C\eta^{3/2}\|u\| \frac{t^{1/2}}{\varepsilon} \exp\left\{-\frac{t}{\varepsilon^2}\left(\frac{1}{2} - \eta - \omega\varepsilon\right)\right\}. \tag{3.32}$$

We prove that

$$\mathfrak{S}(t;\varepsilon)u \to S(t)u \tag{3.33}$$

uniformly in $t \geq t(\varepsilon)$ uniformly with respect to u in bounded sets of E in exactly the same way used for $\mathfrak{R}$; details are omitted.

After an elementary estimation of the first term in (2.5) the proof of the following result is complete:

THEOREM 3.6. <u>Let</u> $u_0(\varepsilon), u_1(\varepsilon) \in E$ <u>be such that</u>

$$u_0(\varepsilon) \to v, \quad \varepsilon^2 u_1(\varepsilon) \to u_0 - v \qquad (\varepsilon \to 0), \tag{3.34}$$

<u>and let</u> $u(t;\varepsilon)$ <u>be the generalized solution of</u> (3.1), $t(\varepsilon) > 0$ <u>a number such that</u> (3.20) <u>holds. Then</u>

$$u(\hat{t};\varepsilon) \to u(\hat{t}) \tag{3.35}$$

uniformly in compacts of $t \geq t(\varepsilon)$ where $u(\hat{t})$ is the generalized solution of (3.2) with $u(0) = u_0$. The convergence is uniform with respect to u_0, v if $\|u_0\|$, $\|v\|$ are bounded.

REMARK 3.7. Obviously, uniform convergence in $t \geq 0$ cannot be expected since in general $u_0(\varepsilon)$ does not converge to u_0 as $\varepsilon \to 0$, thus there is a "boundary layer" near zero where $u(\hat{t};\varepsilon)$ is not a good approximation to $u(\hat{t})$ (see KEVORKIAN-COLE [1981:1] for a thorough treatment of the one dimensional case). Note also that uniform convergence of $e^{-\omega^2 t}u(t;\varepsilon)$ to $e^{-\omega^2 t}u(t)$ in $t \geq t(\varepsilon)$ cannot be assured even in the scalar case. To see this, let $\xi(\hat{t};\varepsilon)$ be the solution of the initial value problem (3.2) with $\xi'(0;\varepsilon) = 0$. Then, since $\lambda^+(\varepsilon), \lambda^-(\varepsilon) < \omega^2$ we have $e^{-\omega^2 t}\xi(t;\varepsilon) \to 0$ as $t \to \infty$, whereas $e^{-\omega^2 t}e^{\omega^2 t} = 1$, $e^{\omega^2 t}$ the solution of the same equation with $\varepsilon = 0$. However, an examination of the proof of Theorem 3.2 shows that $u(\hat{t};\varepsilon) \to u(\hat{t})$ uniformly in $t \geq t(\varepsilon)$ in the particular case $\omega = 0$ (in fact, there is no need to assume the sequence $\{t_n\}$ bounded). We state this special result.

THEOREM 3.8. Let the assumptions of Theorem 3.6 hold. Assume in addition that $\omega = 0$ in (2.17) (i.e. assume that A generates a uniformly bounded cosine function $C(\hat{t})$). Then the convergence in (3.35) is uniform in $t \geq t(\varepsilon)$.

REMARK 3.9. Condition (3.20) is necessary in order that $u(t;\varepsilon) \to u(t)$ uniformly in $t \geq t(\varepsilon)$, at least without additional assumptions on the initial data (see §VI.5). This can be seen as follows. Consider first the one dimensional case (3.2) with, say, $\xi(0,\varepsilon) = 0$, $\xi'(0,\varepsilon) = \varepsilon^{-2}$. For each $\varepsilon > 0$ let $t(\varepsilon)$ be such that $t(\varepsilon)/\varepsilon^2 \to a$, where $0 < a < \infty$. Then we see from (3.3) that $\xi(t(\varepsilon),\varepsilon) \to 1 - e^{-a}$, although $\xi(t;\varepsilon) \to e^{\omega^2 t}$ in the sense of Theorem 3.6. However, uniform convergence in $t \geq 0$ can be obtained if we assume that $\varepsilon^{-2}\xi'(0,\varepsilon) \to 0$. As we shall see in §VI.5 this result has an infinite dimensional counterpart (Theorem 5.5). However, the following example (where $u_1(\varepsilon) = 0$) shows that convergence, although uniform in $t \geq 0$, may not be uniform in $u_0 = \lim u_0(\varepsilon)$ for $\|u_0\| \leq 1$ if condition (3.20) is violated.

EXAMPLE 3.10. Consider the space $E = \ell^2$ of all complex sequences

$u = \{u_n\} = \{u_n ; n \geq 1\}$ with

$$\|u\|^2 = \|\{u_n\}\|^2 = \sum |u_n|^2 < \infty .$$

We check that E is a Hilbert space. The operator A is defined by

$$Au = A\{u_n\} = \{-n^2 u_n\}, \tag{3.36}$$

and generates the strongly continuous, uniformly bounded cosine function

$$C(t)\{u_n\} = \{(\cos nt)u_n\}.$$

The (generalized) solution of the initial value problem (2.1) for A with initial conditions $u_0(\varepsilon) = \{u_{0n}(\varepsilon)\}$ and $u_1(\varepsilon) = \{u_{1n}(\varepsilon)\}$ is $u(t;\varepsilon) = \{u_n(t;\varepsilon)\}$ with

$$u_n(t;\varepsilon) = \Phi_{in}(t;\varepsilon)u_{0n}(\varepsilon) + \Psi_{in}(t;\varepsilon)(\varepsilon^2 u_{1n}(\varepsilon)), \tag{3.37}$$

the functions Φ_{in}, Ψ_{in} defined as in (3.3) for $\omega = in$, with the mandatory modifications when the two roots $\lambda_n^+(\varepsilon)$, $\lambda_n^-(\varepsilon)$ of the characteristic polynomial $\varepsilon^2\lambda^2 + \lambda + n^2 = 0$. coalesce.

Let $0 < b < 1$. Take

$$\varepsilon_m = b/2m. \tag{3.38}$$

Then

$$\lambda_m^-(\varepsilon) = (-1 - \sqrt{1 - b^2})/2\varepsilon_m^2, \quad \lambda_m^+(\varepsilon) = (-1 + \sqrt{1 - b^2})/2\varepsilon_m^2.$$

We solve the initial value problem (2.1) for A with $u_1(\varepsilon) = 0$, $u_0(\varepsilon) = u = \{u_n\}$ independent of ε. Note that the solution of (2.2), that is, the semigroup generated by A, is

$$S(t)\{u_n\} = \{e^{-n^2 t}u_n\}. \tag{3.39}$$

Let $\{t_m\}$ be a sequence of positive numbers such that $t_m/\varepsilon_m^2 \not\to \infty$. Then, passing if necessary to a subsequence we may assume that

$$t_m/\varepsilon_m^2 \to a \geq 0 .$$

We have

$$u_n(t_m;\varepsilon_m) = \Phi_{in}(t_m;\varepsilon_m)u$$

$$= \frac{\lambda_n^+(\varepsilon_m)e^{\lambda_n^-(\varepsilon_m)t_m} - \lambda_n^-(\varepsilon_m)e^{\lambda_n^+(\varepsilon_m)t_m}}{\lambda_n^+(\varepsilon_m) - \lambda_n^-(\varepsilon_m)} .$$

Accordingly,

$$\|u(t_m;\varepsilon_m) - u(t_m)\| \geq |u_m(t_m;\varepsilon_m) - u_m(t_m)|$$

$$= \left| \frac{\gamma^+ e^{\gamma^- a} - \gamma^- e^{\gamma^+ a}}{(1 + b^2)^{1/2}} - e^{-ab^2/4} \right| |u_m|(1 + o(1)),$$

where $\gamma^{\pm} = -1 \pm (1 - b^2)^{1/2}$, which inequality precludes uniform convergence for $\|u\|$ bounded if b is adequately chosen, at least if $a \neq 0$.

We note the following reformulation of Theorem 3.6:

THEOREM 3.11. Let $t(\varepsilon) > 0$ be such that (3.20) holds. Then

$$\mathfrak{R}(t;\varepsilon) \to S(t), \ \mathfrak{S}(t;\varepsilon) \to S(t) \tag{3.40}$$

uniformly on compacts of $t \geq t(\varepsilon)$ in the topology of (E). If $\omega = 0$ in (2.17) (that is, if A generates a uniformly bounded cosine function $C(\hat{t})$) the convergence in (3.40) is uniform in $t \geq t(\varepsilon)$.

§VI.4 Convergence of $u'(t;\varepsilon)$ and higher derivatives.

As pointed out in §VI.2, formula (2.14) does not necessarily provide a genuine solution of (2.1): in fact, if $u_0(\varepsilon)$ and $u_1(\varepsilon)$ are arbitrary, $u(\hat{t};\varepsilon)$ may not even be differentiable. However, if $u_0(\varepsilon) = 0$ it follows from (2.4) (or directly from (2.14)) that $u(\hat{t};\varepsilon)$ is continuously differentiable: noting that $I_0'(x) = I_1(x)$ we obtain

$$u'(t;\varepsilon) = \mathfrak{S}'(t;\varepsilon)(\varepsilon^2 u_1(\varepsilon)) = \frac{e^{-t/2\varepsilon^2}}{\varepsilon^2} C(t/\varepsilon)(\varepsilon^2 u_1(\varepsilon))$$

$$+ \frac{te^{-t/2\varepsilon^2}}{2\varepsilon^4} \int_0^{t/\varepsilon} \frac{I_1(((t/\varepsilon)^2 - s^2)^{1/2}/2\varepsilon)}{((t/\varepsilon)^2 - s^2)^{1/2}} C(s)(\varepsilon^2 u_1(\varepsilon))\, ds$$

$$- \frac{e^{-t/2\varepsilon^2}}{2\varepsilon^3} \int_0^{t/\varepsilon} I_0(((t/\varepsilon)^2 - s^2)^{1/2}/2\varepsilon) C(s)(\varepsilon^2 u_1(\varepsilon))\, ds$$

$$= \frac{e^{-t/2\varepsilon^2}}{\varepsilon^2} C(t/\varepsilon)(\varepsilon^2 u_1(\varepsilon)) + \frac{1}{2\varepsilon^2} \int_0^{t/\varepsilon} \varphi(t,s;\varepsilon) C(s)(\varepsilon^2 u_1(\varepsilon))\, ds$$

$$- \frac{1}{2\varepsilon^2} \int_0^{t/\varepsilon} \psi(t,s;\varepsilon) C(s)(\varepsilon^2 u_1(\varepsilon))\, ds =$$

$$= \frac{e^{-t/2\varepsilon^2}}{\varepsilon^2} C(t/\varepsilon)(\varepsilon^2 u_1(\varepsilon)) + \frac{1}{2\varepsilon^2} \mathfrak{R}(t;\varepsilon)(\varepsilon^2 u_1(\varepsilon))$$

$$- \frac{1}{2\varepsilon^2} \mathfrak{S}(t;\varepsilon)(\varepsilon^2 u_1(\varepsilon)) , \qquad (4.1)$$

where the obvious cancelling out of ε^2 and ε^{-2} is to be frowned upon, since $\varepsilon^2 u_1(\varepsilon)$ not $u_1(\varepsilon)$ will tend to a limit later. In view of (3.34) the last two terms of (4.1) are individually divergent as $\varepsilon \to 0$ thus they will have to be jointly estimated. We divide the integrals defining $\mathfrak{R}$ and $\mathfrak{S}$ at $s(\varepsilon)$ as in the previous section with $\eta < 1/2$ to be fixed later. The outer integrals are estimated individually using Lemma 3.3 and Lemma 3.5. We have

$$\frac{1}{2\varepsilon^2} \int_{s(\varepsilon)}^{t/\varepsilon} \varphi(t,s;\varepsilon)\|C(s)u\| \, ds$$

$$\le C \, \frac{\eta^{1/2}}{t} \|u\| \left(\frac{t}{\varepsilon^2}\right)^{3/2} \exp\left\{- \frac{t}{\varepsilon^2}\left(\frac{1}{2} - \eta - \omega\varepsilon\right)\right\} \qquad (4.2)$$

by virtue of (3.17) and of the bound (3.8) for the length of the interval of integration: in the same way, but this time using (3.31) we obtain

$$\frac{1}{2\varepsilon^2} \int_{s(\varepsilon)}^{t/\varepsilon} \psi(t,s;\varepsilon)\|C(s)u\| \, ds$$

$$\le C \, \frac{\eta^{3/2}}{t} \|u\| \left(\frac{t}{\varepsilon^2}\right)^{3/2} \exp\left\{- \frac{t}{2}\left(\frac{1}{2} - \eta - \omega\varepsilon\right)\right\} . \qquad (4.3)$$

We take $\eta < 1/2$ fixed in both inequalities. Estimation of the combined integral in $0 \le s \le s(\varepsilon)$ will require an additional term in the asymptotic series (3.5). Taking now $m = 1$ we obtain

$$I_0(x) = \frac{e^x}{(2\pi x)^{1/2}} \left(1 + \frac{1}{8x} + 0\left(\frac{1}{x^2}\right)\right) \qquad (4.4)$$

noting that $[0,1] = \Gamma(3/2)/\Gamma(-1/2) = -1/4$. On the other hand, $[1,1] = \Gamma(5/2)/\Gamma(1/2) = 3/4$, thus

$$I_1(x) = \frac{e^x}{(2\pi x)^{1/2}} \left(1 - \frac{3}{8x} + 0\left(\frac{1}{x^2}\right)\right). \qquad (4.5)$$

It follows from (4.5) that

$$\varphi(t,s;\varepsilon) = \chi(t,s;\varepsilon) \exp\left\{- \frac{t}{2\varepsilon^2}\left(1 - \left(1 - \left(\frac{\varepsilon s}{t}\right)^2\right)^{1/2}\right)\right\} \qquad (4.6)$$

with

$$\chi(t,s;\varepsilon) = (\pi t)^{-1/2}\left(1 - \left(\frac{\varepsilon s}{t}\right)^2\right)^{-3/4}$$

$$- \frac{3}{4}\,\varepsilon^2\pi^{-1/2}t^{-3/2}\left(1 - \left(\frac{\varepsilon s}{t}\right)^2\right)^{-5/4}\left(1 + 0\left(\frac{\varepsilon^2}{\eta t}\right)\right) ; \qquad (4.7)$$

On the other hand,

$$\psi(t,s;\varepsilon) = \chi(t,s;\varepsilon)\,\exp\left\{- \frac{t}{2\varepsilon^2}\left(1 - \left(1 - \left(\frac{\varepsilon s}{t}\right)^2\right)^{1/2}\right)\right\} \qquad (4.8)$$

with

$$\chi(t,s;\varepsilon) = (\pi t)^{-1/2}\left(1 - \left(\frac{\varepsilon s}{t}\right)^2\right)^{-1/4}$$

$$+ \frac{1}{4}\,\varepsilon^2\pi^{-1/2}t^{-3/2}\left(1 - \left(\frac{\varepsilon s}{t}\right)^2\right)^{-3/4}\left(1 + 0\left(\frac{\varepsilon^2}{\eta t}\right)\right). \qquad (4.9)$$

where due attention has been paid to (3.12) in the range $0 \leq s \leq t(\varepsilon)$. Formulas (4.6) and (4.7) will be modified as follows. We use Taylor's formula of order 1 in the first parenthesis of (4.7):

$$\left(1 - \left(\frac{\varepsilon s}{t}\right)^2\right)^{-3/4} = 1 + \frac{3}{4}\left(\frac{\varepsilon s}{t}\right)^2\left(1 + 0\left(\left(\frac{\varepsilon s}{t}\right)^2\right)\right) \qquad (4.10)$$

In the second parenthesis it is enough to use Taylor's formula only up to order zero:

$$\left(1 - \left(\frac{\varepsilon s}{t}\right)^2\right)^{-5/4} = 1 + 0\left(\left(\frac{\varepsilon s}{t}\right)^2\right) \qquad (4.11)$$

The result is

$$\chi(t,s;\varepsilon) = (\pi t)^{-1/2}\left\{1 + \frac{3}{4}\left(\frac{\varepsilon s}{t}\right)^2\left(1 + 0\left(\left(\frac{\varepsilon s}{t}\right)^2\right)\right)\right\}$$

$$- \frac{3}{4}\,\varepsilon^2\pi^{-1/2}t^{-3/2}\left(1 + 0\left(\left(\frac{\varepsilon s}{t}\right)^2\right)\right)\left(1 + 0\left(\frac{\varepsilon^2}{\eta t}\right)\right) \qquad (4.12)$$

The treatment of (4.9) is similar: the analogues of (4.10) and (4.11) are

$$\left(1 - \left(\frac{\varepsilon s}{t}\right)^2\right)^{-1/4} = 1 + \frac{1}{4}\left(\frac{\varepsilon s}{t}\right)^2\left(1 + 0\left(\left(\frac{\varepsilon s}{t}\right)^2\right)\right)$$

and

$$\left(1 - \left(\frac{\varepsilon s}{t}\right)^2\right)^{-3/4} = 1 + 0\left(\left(\frac{\varepsilon s}{t}\right)^2\right),$$

thus the function χ in (4.9) can be expressed as follows:

$$\chi(t,s;\varepsilon) = (\pi t)^{-1/2}\left\{1 + \frac{1}{4}\left(\frac{\varepsilon s}{t}\right)^2\left(1 + 0\left(\left(\frac{\varepsilon s}{t}\right)^2\right)\right)\right\}$$

$$+ \frac{1}{4}\varepsilon^2\pi^{-1/2}t^{-3/2}\left(1 + 0\left(\left(\frac{\varepsilon s}{t}\right)^2\right)\right)\left(1 + 0\left(\frac{\varepsilon^2}{\eta t}\right)\right). \qquad (4.13)$$

We combine now (4.6)-(4.7) with (4.8)-(4.9), making use of the asymptotic expressions (4.12) and (4.13): the result is

$$\frac{1}{2\varepsilon^2}(\varphi(t,s;\varepsilon) - \psi(t,s;\varepsilon))$$

$$= \gamma(t,s;\varepsilon)\exp\left\{-\frac{t}{2\varepsilon^2}\left(1 - \left(1 - \left(\frac{\varepsilon s}{t}\right)^2\right)^{1/2}\right)\right\}, \qquad (4.14)$$

with

$$\gamma(t,s;\varepsilon) = \frac{1}{4}s^2\pi^{-1/2}t^{-5/2}\left(1 + 0\left(\left(\frac{\varepsilon s}{t}\right)^2\right)\right)$$

$$- \frac{1}{2}\pi^{-1/2}t^{-3/2}\left(1 + 0\left(\left(\frac{\varepsilon s}{t}\right)^2\right)\right)\left(1 + 0\left(\frac{\varepsilon^2}{\eta t}\right)\right). \qquad (4.15)$$

The next task is to obtain an estimate for the function in (4.14)-(4.15) which is uniform with respect to s in $0 \leq s \leq t(\varepsilon)$ and which is at the same time independent of ε, $t > 0$. This is easily done keeping in mind the comments after (3.6); all we have to do is to replace the remainder

$$1 + 0\left(\frac{\varepsilon^2}{\eta t}\right)$$

by

$$1 + 0\left(\left(1 + \frac{\eta t}{\varepsilon^2}\right)^{-1}\right)$$

(which amounts to replacing $0(x^{-\alpha})$ by $0((1 + x)^{-\alpha})$ in (3.5)) and then note that this last expression remains bounded for all values of ε, $t > 0$. The result of these manipulations is the bound

$$\varepsilon^{-2}|\varphi(t,s;\varepsilon) - \psi(t,s;\varepsilon)| \leq C(t^{-3/2} + s^2t^{-5/2})e^{-s^2/4t} \qquad (4.16)$$

in $0 \leq s \leq t(\varepsilon)$, where C does not depend on ε or t; in the exponent we take advantage of the inequality $(1 - (\varepsilon s/t)^2)^{1/2} \leq 1 - (\varepsilon s/t)^2/2$.

To use (4.16) in the estimation of (4.1) we take profit of the well known formula.

$$\int_0^\infty e^{-s^2/4t}\cosh \omega s \, ds = (\pi t)^{1/2}e^{\omega^2 t} \qquad (4.17)$$

(which was already used in §VI.2) and of its differentiated version

$$\int_0^\infty s^2 e^{-s^2/4t} \cosh \omega s \, ds$$

$$= 2\pi^{1/2} t^{3/2} e^{\omega^2 t} + 4\pi^{1/2} t^{5/2} \omega^2 e^{\omega^2 t}, \tag{4.18}$$

both valid for $t > 0$.

To complete the estimation of (4.1) it only remains to dispose of the first (nonintegral) term. This is done observing that

$$\frac{e^{-t/2\varepsilon^2}}{\varepsilon^2} \|C(t/\varepsilon)u\|$$

$$\leq \frac{C_0}{t} \cdot \frac{t}{\varepsilon^2} \exp\left\{- \frac{t}{2\varepsilon^2}\left(1 - 2\omega\varepsilon\right)\right\} \leq C/t \tag{4.19}$$

if $0 < \varepsilon < 1/4\omega$.

Taking $\eta < 1/2$ fixed in (4.12), (4.13) and their consequences (4.14), (4.15) and (4.16) we obtain from (4.17) and (4.18):

THEOREM 4.1. <u>There exists a constant</u> C <u>independent of</u> $\omega \geq 0$ <u>and of</u> ε, t $(0 < \varepsilon < 1/4\omega,\ t > 0)$ <u>such that</u>

$$\|\mathfrak{S}'(t;\varepsilon)\| \leq C(\omega^2 + 1/t)e^{\omega^2 t} . \tag{4.20}$$

The following result establishes that $\mathfrak{S}'(\hat{t};\varepsilon)$ is an approximation of the derivative $S'(\hat{t})$ of the semigroup $S(\hat{t})$.

THEOREM 4.2. <u>Let</u> $u_0(\varepsilon) = 0$ <u>for all</u> $\varepsilon > 0$, $u_1(\varepsilon) \in E$ <u>such that</u>

$$\varepsilon^2 u_1(\varepsilon) \to u_0 \qquad (\varepsilon \to 0), \tag{4.21}$$

$u(t;\varepsilon)$ <u>the generalized solution of</u> (2.1), $t(\varepsilon) > 0$ <u>such that</u> (3.20) <u>holds. Then</u>

$$tu'(t;\varepsilon) \to tu'(t) \tag{4.22}$$

<u>uniformly on compacts of</u> $t \geq t(\varepsilon)$, <u>where</u> $u(t)$ <u>is the generalized solution of</u> (2.2) <u>with</u> $u(0) = u_0$. <u>The convergence is uniform with respect to</u> u_0 <u>if</u> $\|u_0\|$ <u>is bounded</u>.

The proof is essentially similar to that of Theorem 3.6; assuming that (4.22) fails to hold there exists a sequence $\{\varepsilon_n\}$, a bounded sequence $\{t_n\}$ and a bounded sequence $\{u_n\}$ in E such that

$t_n/\varepsilon_n^2 \to \infty$ and $t_n\|\mathfrak{S}'(t_n;\varepsilon_n)u_n - S'(t_n)u_n\| \geq \delta > 0$. We choose η_n according to (3.21). An obvious differentiation under the integral sign shows that

$$S'(t)u = \frac{1}{\pi^{1/2}}\int_0^\infty \left(\frac{s^2}{4t^{5/2}} - \frac{1}{2t^{3/2}}\right) e^{-s^2/4t}\, C(s)u\, ds. \qquad (4.23)$$

We have

$$t_n\| \mathfrak{S}'(t_n;\varepsilon_n)u_n - S'(t_n)u_n\| \leq \frac{t_n}{\varepsilon_n^2}\, e^{-t_n/2\varepsilon^2}\, \|C(t)u_n\|$$

$$+ \left\|\frac{t_n}{\varepsilon_n^2}\left(\mathfrak{R}(t_n;\varepsilon_n)u_n - \mathfrak{S}(t_n;\varepsilon_n)u_n\right) - t_nS'(t_n)u_n\right\| \leq \frac{t_n}{\varepsilon_n^2}\, e^{-t_n/2\varepsilon_n^2}\|C(t)u_n\|$$

$$+ \int_0^{s(\varepsilon_n)} t_n|\eta(t_n,s;\varepsilon_n)|\ \|C(s)u_n\|ds$$

$$+ \frac{t_n}{2\varepsilon_n^2}\int_{s(\varepsilon_n)}^{t_n/s_n} \varphi(t_n,s;\varepsilon_n)\|C(s)u_n\|\, ds + \frac{t_n}{2\varepsilon_n^2}\int_{s(\varepsilon_n)}^{t_n/s_n} \psi(t_n,s;\varepsilon_n)\|C(s)u_n\|\, ds$$

$$+ \frac{t_n}{\pi^{1/2}}\int_{s(\varepsilon_n)}^{\infty} \left(\frac{s^2}{4t_n^{5/2}} + \frac{1}{2t_n^{3/2}}\right) e^{-s^2/4t_n}\|C(s)u_n\|\, ds\,, \qquad (4.24)$$

where

$$\eta(t,s;\varepsilon) = \frac{1}{2\varepsilon^2}\left(\varphi(t,s;\varepsilon) - \psi(t,s;\varepsilon)\right) - \frac{1}{\pi^{1/2}}\left(\frac{s^2}{4t^{5/2}} - \frac{1}{2t^{3/2}}\right) e^{-s^2/4t}\,.$$

The first term in (4.24) is immediately seen to tend to zero using (2.17) for $\|C(t)u\|$. In the first integral we use the asymptotic development (4.14)-(4.15) to show that $\eta(t_n,s;\varepsilon_n) \to 0$ in $0 \leq s \leq s(\varepsilon)$ and the estimate (4.16); the integral tends to zero then by the dominated convergence theorem. The second and third integrals are taken care of by (4.2) and (4.3) respectively. Finally, the fourth integral is seen to tend to zero by means of the change of variables $t_n^{-1/2}s = \sigma$, keeping in mind that the sequence $\{t_n\}$ is bounded: in fact,

$$t_n\int_{s(\varepsilon_n)}^{\infty} \left(\frac{s^2}{4t_n^{5/2}} + \frac{1}{2t_n^{3/2}}\right) e^{-s^2/4t_n + \omega s}\, ds$$

$$= \int_{t_n^{-1/2}s(\varepsilon_n)} \left(\frac{\sigma^2}{4} + \frac{1}{2}\right) e^{-\sigma^2/4 + \omega t_n^{1/2}\sigma}\, d\sigma\,, \qquad (4.25)$$

where $t_n^{-1/2}s(\varepsilon_n) \to \infty$ (see (3.24)). We obtain then a contradiction and thus complete the proof of Theorem 4.2.

In the case $\omega = 0$, (4.3) tends to zero as $t_n/\varepsilon_n^2 \to \infty$ even if the t_n are unbounded. Hence we may improve Theorem 4.2 as follows:

THEOREM 4.3. Let the assumptions of Theorem 4.2 hold. Assume in addition that $\omega = 0$ in (2.17) (that is, that A generates a uniformly bounded cosine function $C(\hat{t})$). Then the convergence in (4.22) is uniform in $t \geq t(\varepsilon)$.

Results of the type of Lemma 4.1 and Theorems 4.2 and 4.3 can be obtained for higher derivatives; we limit ourselves to $\mathfrak{S}''(t;\varepsilon)$. Since $\mathfrak{S}(\hat{t};\varepsilon)u$ is not twice continuously differentiable for all $u \in E$, however, the estimates will refer not to $\mathfrak{S}''(t;\varepsilon)$ itself but to the "mollified" operator $\mathfrak{S}''(t;\varepsilon)(\varepsilon^{-1}R(\varepsilon^{-1};A))$.

THEOREM 4.4. There exists a constant C independent of $\omega > 0$ and of t,ε $(t > 0,\ 0 < \varepsilon < 1/4\omega)$ such that

$$\|\mathfrak{S}''(t;\varepsilon)(\varepsilon^{-1}R(\varepsilon^{-1};A))\| \leq C(\omega^4 + \omega^2/t + 1/t^2)e^{\omega^2 t}. \tag{4.26}$$

The proof is straightforward but tedious. Assume for the moment that $u \in D(A)$, so that $\mathfrak{S}''(t;\varepsilon)u$ exists; an explicit formula for it can be obtained from (4.1):

$$\begin{aligned}
\mathfrak{S}''(t;\varepsilon)u &= \frac{e^{-t/2\varepsilon^2}}{\varepsilon^3} C'(t/\varepsilon)u - \frac{e^{-t/2\varepsilon^2}}{2\varepsilon^4} C(t/\varepsilon)u + \frac{te^{-t/2\varepsilon^2}}{8\varepsilon^6} C(t/\varepsilon)u \\
&+ \frac{e^{-t/2\varepsilon^2}}{2\varepsilon^4} \int_0^{t/\varepsilon} \frac{I_1(((t/\varepsilon)^2 - s^2)^{1/2}/2\varepsilon)}{((t/\varepsilon)^2 - s^2)^{1/2}} C(s)u\,ds \\
&- \frac{te^{-t/2\varepsilon^2}}{4\varepsilon^6} \int_0^{t/\varepsilon} \frac{I_1(((t/\varepsilon)^2 - s^2)^{1/2}/2\varepsilon)}{((t/\varepsilon)^2 - s^2)^{1/2}} C(s)u\,ds \\
&+ \frac{t^2e^{-t/2\varepsilon^2}}{4\varepsilon^7} \int_0^{t/\varepsilon} \frac{I_1'(((t/\varepsilon)^2 - s^2)^{1/2}/2\varepsilon)}{(t/\varepsilon)^2 - s^2} C(s)u\,ds \\
&- \frac{t^2e^{-t/2\varepsilon^2}}{2\varepsilon^6} \int_0^{t/\varepsilon} \frac{I_1(((t/\varepsilon)^2 - s^2)^{1/2}/2\varepsilon)}{((t/\varepsilon)^2 - s^2)^{3/2}} C(s)u\,ds \\
&- \frac{e^{-t/2\varepsilon^2}}{2\varepsilon^4} C(t/\varepsilon)u +
\end{aligned}$$

$$+ \frac{e^{-t/2\varepsilon^2}}{4\varepsilon^5} \int_0^{t/\varepsilon} I_0(((t/\varepsilon)^2 - s^2)^{1/2}/2\varepsilon) C(s)u \, ds$$

$$- \frac{te^{-t/2\varepsilon^2}}{4\varepsilon^6} \int_0^{t/\varepsilon} \frac{I_1(((t/\varepsilon)^2 - s^2)^{1/2}/2\varepsilon)}{((t/\varepsilon)^2 - s^2)^{1/2}} C(s)u \, ds, \qquad (4.27)$$

where terms are grouped together as they appear in differentiating (4.1). Note also that the third and fourth integrals are individually divergent and must be combined into one. We take a look first at the terms that lay outside of integrals. For the second we have

$$\frac{e^{-t/2\varepsilon^2}}{2\varepsilon^4} \|C(t/\varepsilon)u\| \leq \frac{C}{t^2} \|u\| \left(\frac{t}{2\varepsilon^2}\right)^2 \exp\left\{- \frac{t}{2\varepsilon^2} \left(1 - 2\omega\varepsilon\right)\right\}, \qquad (4.28)$$

and the same estimate obtains for the third and the fourth, so that they satisfy (4.26) even without the intercession of the mollifying operator $\varepsilon^{-1}R(\varepsilon^{-1};A)$.[4] For the first term we note that if $v \in D(A)$ then $C(\hat{t})v$ is continuously differentiable with $C''(t)v = C(t)Av$, hence $C'(t)v = S(t)Av$ and we have

$$\frac{e^{-t/2\varepsilon^2}}{\varepsilon^3} \|C'(t/\varepsilon)(\varepsilon^{-1}R(\varepsilon^{-1};A)u)\| = \frac{e^{-t/2\varepsilon^2}}{\varepsilon^4} \|S(t/\varepsilon)AR(\varepsilon^{-1};A)u\|. \qquad (4.29)$$

Since $\omega > 0$, $\|S(t)\| \leq C \exp(\omega t)$[5] and the right hand side of (4.29) can be estimated in the same way as (4.28).

To estimate the six integrals in (4.27) we divide the domain of integration at $s = s(\varepsilon)$ given by (3.7), with $\eta < 1/2$ to be specified later. For the first outer integral we take advantage of the estimate (3.17) for $\varphi(t,s;\varepsilon)$, divided by $t\varepsilon^2$; for the interval of integration we use (3.8). The result is a bound of the form

$$C\eta^{1/2}\|u\|t^{-1/2}\varepsilon^{-3}e^{-t/2\varepsilon^2}e^{\eta t/\varepsilon^2}e^{\omega t/\varepsilon}$$

$$= C\frac{\eta^{1/2}}{t^2} \|u\| \left(\frac{t}{\varepsilon^2}\right)^{3/2} \exp\left\{- \frac{t}{2\varepsilon^2} \left(\frac{1}{2} - \eta - \omega\varepsilon\right)\right\}. \qquad (4.30)$$

The second, fifth and sixth integrals are treated in the same way: in all cases, due to the additional factor t/ε^2 we end up with an estimate of the form

$$C\frac{\eta^{1/2}}{t^2} \|u\| \left(\frac{t}{\varepsilon^2}\right)^{5/2} \exp\left\{- \frac{t}{2\varepsilon^2} \left(\frac{1}{2} - \eta - \omega\varepsilon\right)\right\}. \qquad (4.31)$$

As pointed out after (4.27) the third and fourth integrals must be combined into one to avoid divergence at $s = t/\varepsilon$ (in fact, they are written separately only for typographical reasons). The basis of the resulting estimation will be the asymptotic series for the function $Q(x) = x^{-1}(x^{-1}I_1(x))'$ obtained from (3.6): we deduce from it that

$$|Q(x)| = \left|\frac{1}{x}\left(\frac{I_1(x)}{x}\right)'\right| \leq C\,\frac{e^x}{(3+x)^{5/2}} \quad (x \geq 0)\,. \tag{4.32}$$

The combined integrand of the fourth and fifth integral (including factors outside of the integral) is

$$\frac{te^{-t/2\varepsilon^2}}{2\varepsilon^4}\,\frac{d}{dt}\left(\frac{I_1(((t/\varepsilon)^2 - s^2)^{1/2}/2\varepsilon)}{((t/\varepsilon)^2 - s^2)^{1/2}}\right)C(s)u$$

$$= \frac{t^2e^{-t/2\varepsilon^2}}{16\varepsilon^9}\,Q(((t/\varepsilon)^2 - s^2)^{1/2}/2\varepsilon)C(s)u\,.$$

In view of (3.30) we have

$$\frac{t^2e^{-t/2\varepsilon^2}}{16\varepsilon^9}\,|Q(((t/\varepsilon)^2 - s^2)^{1/2}/2\varepsilon)|$$

$$\leq C\rho(t,s;\varepsilon)\exp\left\{-\frac{t}{2\varepsilon^2}\left(1 - \left(1 - \left(\frac{\varepsilon s}{t}\right)^2\right)^{1/2}\right)\right\}, \tag{4.33}$$

where

$$\rho(t,s;\varepsilon) = t^{-1/2}\varepsilon^{-4}\left(\frac{6\varepsilon^2}{t} + \left(1 - \left(\frac{\varepsilon s}{t}\right)^2\right)^{1/2}\right)^{-5/2}. \tag{4.34}$$

Since $(3 + x)^{-5/2}e^x$ is increasing we can bound the right hand side of (4.34) by its value at $s = s(\varepsilon)$ subsequently deleting the factor $6\varepsilon^2/t$ from the outer parenthesis. The result is an upper bound for the combined integrand of the form

$$C\eta^{-5/2}\|u\|t^{-1/2}\varepsilon^{-4}e^{-t/2\varepsilon^2}e^{\eta t/\varepsilon^2}e^{\omega t/\varepsilon} \qquad (s(\varepsilon) \leq s \leq t/\varepsilon)\,.$$

Therefore, the integral can be bounded by the following expression:

$$C\,\frac{\eta^{-1/2}}{t^2}\,\|u\|\left(\frac{t}{\varepsilon^2}\right)^{5/2}\exp\left\{-\frac{1}{2\varepsilon^2}\left(\frac{1}{2} - \eta - \omega\varepsilon\right)\right\}. \tag{4.35}$$

This completes the consideration of the outer integrals.

We look at the inner integrals. We begin by grouping them into the integral below:

$$\int_0^{s(\varepsilon)} \Theta(t,s;\varepsilon)C(s)u\,ds. \tag{4.36}$$

Using the asymptotic developments (3.5) for I_0, I_1 and I_1' of order $m = 1$ in the first and fourth integrals and of order $m = 2$ in the rest we obtain for Θ an expression of the form

$$\Theta(t,s;\varepsilon) = \chi(t,s;\varepsilon)\exp\left\{-\frac{t}{2\varepsilon^2}\left(1-\left(1-\left(\frac{\varepsilon s}{t}\right)^2\right)^{1/2}\right)\right\}, \tag{4.37}$$

with χ a linear combination of terms of the form

$$\varepsilon^{-2j}t^{-\alpha_j}\left(1-\left(\frac{\varepsilon s}{t}\right)^2\right)^{-\beta_j}\left(1+O\left(\frac{\varepsilon^2}{\eta t}\right)\right) \tag{4.38}$$

with $j = 2,1,0$, $\alpha_j,\beta_j > 0$. We then use Taylor's formula of order $2j$ for each term $(1-(\varepsilon s/t)^2)^{-\beta_j}$, ending up with the following expression for χ:

$$\chi(t,s;\varepsilon) = \sum_{j=1}^{2}\varepsilon^{-2j}\chi_j(t,s) + \chi_0(t,s)\left(1+O\left(\left(\frac{\varepsilon s}{t}\right)^2\right)\right)\left(1+O\left(\frac{\varepsilon^2}{\eta t}\right)\right) \tag{4.39}$$

where each $\chi_j(t,s)$ is independent of ε (in fact, χ_j is a finite linear combination of terms of the form $s^{\alpha}t^{-\beta}$ with $\alpha,\beta \geq 0$). We then fix $t > 0$ and apply formula (4.27) in the space $E = \mathbb{C}$ to the cosine function $C(\hat{s}) = \cos\sigma\hat{s}$, where σ is a real parameter. Naturally, the result must be the second derivative of the solution of

$$\varepsilon^2\xi''(t;\varepsilon) + \xi'(t;\varepsilon) = -\sigma^2\xi(t;\varepsilon)\,, \tag{4.40}$$

with initial conditions

$$\xi(t,0) = 0,\ \xi'(t;0) = \varepsilon^{-2}, \tag{4.41}$$

hence

$$\xi(t;\varepsilon) = \Psi_{1\sigma}(t;\varepsilon)\,,$$

where $\Psi_{\omega}(t;\varepsilon)$ is the function defined in (3.3). We check easily that

$$\xi''(t;\varepsilon) \to \sigma^4 e^{-\sigma^2 t}. \tag{4.42}$$

Since $t > 0$ is being kept fixed, it follows from (4.30), (4.31) and (4.35) that the outer integrals tend to zero as $\varepsilon \to 0$; we obtain then making use of (4.28) and following comments that

$$\lim_{\varepsilon\to 0}\int_0^{s(\varepsilon)} \Theta(t,s;\varepsilon)\cos\sigma s\,ds = \sigma^4 e^{-\sigma^2 t}. \tag{4.43}$$

We try next to identify the different terms in (4.39). Observe that $\varepsilon^2/\eta t \to 0$ and that $\varepsilon s/t \to 0$ in the interval $0 < s < t(\varepsilon)$. On the other hand, on account of (4.37) and (4.39), of the form of each χ_j, estimating the exponent in (4.37) in the same way used to obtain (3.16) and making use of the dominated convergence theorem we deduce from (4.39) that

$$\int_0^\infty \chi_j(t,s)e^{-s^2/4t} \cos \sigma s \, ds = \begin{cases} 0 & \text{if} \quad j = 1,2 \\ \sigma^4 e^{-\sigma^2 t} & \text{if} \quad j = 0. \end{cases} \tag{4.44}$$

By uniqueness of Fourier cosine transforms we obtain that $\chi_j(t,s) = 0$ for $j = 1,2$ and

$$e^{-s^2/4t}\chi_0(t,s) = \left(\frac{\partial}{\partial t}\right)^2 \frac{1}{(\pi t)^{1/2}} e^{-s^2/4t}$$

$$= \frac{3}{4\pi^{1/2}t^{5/2}} e^{-s^2/4t} - \frac{3s^2}{4\pi^{1/2}t^{7/2}} e^{-s^2/4t}$$

$$+ \frac{s^4}{16\pi^{1/2}t^{9/2}} e^{-s^2/4t}. \tag{4.45}$$

Replacing back in (4.39), we see that the sum reduces to a unique term,

$$\chi(t,s;\varepsilon) = \chi_0(t,s)\left(1 + 0\left(\left(\frac{\varepsilon s}{t}\right)^2\right)\right)\left(1 + 0\left(\frac{\varepsilon^2}{\eta t}\right)\right). \tag{4.46}$$

The estimate derived from (4.46) is not good enough for the proof of Lemma 4.4 since ε^2/t need not remain bounded. To obtain the required bound we work with the asymptotic development (3.5) minding the comments after (3.6) on obtention of bounds, as in the argument leading to (4.16). The estimate obtained is of the form

$$|\chi(t,s,\varepsilon)| \leq C(t^{-5/2} + s^2t^{-7/2} + s^4t^{-9/2}) \tag{4.47}$$

in $0 \leq s \leq t(\varepsilon)$, where C does not depend on ε or t. Using the inequality $(1 - (\varepsilon s/t)^2)^{1/2} \leq 1 - (\varepsilon s/t)^2/2$ in the exponent as we did in obtaining (4.16) we deduce that

$$|\Theta(t,s;\varepsilon)| \leq C(t^{-5/2} + s^2t^{-7/2} + s^4t^{-9/2})e^{-s^2/4t} \tag{4.48}$$

in $0 \leq s \leq t(\varepsilon)$, where C is independent of ε and t.

To complete the proof of Lemma 4.4 we use (4.18), differentiated once again:

$$\int_0^\infty s^4 e^{-s^2/4t} \cosh \omega s \, ds = 12\pi^{1/2} t^{5/2} e^{\omega^2 t} + 48\pi^{1/2} t^{7/2} \omega^2 e^{\omega^2 t}$$

$$+ 16\pi^{1/2} t^{9/2} \omega^4 e^{\omega^2 t} \tag{4.49}$$

REMARK 4.5. The case $\omega = 0$, excluded from Lemma 4.4 can be incorporated noting that the hypotesis $\omega > 0$ was only used to deduce that $\|S(t)\| \leq C \exp \omega t$, which inequality appears in the estimation (4.29). If $\omega = 0$ then we can only assert in general that $\|S(t)\| \leq Ct$, thus the right-hand side of (4.29) is bounded as follows:

$$\frac{e^{-t/2\varepsilon^2}}{2\varepsilon^4} \|S(t/\varepsilon)AR(\varepsilon^{-1};A)u\| \leq \frac{C'}{t} \|u\| \left(\frac{t}{\varepsilon^2}\right)^2 e^{-t/2\varepsilon^2} \leq \frac{C}{t} \|u\| . \tag{4.50}$$

Accordingly, the following estimate holds for $\mathfrak{S}''(t;\varepsilon)$:

$$\|\mathfrak{S}''(t;\varepsilon)(\varepsilon^{-1}R(\varepsilon^{-1};A))\| \leq C(1/t + 1/t^2) \quad (t > 0). \tag{4.51}$$

We can state using the preceding arguments a convergence result for $\mathfrak{S}''(t;\varepsilon)(\varepsilon^{-1}R(\varepsilon^{-1};A))$ whose proof is very much the same as that of Theorem 4.2. We include in the statement below the corresponding result for $\mathfrak{S}'(t;\varepsilon)$ which is nothing but a reformulation of Theorem 4.2:

THEOREM 4.6. <u>Let</u> $t(\varepsilon) > 0$ <u>be such that</u> (3.20) <u>holds. Then</u>

$$t\,\mathfrak{S}'(t;\varepsilon) \to tS'(t), \tag{4.52}$$

$$t^2\mathfrak{S}''(t;\varepsilon)(\varepsilon^{-1}R(\varepsilon^{-1};A)) \to t^2S''(t), \tag{4.53}$$

<u>uniformly on compacts of</u> $t \geq t(\varepsilon)$ <u>in the topology of</u> (E).

Obviously, results of the type of Lemma 4.4 and Theorem 4.6 can be obtained for derivatives of any order of $\mathfrak{S}$. We omit the details.

In a sense, the results above do not tell the whole story about $\mathfrak{S}''$ since the smoothing operator $\varepsilon^{-1}R(\varepsilon^{-1};A)$ only plays a role in the first term on the right hand side of (4.27); besides, the first term also makes it necessary to separate the case $\omega = 0$ for estimation purposes. A statement on $\mathfrak{S}''(\hat{t};\varepsilon)$ that avoid these inconvenients is:

THEOREM 4.7 (a) There exists a constant C independent of $\omega \geq 0$ and of t, ε $(t > 0,\ 0 < \varepsilon < 1/4\omega)$ such that

$$\|\mathfrak{S}''(t;\varepsilon)u - \varepsilon^{-3}e^{-t/2\varepsilon^2}C'(t/\varepsilon)u\| \leq C(\omega^4 + \omega^2/t + 1/t^2)\|u\| \quad (u \in D(A)). \tag{4.54}$$

(b) We have

$$t^2\left(\mathfrak{S}''(t;\varepsilon)u - \varepsilon^{-3}e^{-t/2\varepsilon^2}C'(t/\varepsilon)u\right) \to t^2S''(t)u \tag{4.55}$$

uniformly on compacts of $t \geq t(\varepsilon)$ uniformly with respect to $u \in D(A) \cap \mathfrak{B}$, where $\mathfrak{B}$ is any bounded subset in E.

§VI.5 The homogeneous equation. Rates of convergence.

We show in this section that if there is no "crossover" of initial conditions (i.e. if we have

$$u_0(\varepsilon) \to u_0,\ \varepsilon^2 u_1(\varepsilon) \to 0 \quad \text{as} \quad \varepsilon \to 0 \tag{5.1}$$

rather than (3.34)) then $u(\hat{t};\varepsilon)$ converges to $u(\hat{t})$ uniformly in $t \geq 0$, with precise rates of convergence if $u_0 \in D(A)$ or to certain subspaces intermediate between $D(A)$ and E. In contrast with the results of §VI.3 and §VI.4, convergence will not be uniform with respect to u even if $\|u\|$ is bounded.

Let $\mathfrak{C}(t;\varepsilon)$ be the operator acting on the initial condition $u_0(\varepsilon)$ in (2.14), i.e.

$$\mathfrak{C}(t;\varepsilon) = e^{-t/2\varepsilon^2}\, C(t/\varepsilon) + \frac{1}{2}\,\mathfrak{R}(t;\varepsilon) + \frac{1}{2}\,\mathfrak{S}(t;\varepsilon). \tag{5.2}$$

If $u \in D(A)$ both $\mathfrak{R}(\hat{t};\varepsilon)u$ and $\mathfrak{S}(\hat{t};\varepsilon)u$ are twice continuously differentiable, thus so is $u(\hat{t};\varepsilon) = \mathfrak{C}(\hat{t};\varepsilon)u$. The derivative $v(\hat{t};\varepsilon) = u'(t;\varepsilon)$ is a generalized solution of (2.1) with initial conditions $v(0;\varepsilon) = u'(0;\varepsilon) = 0$ and $v'(0;\varepsilon) = u''(0;\varepsilon) = \varepsilon^{-2}Au$. Hence, by uniqueness, we must have

$$u'(t;\varepsilon) = \mathfrak{C}'(t;\varepsilon)u = \mathfrak{S}(t;\varepsilon)Au. \tag{5.3}$$

On the other hand, we may write (4.1) in the form $\mathfrak{S}'(t;\varepsilon) = \varepsilon^{-2}\mathfrak{C}(t;\varepsilon) - \varepsilon^{-2}\,\mathfrak{S}(t;\varepsilon)$, hence

$$\mathfrak{S}(t;\varepsilon)u = \mathfrak{C}(t;\varepsilon)u - \varepsilon^2\,\mathfrak{S}'(t;\varepsilon)u\,. \tag{5.4}$$

Applying this equality to an element of the form Au and using (5.3) we obtain

$$\mathfrak{C}'(t;\varepsilon)u = A\mathfrak{C}(t;\varepsilon)u - \varepsilon^2\mathfrak{S}'(t;\varepsilon)Au . \tag{5.5}$$

so that $\mathfrak{C}(\hat{t};\varepsilon)u$ is a genuine solution of the nonhomogeneous first order equation (2.2). Consequently, the **variation-of-constants** formula (I.5.3) applies and we have

$$\mathfrak{C}(t;\varepsilon)u - S(t)u = -\varepsilon^2 \int_0^t S(t-s)\mathfrak{S}'(s;\varepsilon)Au\, ds = -\varepsilon^2\mathfrak{H}(t;\varepsilon)\, Au . \tag{5.6}$$

THEOREM 5.1.

$$\|\mathfrak{H}(t;\varepsilon)\| \le C_0\Theta_\omega(t;\varepsilon) = C_0\int_0^t e^{\omega^2(t-s)}\Psi_\omega'(s;\varepsilon)\, ds \quad (t \ge 0) , \tag{5.7}$$

__where__ C_0, ω __are the constants in__ (2.17) __and__ Ψ_ω __is the function in__ (3.3).

__Proof__: It will be divided in several steps. If $u \in E$,

$$R(\lambda;A)u = \int_0^\infty e^{-\lambda t}S(t)u\, dt \qquad (\lambda > \omega) \tag{5.8}$$

(see (I.3.8)). On the other hand, if $u \in D(A)$,

$$\mathfrak{L}(\lambda)u = \int_0^\infty e^{-\lambda t}\mathfrak{S}(t;\varepsilon)u\, dt = \frac{1}{\lambda}\int_0^\infty e^{-\lambda t}\mathfrak{S}'(t;\varepsilon)u\, dt = \frac{1}{\varepsilon^2\lambda^2}u$$

$$+ \frac{1}{\lambda^2}\int_0^\infty e^{-\lambda t}\mathfrak{S}''(t;\varepsilon)u\, dt = \frac{1}{\varepsilon^2\lambda^2}u + \frac{1}{\varepsilon^2\lambda^2}A\mathfrak{L}(\lambda)u - \frac{1}{\varepsilon^2\lambda}\mathfrak{L}(\lambda)u \qquad (\lambda > \omega).$$

so that $(\varepsilon^2\lambda^2 I + \lambda - A)\mathfrak{L}(\lambda)u = u$ and we deduce using denseness of $D(A)$ that

$$\mathfrak{L}(\lambda) = R(\varepsilon^2\lambda^2 + \lambda;A) \qquad (\lambda > \omega^2). \tag{5.9}$$

Accordingly,

$$\lambda R(\varepsilon^2\lambda^2 + \lambda;A)u = \int_0^\infty e^{-\lambda t}\mathfrak{S}'(t;\varepsilon)u\, dt \qquad (\lambda > \omega^2,\ u \in E). \tag{5.10}$$

We use now (II.2.11):

$$\lambda R(\lambda^2;A)u = \int_0^\infty e^{-\lambda t}C(t)u\, dt \qquad (\lambda > \omega,\ u \in E). \tag{5.11}$$

Making use of (5.11) and of the cosine functional equation (II.3.1) for $C(t)$ we obtain

$$2\mu\nu\, R(\mu^2;A)R(\nu^2;A)u$$

$$= \int_0^\infty\int_0^\infty e^{-(\mu s+\nu t)}(C(s+t) + C(s-t))u\, ds dt \qquad (\mu,\nu > \omega). \tag{5.12}$$

Taking advantage of the convolution theorem in the definition of $\mathfrak{H}(\hat{t};\varepsilon)$ in (5.6) we deduce, making use of (5.8) and (5.10) that

$$\lambda R(\lambda;A)R(\varepsilon^2\lambda^2 + \lambda;\ A)u = \int_0^\infty e^{-\lambda t}\mathfrak{H}(t;\varepsilon)u\ dt \qquad (\lambda > \omega^2,\ u \in E). \quad (5.13)$$

By virtue of (5.12) we may also write

$$\lambda R(\lambda;A)R(\varepsilon^2\lambda^2 + \lambda;\ A)u =$$

$$\int_0^\infty\int_0^\infty h(t,s,\lambda;\varepsilon)(C(s + t) + C(s - t))u\ dsdt \qquad (\lambda > \omega^2), \quad (5.14)$$

with

$$h(t,s,\lambda;\varepsilon) = \frac{1}{2}\, g(\lambda;\varepsilon)\exp(-g_1(\lambda;\varepsilon)s - g_2(\lambda;\varepsilon)t)$$

$$= \frac{1}{2}\,(\varepsilon^2\lambda + 1)^{-1/2}\exp\,(-\lambda^{1/2}s - (\varepsilon^2\lambda^2 + \lambda)^{1/2}t). \quad (5.15)$$

Consider the scalar cosine function

$$C(t) = \cosh \omega\hat{t} \qquad (-\infty < t < \infty)\ . \quad (5.16)$$

Here we have

$$S(\hat{t}) = e^{\omega^2 t} \qquad (t \geq 0)\ ,$$

and

$$\mathfrak{S}(t;\varepsilon) = \Psi_\omega(t;\varepsilon),$$

Ψ_ω as defined in (3.3); accordingly it follows from (5.7) that

$$\mathfrak{H}(t;\varepsilon) = \Theta_\omega(t;\varepsilon)\ .$$

Applying formulas (5.13) and (5.14) we obtain

$$\int_0^\infty e^{-\lambda t}\Theta_\omega(t;\varepsilon)\ dt$$

$$= \int_0^\infty h(t,s,\lambda,\varepsilon)(\cos \omega(s + t) - \cos \omega(s - t))\ dsdt\ .$$

Let now u be an arbitrary element of E, u^* an arbitrary element of the dual space E^* with $\|u^*\| = \|u\| = 1$, and consider the scalar function

$$r(t;\varepsilon) = c_0\Theta_\omega(t;\varepsilon) - \langle u^*,\mathfrak{H}(t;\varepsilon)u\rangle\ . \quad (5.17)$$

According to the previous arguments,

$$\int_0^\infty e^{-\lambda t} r(t;\varepsilon)\, dt = \int_0^\infty \int_0^\infty h(t,s,\lambda;\varepsilon)(k(s + t) - k(s - t))\, dsdt \tag{5.18}$$

where

$$k(s) = C_0 \cosh \omega s - \langle u^*, C(s)u\rangle . \tag{5.19}$$

Obviously,

$$k(s) \geq 0 \quad (-\infty < s < \infty) . \tag{5.20}$$

Let $\ell(\hat{\lambda})$ be a function defined and infinitely differentiable in $\lambda \geq 0$. We say that ℓ is <u>alternating</u> (in $t \geq 0$) if

$$(-1)^n \ell^{(n)}(\lambda) \geq 0 \quad (\lambda \geq 0,\ n = 0,1,\ldots) . \tag{5.21}$$

We define correspondingly alternating functions in $t \geq a$.

It is obvious that the sum of two alternating functions and the product of an alternating function by a nonnegative constant is alternating. More generally, it follows from Leibniz's formula that <u>the product of two alternating functions is alternating</u>.

LEMMA 5.2. *Let* $m(\hat{\lambda})$ *be a function such that* $m'(\hat{\lambda})$ *is alternating. Then*

$$\ell(\hat{\lambda}) = e^{-m(\hat{\lambda})} \tag{5.22}$$

is alternating.

Proof: Obviously, it is enough to show that each summand in the derivative of order $n \geq 1$ of $\ell(\hat{\lambda})$ is of the form

$$(-1)^k m^{(j)}(\lambda) \ldots m^{(p)}(\lambda) e^{-m(\lambda)} \tag{5.23}$$

with $j,\ldots,p \geq 1$ and

$$(-1)^n = (-1)^{k+(j-1)+\ldots+(p-1)} . \tag{5.24}$$

This statement is obvious for $n = 1$; assuming it is true for n, its validity for $n + 1$ follows from Leibniz's formula.

LEMMA 5.3. *Let* $\varepsilon > 0$,

$$m(\lambda) = (\varepsilon^2\lambda^2 + \lambda)^{1/2} \quad (\lambda \geq 0) . \tag{5.25}$$

Then $m'(\lambda)$ *is alternating.*

The proof is left to the reader (Exercise 1).

LEMMA 5.4. Let $f(\hat{t})$ be continuous in $t \geq 0$, $f(t) = 0(\exp \alpha t)$ as $t \to \infty$. Assume the Laplace transform $\mathcal{L}f(\hat{\lambda})$ is alternating in $\lambda \geq a$. Then

$$f(t) \geq 0 \quad (t \geq 0). \tag{5.26}$$

The proof is an immediate consequence of Lemma I.3.2 (see (I.3.14)).

End of proof of Theorem 5.1. We go back to (5.18). The definition (5.15) of the function $h(t,s,\lambda;\varepsilon)$, Lemma 5.3, Lemma 5.2 and the comments preceding it show that h, as a function of λ, is alternating for any $s,t \geq 0$, $\varepsilon > 0$. Since the function $k(s)$ defined in (5.19) is nonnegative, it follows from (5.18) that the Laplace transform of $r(\hat{t};\varepsilon)$ is alternating. Thus, by Lemma 5.4, $r(t;\varepsilon) \geq 0$ $(t \geq 0,\ \varepsilon > 0)$. Taking into account the arbitrariness of u and u^*, (5.7) follows, completing the proof of Theorem 5.1.

In all of the results that follow $u(\hat{t};\varepsilon)$ (resp. $u(\hat{t})$) is the solution of the homogeneous initial value problem (2.1) (resp. (2.2)).

THEOREM 5.5. Let $u_0 \in D(A)$. Then

$$\|u(t;\varepsilon) - u(t)\| \leq C_0\varepsilon^2\Theta_\omega(t;\varepsilon)\|Au_0\| + C_0\Phi_\omega(t;\varepsilon)\|u_0(\varepsilon) - u_0\|$$

$$+ C_0\varepsilon^2\Psi_\omega(t;\varepsilon)\|u_1(\varepsilon)\| \quad (t \geq 0,\ \varepsilon > 0). \tag{5.27}$$

The proof results noting that

$$u(t;\varepsilon) - u(t) = \mathfrak{C}(t;\varepsilon)u_0 - S(t)u_0 + \mathfrak{C}(t;\varepsilon)(u_0(\varepsilon) - u_0)$$

$$+ \mathfrak{S}(t;\varepsilon)(\varepsilon^2 u_1(\varepsilon)), \tag{5.28}$$

and applying (5.6) and (5.7) to the first term on the right hand side of (5.28): to estimate the other summands we use (3.4) which implies (taking $u_0(\varepsilon) = 0$ or $u_1(\varepsilon) = 0$ alternately)

$$\|\mathfrak{C}(t;\varepsilon)\| \leq C_0\Phi_\omega(t;\varepsilon),\ \|\mathfrak{S}(t;\varepsilon)\| \leq C_0\Psi_\omega(t;\varepsilon) \quad (t \geq 0,\ \varepsilon > 0). \tag{5.29}$$

We obtain a simpler but less precise bound noting that $\Phi_\omega(t;\varepsilon)$, $\Psi_\omega(t,\varepsilon) \leq e^{\omega^2 t}$ (Lemma 3.1) and integrating (5.7) by parts; it results that $\Theta_\omega(t;\varepsilon) \leq (1 + \omega^2 t)e^{\omega^2 t}$ so that (5.27) becomes

$$\|u(t;\varepsilon) - u(t)\| \leq C_0\varepsilon^2(1 + \omega^2 t)e^{\omega^2 t}\|Au_0\| + C_0 e^{\omega^2 t}\|u_0(\varepsilon) - u_0\| +$$

$$+ C_0\varepsilon^2 e^{\omega^2 t}\|u_1(\varepsilon)\| \qquad (t \geq 0,\ \varepsilon > 0) . \tag{5.30}$$

Theorem 5.5 implies that when $u_0 \in D(A)$ we have

$$\|u(t;\varepsilon) - u(t)\| = 0(\varepsilon^2) \tag{5.31}$$

uniformly on compacts of $t \geq 0$ if

$$\|u_0(\varepsilon) - u_0\| = 0(\varepsilon^2) \quad \text{and} \quad \|u_1(\varepsilon)\| = 0(1). \tag{5.32}$$

Estimates of the same sort can be easily obtained for the derivative $u'(t;\varepsilon)$ if $u_0 \in D(A^2)$ and $u_0(\varepsilon) \in D(A)$. In fact, $v(\hat{t};\varepsilon) = u'(\hat{t};\varepsilon)$ is the solution of the initial value problem (2.1) with

$$v(0;\varepsilon) = u'(0;\varepsilon) = u_1(\varepsilon),\ v'(0,\varepsilon) = u''(0;\varepsilon) = \varepsilon^{-2}(Au_0(\varepsilon) - u_1(\varepsilon)). \tag{5.33}$$

On the other hand, $v(\hat{t}) = u'(\hat{t})$ is the solution of (2.2) with

$$v(0) = u'(0) = Au_0. \tag{5.34}$$

Accordingly, we have

THEOREM 5.6. <u>Assume that</u> $u_0 \in D(A^2)$ <u>and</u> $u_0(\varepsilon) \in D(A)$. <u>Then</u>

$$\begin{aligned} \|u'(t;\varepsilon) - u'(t)\| &\leq C_0\varepsilon^2\Theta_\omega(t;\varepsilon)\|A^2u_0\| + C_0\Phi_\omega(t;\varepsilon)\|u_1(\varepsilon) - Au_0\| \\ &\quad + C_0\Psi_\omega(t;\varepsilon)\|u_1(\varepsilon) - Au_0(\varepsilon)\| \\ &\leq C_0\varepsilon^2(1 + \omega^2 t)e^{\omega^2 t}\|A^2u_0\| \\ &\quad + C_0 e^{\omega^2 t}(\|u_1(\varepsilon) - Au_0\| + \|u_1(\varepsilon) - Au_0(\varepsilon)\|) \\ &\qquad\qquad (t \geq 0) . \end{aligned} \tag{5.35}$$

It follows from this result that if $u_0 \in D(A^2)$ and $u_0(\varepsilon) \in D(A)$ then

$$\|u'(t;\varepsilon) - u'(t)\| = 0(\varepsilon^2) \tag{5.36}$$

uniformly on compacts of $t \geq 0$ if

$$\|u_1(\varepsilon) - Au_0\| = 0(\varepsilon^2) \quad \text{and} \quad \|u_1(\varepsilon) - Au_0(\varepsilon)\| = 0(\varepsilon^2), \tag{5.37}$$

or, equivalently, if

$$\|u_1(\varepsilon) - Au_0\| = 0(\varepsilon^2) \quad \text{and} \quad \|Au_0(\varepsilon) - Au_0\| = 0(\varepsilon^2). \tag{5.38}$$

Theorems 5.5 and 5.6 are easily seen to imply convergence results valid for arbitrary initial conditions.

THEOREM 5.7. *Let* $u(\hat{t};\varepsilon)$ (*resp.* $u(\hat{t})$) *be the generalized solution of* (2.1) (*resp.* (2.2) *with* $u_0 \in E$ *arbitrary). Assume that*

$$u_0(\varepsilon) \to u_0, \quad \varepsilon^2 u_1(\varepsilon) \to 0 \quad \text{as} \quad \varepsilon \to 0. \tag{5.39}$$

Then

$$u(t;\varepsilon) \to u(t) \quad \text{as} \quad \varepsilon \to 0 \tag{5.40}$$

uniformly on compacts of $t \geq 0$.

Proof. Pick $\delta > 0$ and choose $\bar{u} \in D(A)$ with $\|\bar{u} - u_0\| \leq \delta$. Let $\bar{u}(\hat{t})$ be the solution of the initial value problem (2.2) with $\bar{u}(0) = \bar{u}$. Applying Theorem 5.5 and inequalities (5.29), (5.30), we obtain

$$\begin{aligned}
\|u(t;\varepsilon) - u(t)\| &\leq \|u(t;\varepsilon) - \bar{u}(t)\| + \|\bar{u}(t) - u(t)\| \\
&\leq C_0\varepsilon^2(1 + \omega^2 t)e^{\omega^2 t}\|A\bar{u}\| + C_0 e^{\omega^2 t}\|u_0(\varepsilon) - \bar{u}\| \\
&\quad + C_0\varepsilon^2 e^{\omega^2 t}\|u_1(\varepsilon)\| + C_0 e^{\omega^2 t}\|\bar{u} - u_0\| \\
&\leq C_0\varepsilon^2(1 + \omega^2 t)e^{\omega^2 t}\|A\bar{u}\| + C_0 e^{\omega^2 t}\|u_0(\varepsilon) - u_0\| \\
&\quad + C_0\varepsilon^2 e^{\omega^2 t}\|u_1(\varepsilon)\| + C_0\delta\, e^{\omega^2 t}.
\end{aligned} \tag{5.41}$$

Taking $\varepsilon > 0$ sufficiently small we can obviously make the right hand side $\leq 2C_0\,\delta\, e^{\omega^2 a}$ in $0 \leq t \leq a$, $a > 0$. This ends the proof.

Concerning derivatives, we have

THEOREM 5.8. *Let* $u(t;\varepsilon)$, $u(t)$ *be as in Theorem* 5.7. *Assume that* $u_0, u_0(\varepsilon) \in D(A)$ *and*

$$Au_0(\varepsilon) \to Au, \quad u_1(\varepsilon) \to Au_0 \quad \text{as} \quad \varepsilon \to 0. \tag{5.42}$$

Then

$$u'(t;\varepsilon) \to u'(t) \quad \text{as} \quad \varepsilon \to 0 \tag{5.43}$$

uniformly on compacts of $t \geq 0$.

The proof follows the lines of that of the previous result. Let $\delta > 0$, and choose $\bar{u} \in D(A^2)$ such that $\|A\bar{u} - Au_0\| \leq \delta$. Then, if $\bar{u}(\hat{t})$ is again the solution of the initial value problem (2.2) with $\bar{u}(0) = u_0$ we apply (5.29) and (5.35), obtaining

$$\|u'(t;\varepsilon) - u'(t)\| \le \|u'(t;\varepsilon) - \bar{u}'(t)\| + \|\bar{u}'(t) - u'(t)\|$$

$$\le C_0\varepsilon^2(1 + \omega^2 t)e^{\omega^2 t}\|A^2\bar{u}\|$$

$$+ C_0 e^{\omega^2 t}(\|u_1(\varepsilon) - A\bar{u}\| + \|u_1(\varepsilon) - Au_0(\varepsilon)\|) + \|S(t)(A\bar{u} - Au_0)\|$$

$$\le C_0\varepsilon^2(1 + \omega^2 t)e^{\omega^2 t}\|A^2\bar{u}\|$$

$$+ C_0 e^{\omega^2 t}(\|u_1(\varepsilon) - A\bar{u}_0\| + \|u_1(\varepsilon) - Au_0(\varepsilon)\|)$$

$$+ C_0\delta\ e^{\omega^2 t} \qquad (t \ge 0,\ \varepsilon > 0). \tag{5.44}$$

This completes the proof.

REMARK 5.9. Convergence in (5.31) and (5.36) is uniform in $t \ge 0$ (rather than just uniform on compacts of $t \ge 0$) if $\omega = 0$. Of course, the same observation applies to all the other results in this section. For easy reference later we collect these particular cases of Theorems 5.5, 5.6 5.7 and 5.8 under a single heading.

THEOREM 5.10. _Assume that_ A _generates a strongly continuous cosine function_ $C(\hat{t})$ _with_

$$\|C(t)\| \le C_0 \qquad (-\infty < t < \infty).$$

Let $u(\hat{t};\varepsilon)$ _be the generalized solution of the initial value problem_ (2.1), $u(\hat{t})$ _the generalized solution of the initial value problem_ (2.2). _Then_, (a) _if_ $u_0 \in D(A)$,

$$\|u(t;\varepsilon) - u(t)\| \le C_0\varepsilon^2(\|Au_0\| + \|u_1(\varepsilon)\|)$$

$$+ C_0\|u_0(\varepsilon) - u_0\| \qquad (t \ge 0,\ \varepsilon > 0). \tag{5.45}$$

(b) _If_ $u_0 \in D(A^2)$ _and_ $u_0(\varepsilon) \in D(A)$ _then_

$$\|u'(t;\varepsilon) - u'(t)\| \le C_0\varepsilon^2\|A^2u_0\|$$

$$+ C_0(\|u_1(\varepsilon) - Au_0\| + \|u_1(\varepsilon) - Au_0(\varepsilon)\|)$$

$$(t \ge 0,\ \varepsilon > 0)\ . \tag{5.46}$$

(c) _If_ u_0, $u_0(\varepsilon), u_1(\varepsilon)$ _are arbitrary elements of_ E, _then_

$$u(t;\varepsilon) \to u(t) \ \ \text{as} \ \ \varepsilon \to 0 \tag{5.47}$$

uniformly in $t \ge 0$ _if_

$$u_0(\varepsilon) \to u_0, \ \varepsilon^2 u_1(\varepsilon) \to 0 \ \underline{\text{as}} \ \varepsilon \to 0. \tag{5.48}$$

(d) <u>If</u> $u_0, u_0(\varepsilon) \in D(A)$ <u>then</u>

$$u'(t;\varepsilon) \to u'(t) \ \underline{\text{as}} \ \varepsilon \to 0 \tag{5.49}$$

<u>uniformly in</u> $t \geq 0$ <u>if</u>

$$Au_0(\varepsilon) \to Au, \ u_1(\varepsilon) \to Au_0 \ \underline{\text{as}} \ \varepsilon \to 0. \tag{5.50}$$

REMARK 5.11. The results in this section do not supersede those in §VI.4 (in particular, Theorem 3.6 on convergence of $u(t;\varepsilon)$ and Theorem 4.2 on convergence of the derivative $u'(t;\varepsilon)$). One reason is that "crossover" of initial conditions (see (3.34)) is ruled out here, although this will be remedied in a sense in §VI.8 by means of correction terms to straighten out convergence near zero. Another, and far more important reason, is that the convergence results in §VI.3 and §VI.4 are <u>uniform with respect to bounded initial conditions</u> (see the statement of Theorem 3.6); more precisely, the results are on convergence of $\mathfrak{C}(\hat{t};\varepsilon), \mathfrak{S}(\hat{t};\varepsilon), \mathfrak{S}'(\hat{t};\varepsilon)$ <u>in the norm of</u> (E), <u>that is, in the uniform topology of operators</u>. As the theorems in this section make clear, the price we pay for convergence in the norm of (E) is loss of convergence in an "initial layer" $0 \leq t \leq t(\varepsilon)$, even if crossover of initial conditions is avoided (see Example 3.10).

We note finally that the arguments in §VI.4 yield bounds for $\mathfrak{S}'(t;\varepsilon)$ and $\mathfrak{S}''(t;\varepsilon)$ (see Theorems 4.1 and 4.4) not amenable to the methods of the present section.

It is natural to ask whether we may obtain convergence of order $\varepsilon^{2\alpha}$, $0 < \alpha < 1$ for u_0 in spaces intermediate between E and D(A) ($D(A^2)$ for the derivative). This question has several equally natural answers, one of which we present here; another can be found in the author [1983:4] and a third, more useful in practice, will be examined in §VI.9.

We introduce the spaces $\mathfrak{H}_\alpha$ $(0 \leq \alpha \leq 1)$ consisting of all $u \in E$ such that

$$\|AS(t)u\| = O(t^{\alpha-1}) \ \text{as} \ t \to 0+. \tag{5.51}$$

Equivalent characterizations of the spaces $\mathfrak{H}_\alpha$ are given in BUTZER-BERENS [1967:1, pp. 111 and 115]: if $0 \leq \alpha \leq 1$, $u \in \mathfrak{H}_\alpha$ if and only if

$$\|S(t)u - u\| = O(t^{\alpha}) \quad \text{as} \quad t \to 0+, \tag{5.52}$$

(equivalently, if $S(\hat{t})u$ is locally Hölder continuous with exponent α). We have $\mathfrak{H}_1 = D(A)$ when E is reflexive; in the general case only $\mathfrak{H}_1 \supseteq D(A)$ can be assured.

We use now (4.23) to estimate $S'(t)$:

$$\|S'(t)\| = \|AS(t)\| \le \frac{C_0}{\pi^{1/2}} \int_0^{\infty} \left(\frac{s^2}{4t^{5/2}} + \frac{1}{2t^{3/2}} \right) e^{-s^2/4t} \cosh \omega s \, ds$$

$$\le C(\omega^2 + 1/t)e^{\omega^2 t} \qquad (t \ge 0). \tag{5.53}$$

Let $u \in \mathfrak{H}_{\alpha}$. Define

$$|u|_{\alpha} = \|u\| + \sup_{t \ge 0} (\omega^2 + 1/t)^{\alpha - 1} e^{-\omega^2 t} \|AS(t)u\|. \tag{5.54}$$

Obviously, $|\cdot|_{\alpha}$ is a norm in $\mathfrak{H}_{\alpha}$ (which, incidentally, makes $\mathfrak{H}_{\alpha}$ a Banach space).

The following two results are formal counterparts of Theorem 5.5 and 5.6.

THEOREM 5.12. _Let_ $u_0 \in \mathfrak{H}_{\alpha}$ _with_ $0 < \alpha < 1$, _and let_ $\beta = \max(\alpha, 1 - \alpha)$. _Then_

$$\|u(t;\varepsilon) - u(t)\| \le C\varepsilon^{2\alpha}(1 + \omega^2 t)^{\beta} e^{\omega^2 t} |u_0|_{\alpha} + C_0 e^{\omega^2 t} \|u_0(\varepsilon) - u_0\|$$

$$+ C_0 \varepsilon^{2\alpha} e^{\omega^2 t} (\varepsilon^{2(1-\alpha)} \|u_1(\varepsilon)\|) \quad (t \ge 0). \tag{5.55}$$

THEOREM 5.13. _Assume that_ $u_0, u_0(\varepsilon) \in D(A)$ _and_ $Au_0 \in \mathfrak{H}_{\alpha}$, $0 < \alpha < 1$, _and let_ β _be as in Theorem_ 5.12. _Then_

$$\|u'(t;\varepsilon) - u'(t)\| \le C\varepsilon^{2\alpha}(1 + \omega^2 t)^{\beta} e^{\omega^2 t} |Au_0|_{\alpha}$$

$$+ C_0 e^{\omega^2 t} (\|u_1(\varepsilon) - Au_0\| + \|u_1(\varepsilon) - Au_0(\varepsilon)\|)$$

$$(t \ge 0). \tag{5.56}$$

The proof of both results is based on a different estimation of the operator $\mathfrak{H}$. Assume first that $u \in D(A)$; since S and $\mathfrak{S}'$ commute with A and with each other we can write

$$\mathfrak{H}(t;\varepsilon)u = \int_0^t \mathfrak{S}'(t - s;\varepsilon) AS(s)u \, ds. \tag{5.57}$$

It follows from (4.1) that

$$\|\mathfrak{S}'(t;\varepsilon)\| \leq C\varepsilon^{-2}e^{\omega^2 t} \qquad (t \geq 0). \tag{5.58}$$

Take the $(1-\alpha)$-th power of both sides, take the α-th power of both sides of (4.20) and multiply the inequalities thus obtained term by term. The result is

$$\|\mathfrak{S}'(t;\varepsilon)\| \leq C\varepsilon^{-2(1-\alpha)}(\omega^2 + 1/t)^{\alpha}e^{\omega^2 t} \qquad (t \geq 0). \tag{5.59}$$

Hence, if $u \in \mathcal{H}_\alpha$,

$$\|\mathfrak{H}(t;\varepsilon)u\| \leq C\varepsilon^{-2(1-\alpha)}|u|_\alpha e^{\omega^2 t}\int_0^t (\omega^2 + 1/(t-s))^{\alpha}(\omega^2 + 1/s)^{1-\alpha}\, ds, \tag{5.60}$$

whence the estimates (5.55) and (5.56) result.

REMARK 5.14. Direct verification that an element $u \in E$ belongs to $\mathcal{H}_\alpha$ may be difficult. Sufficient conditions can be obtained using the theory of fractional powers of infinitesimal generators touched upon in Chapter III. In fact, we have

LEMMA 5.15. *Let* $b > \omega$. *Then*

$$D((b^2I - A)^{\alpha}) \subseteq \mathcal{H}_\alpha \qquad (0 < \alpha \leq 1). \tag{5.61}$$

The result is a consequence of Exercises 3 and 7 in Chapter III. In fact, it follows from Exercise 3 that $u \in D((bI - A)^{\alpha})$ if and only if $S(\hat{t})u$ is continuously differentiable of order $\alpha \geq 0$. On the other hand, it has been proved in Exercise that if $S(t)u$ is continuously differentiable of order α $(0 < \alpha \leq 1)$ then $S(\hat{t})u$ is Hölder continuous with exponent α in $t \geq 0$, so that

$$\|S(t)u - u\| = \|S(t)u - S(0)u\| = o(t^{\alpha})$$

as $t \to 0+$, thus $u \in \mathcal{H}_\alpha$.

§VI.6. Singular integrals of Hilbert space-valued functions and applications to inhomogeneous first order equations.

We prove in this section a result on the initial value problem

$$u'(t) = Au(t) + f(t) \quad (t \geq 0), \qquad u(0) = 0. \tag{6.1}$$

valid whenever A generates an analytic semigroup in a Hilbert space H. The basis of the argument (as well as of the corresponding theorem for the initial value problem (2.1)) is the following result on singular integrals of Hilbert space-valued functions.

THEOREM 6.1. *Let* H *be a Hilbert space,* $K(\hat{t})$ *a function with values in* (H) *defined in* $-\infty < t < \infty$ *and such that* (a)

$$K(\hat{t}) \in L^1(-\infty,\infty;H) \cap L^2(-\infty,\infty;H). \tag{6.2}$$

(b) *The Fourier transform* $\tilde{K}(\hat{\sigma})$ *satisfies*

$$\|\tilde{K}(\sigma)\| \leq B \text{ a.e. in } -\infty < \sigma < \infty. \tag{6.3}$$

(c)

$$\int_{|s| \geq 2|t|} \|K(s - t) - K(s)\| \, ds \leq B \qquad (-\infty < t < \infty). \tag{6.4}$$

Let $1 < p < \infty$. *For* $f \in L^p(-\infty,\infty;H)$ *define*

$$(\mathcal{K}f)(t) = \int_{-\infty}^{\infty} K(t - s)f(s) \, ds . \tag{6.5}$$

Then there exists a constant C *depending only on* p *and* B *such that*

$$\|\mathcal{K}f\|_p \leq C\|f\|_p , \tag{6.6}$$

where $\|\cdot\|_p$ *indicates the norm of* $L^p(-\infty,\infty;H)$.

The proof (of a more general theorem) can be found in STEIN [1970:1]; the result for scalar functions is in pp. 29 and 34, while the extension to vector-valued function is in pp. 46-48. We note that, since $K(\hat{t}) \in L^1(-\infty,\infty;H)$, (6.6) follows from (6.5) and Young's inequality (STEIN [1970:1]). However, what is significant in Theorem 6.1 is that C does not depend, among other things, of the L^1 norm of $K(\hat{t})$ but only on p and B. This will be decisive below.

THEOREM 6.2. *Assume that* $A \in \mathfrak{G}(\varphi)$, $0 < \varphi < \pi/2$ *(that is, that* A *generates an analytic semigroup) in the Hilbert space* H. *Let* $T > 0$, $1 < p < \infty$, $f(\hat{t}) \in L^p(0,T;H)$ *and* $u(\hat{t})$ *the (generalized) solution of the initial value problem* (6.1). *Then* $u(\hat{t})$ *has a derivative* $u'(t) \in L^p(0,T;H)$ *in* $0 \leq t \leq T$ *(in the sense that there exists* $u'(\hat{t}) \in L^p(0,T;H)$ *such that*

$$u(t) = \int_0^t u'(s) \, ds \qquad (0 \leq t \leq T)) . \tag{6.7}$$

Moreover, $u(t) \in D(A)$ *a.e. in* $0 \leq t \leq T$ *and the equation* (6.1) *is satisfied a.e. Finally, there exists a constant* C *depending only on* p *(and of course on* A) *such that*

$$\|u'(\hat{t})\|_p \leq C\|f\|_p . \tag{6.8}$$

The proof naturally will depend on Theorem 6.1 and (a variant of) the representation

$$u'(t) = \int_0^t S'(t - s)f(s)\, ds \tag{6.9}$$

that follows from (I.5.3). The hypotesis less easy to verify will be (c). For this we shall use the following result.

LEMMA 6.3. *Let* $K(\hat{t})$ *be a function with values in a Banach space* E. *Assume that* $K(\hat{t})$ *is continuously differentiable for* $t \neq 0$ *and that*

$$\|K'(t)\| \leq B'|t|^{-2} \quad (t \neq 0). \tag{6.10}$$

Then $K(\hat{t})$ *satisfies* (6.4) (*with* $B' = (2\log 2)B'$).

Proof: Let $t > 0$. We have

$$\|K(s - t) - K(s)\| \leq \int_{s-t}^{s} \|K'(\sigma)\|\, d\sigma$$

$$\leq B' \int_{s-t}^{s} \frac{d\sigma}{\sigma^2} = B'\left(\frac{1}{s - t} - \frac{1}{s}\right) \quad (s > t),$$

so that

$$\int_{2t}^{\infty} \|K(s - t) - K(s)\|\, ds \leq B' \int_{2t}^{\infty}\left(\frac{1}{s - t} - \frac{1}{s}\right) ds = (\log 2)B'.$$

We use a similar argument in the range $s < t$.

The estimation runs in exactly the same way when $t < 0$ and we omit the details.

Proof of Theorem 6.2. We say that an H-valued function $f(\hat{t})$ defined in $0 \leq t \leq T$ belongs to $C(0,T;D(A))$ if $f(t) \in D(A)$ and $f(\hat{t})$, $Af(\hat{t})$ are continuous. We have already proved (Lemma I.5.1 (a)) that if $f(\hat{t}) \in C(0,T;D(A))$ then the generalized solution $u(\hat{t})$ of (6.1) defined by

$$u(t) = \int_0^t S(t - s)f(s)\, ds \quad (0 \leq t \leq T) \tag{6.11}$$

is actually a genuine solution of (6.1); in particular,

$$u'(t) = \int_0^t S'(t-s)f(s)\,ds + f(t) = A\int_0^t S(t-s)f(s)\,ds + f(t)$$
$$= \int_0^t S(t-s)Af(s)\,ds + f(t) \qquad (0 \le t \le T) \tag{6.12}$$

Let $f(t)$ be a function in $C(0,T;D(A))$. Assume that $\mu > \omega$, where ω is a constant such

$$\|S(t)\| \le C_0 e^{\omega t} \qquad (t \ge 0) \tag{6.13}$$

for some C_0. Then $f_\mu(t) = \mu R(\mu;A)f(\hat{t})$ belongs to $C(0,T;D(A))$ so that the previous considerations apply. We write (6.12) for $f = f_\mu$:

$$u_\mu'(t) = \int_0^t S'(t-s)\mu R(\mu;A)f(s)\,ds + \mu R(\mu;A)f(t)$$
$$= A\int_0^t S(t-s)\mu R(\mu;A)f(s)\,ds + \mu R(\mu;A)f(t)$$
$$= \int_0^t S(t-s)\mu R(\mu;A)Af(s)\,ds + \mu R(\mu;A)f(t)\ . \tag{6.14}$$

Let $\rho > \omega, \omega$ the constant in (6.13). Consider the (H)-valued function

$$K_{\mu,\rho}(t) = e^{-\rho t}S'(t)\mu R(\mu;A) = e^{-\rho t}S(t)\mu AR(\mu;A) \tag{6.15}$$

for $t \ge 0$; for $t < 0$ we set $K_{\mu,\rho}(t) = 0$. Obviously, $K_{\mu,\rho}(t)$ is infinitely differentiable in $t > 0$ (see §III.7). Using the second expression for $K_{\mu,\rho}(\hat{t})$ in (6.15) we see that

$$\|K_{\mu,\rho}(t)\| \le C_\mu e^{-(\rho-\omega)t} \qquad (t \ge 0)\ . \tag{6.16}$$

It follows that $K_{\mu,\rho}(\hat{t}) \in L^r(-\infty,\infty;H)$ for all r, $1 \le r \le \infty$, in particular for $r = 1, 2$ as required in (a) of Theorem 6.1.

We have

$$(2\pi)^{-1/2}\tilde{K}_{\mu,\rho}(\sigma) = \int_0^\infty e^{i\sigma t}K_{\mu,\rho}(t)\,dt$$
$$= \int_0^\infty e^{-(\rho-i\sigma)t}S'(t)\mu R(\mu;A)\,dt$$
$$= -\mu R(\mu;A) + (\rho - i\sigma)\int_0^\infty e^{-(\rho-i\sigma)t}S(t)\mu R(\mu;A)dt$$
$$= -\mu R(\mu;A) + (\rho - i\sigma)R(\rho - i\sigma;A)\mu R(\mu;A)\ . \tag{6.17}$$

It follows from the first inequality (I.3.9) that

$$\|\mu R(\mu;A)\| \leq \mu C_0(\mu - \omega)^{-1} \qquad (\mu > \omega), \tag{6.18}$$

and from (III.7.6) in Theorem III.7.1 that, if ρ is sufficiently large,

$$\|(\rho - i\sigma)R(\rho - i\sigma)\| \leq C \qquad (-\infty < \sigma < \infty),$$

so that (6.17) implies that (6.3) in Theorem 6.1 is satisfies with a constant B <u>that does not depend on</u> μ (say, in $\mu \geq \rho$). To handle (c) we use the following estimate for the derivatives of an analytic semigroup, which is a consequence of the results in Chapter III (see formula (III.7.11)):

$$\|S^{(m)}(t)\| = \|A^m S(t)\| \leq C_m t^{-m} e^{\omega' t} \qquad (t > 0), \tag{6.19}$$

where C_m is a positive constant and $\omega' > \omega$, ω the constant in (6.13). A moment's reflection shows that it is possible to deduce from this inequality, from (6.18) and from the first expression in (6.15) for $K_{\mu,\rho}(t)$ that

$$\|K'_{\mu,\rho}(t)\| \leq C|t|^{-2} \qquad (t \neq 0),$$

thus we can apply Lemma 6.3 to show that $K_{\mu,\rho}$ satisfies as well (6.4). We have thus verified all the hypoteses of Theorem 6.1 for the kernel $K_{\mu,\rho}(\hat{t})$ <u>with a constant</u> B <u>that does not depend on</u> μ. Accordingly, there exists a constant C depending on p but not on μ such that if

$$u(t) = \int_0^t K_{\mu,\rho}(t - s)f(s)\,ds = e^{-\rho t}\int_0^t e^{\rho s}S'(t - s)\mu R(\mu;A)f(s)\,ds \tag{6.20}$$

for $f \in L^p(0,\infty;H)$, then

$$\|u\|_{L^p(0,\infty;H)} \leq C\|f\|_{L^p(0,\infty;H)}. \tag{6.21}$$

Now, if $T > 0$ and α is arbitrary, there obviously exist constants $M > m > 0$ (depending only on p,α,T) such that

$$m\|e^{\alpha\hat{t}}f(\hat{t})\|_{L^p(0,T;H)} \leq \|f(\hat{t})\|_{L^p(0,T;H)}$$

$$\leq M\|e^{\alpha\hat{t}}f(\hat{t})\|_{L^p(0,T;H)} \qquad (f(t) \in L^p(0,T;H)), \tag{6.22}$$

hence, if

$$v(t) = \int_0^t S'(t - s)\mu R(\mu;A)f(s)\,ds$$

for $f \in L^p(0,T;H)$,

$$\|v(\hat{t})\|_{L^p(0,T;H)} \le C\|f\|_{L^p(0,T;H)} . \tag{6.23}$$

We return to (6.14). Keeping in mind that $f(t) \in C(0,T;D(A))$ and making use of the fact that

$$\mu R(\mu;A)u \to u \quad \text{as} \quad \mu \to \infty \tag{6.24}$$

(see Exercise I.17), and of the uniform bound (6.18) we can use the dominated convergence theorem in combination with (6.22) and deduce that if $u(\hat{t})$ is the solution of (6.1) (given by (6.11)) then

$$\|u'(t)\|_{L^p(0,T;H)} \le C\|f\|_{L^p(0,T;H)} . \tag{6.25}$$

Let, finally, $f(\hat{t})$ be an arbitrary function in $L^p(0,T;H)$, $\{f_n(\hat{t})\}$ a sequence in $C(0,T;D(A))$ such that

$$\|f_n - f\|_{L^p(0,T;H)} \to 0 \quad \text{as} \quad n \to \infty$$

and $u_n(\hat{t})$ the solution of (6.1) with $f = f_n$. It results from (6.25) applied to $f_n - f_m$ that $u_n'(\hat{t})$ converges in the L^p norm to a $v(\hat{t}) \in L^p(0,T;H)$. Since $u_n(t) \to u(t)$ uniformly in $0 \le t \le T$, where $u(\hat{t})$ is the weak solution of (6.1), we can take limits in the equality

$$u_n(t) = \int_0^t u_n'(s)ds$$

and conclude that (6.7) holds with $u'(t) = v(t)$. To prove the rest of the claims on $u(\hat{t})$ we write (6.12) for f_n in the form

$$u_n'(t) = Au_n(t) + f_n(t) . \tag{6.26}$$

Since $u_n'(\hat{t}) \to u'(\hat{t})$ in $L^p(0,T;H)$, passing if necessary to a subsequence, we may assume that $u_n'(t) \to u'(t)$ and $f_n(t) \to f(t)$ in a set e of full measure in $0 \le t \le T$. Using the fact that A is closed we obtain taking limits in (6.26) that $u(t) \in D(A)$ and

$$u'(t) = Au(t) + f(t) \qquad (t \in e).$$

This ends the proof of Theorem 6.2.

The next result on the nonhomogeneous equation (6.1) is considerably simpler and not restricted to Hilbert spaces. Given an arbitrary Banach space E, the class $\mathcal{H}_\alpha(0,T;E)$ is defined as follows:

for $0 \leq \alpha \leq 1$, $\mathcal{H}_\alpha(0,T;E)$ consists of all functions f defined and α-Hölder continuous in $0 \leq t \leq T$, and is endowed with the norm

$$|f|_\alpha = \sup_{0\leq t\leq T} \|f(t)\| + \sup_{0\leq t\leq t'\leq T} |t'-t|^{-\alpha}\|f(t')-f(t)\|. \quad (6.27)$$

The spaces $\mathcal{H}_\alpha(0,T;E)$ are Banach spaces.

THEOREM 6.4. <u>Let</u> E <u>be an arbitrary Banach space</u>, $f(\hat{t}) \in \mathcal{H}_\alpha(0,T;E)$, $0 < \alpha \leq 1$, <u>and assume</u> A <u>generates an analytic semigroup</u> $S(\hat{t})$ <u>in</u> E. <u>Then</u> $u(\hat{t})$, <u>the generalized solution of</u> (6.1) <u>is a genuine solution, that is</u>, $u(\hat{t})$ <u>is continuously differentiable in</u> $0 \leq t \leq T$, $u(t) \in D(A)$ <u>and</u> (6.1) <u>is satisfied in</u> $0 \leq t \leq T$. <u>Finally, there exists a constant</u> C_α <u>such that</u>

$$\|u'(t)\| \leq C_\alpha |f(\hat{t})|_\alpha \qquad (0 \leq t \leq T). \quad (6.28)$$

<u>Proof</u>: We use again inequality (6.19), this time for the first derivative $S'(t)$. In view of (6.27) we have

$$\|S'(t-s)(f(s)-f(t))\| \leq C(t-s)^{\alpha-1}|f|_\alpha, \quad (6.29)$$

hence the function

$$v(t) = \int_0^t S'(t-s)(f(s)-f(t))ds + S(t)f(t) \quad (6.30)$$

is well defined in $0 \leq t \leq T$. We show that $v(\hat{t})$ is continuous. Let $0 \leq t < t' \leq T$. Ignoring (as we may) the nonintegral term, we have

$$\|v(t')-v(t)\| \leq \int_t^{t'} \|S(t'-s)(f(s)-f(t'))\|\,ds$$

$$+\int_0^t \|S(t'-s)(f(s)-f(t')) - S(t-s)(f(s)-f(t))\|\,ds. \quad (6.31)$$

In view of (6.29), the first integral is bounded by a constant times

$$\int_t^{t'} (t'-s)^{\alpha-1}\,ds = \frac{1}{\alpha}(t'^\alpha - t^\alpha).$$

The integrand in the second integral tends to zero as $t'-t \to 0$ and can be estimated, using (6.29) again, by a constant times

$$(t'-s)^{\alpha-1} + (t-s)^{\alpha-1} \leq 2(t-s)^{\alpha-1},$$

thus the integral tends to zero due to the dominated convergence theorem. It follows from inequality (6.29) that there exists a constant C_α

such that

$$\|v(t)\| \le C_\alpha |f(\hat{t})|_\alpha . \tag{6.32}$$

Let $u(\hat{t})$ be the weak solution of (6.1). Writing

$$u(t) = \int_0^t S(t - s)(f(s) - f(t))ds + \int_0^t S(t - s)f(t)ds \tag{6.33}$$

we see using once again (6.29) that $u(t) \in D(A)$ and

$$\begin{aligned} Au(t) &= \int_0^t AS(t - s)(f(s) - f(t))ds + S(t)f(t) - f(t) \\ &= \int_0^t S'(t - s)(f(s) - f(t))ds + S(t)f(t) - f(t) , \end{aligned} \tag{6.34}$$

so that

$$v(t) = Au(t) + f(t) \qquad (0 \le t \le T) , \tag{6.35}$$

which shows that $Au(\hat{t})$ is continuous in $0 \le t \le T$. Thus, to complete the proof that $u(\hat{t})$ is a genuine solution of (6.1) we only have to show that $u'(t) = v(t)$ a.e. or

$$u(t) = \int_0^t v(s)ds , \tag{6.36}$$

which, in view of (6.32), also implies (6.27). We can do this as follows. Assuming that $f(\hat{t})$ is continuously differentiable we have, integrating by parts,

$$\begin{aligned} v(t) &= - \int_0^t (D_s S(t - s))(f(s) - f(t))ds + S(t)f(t) \\ &= \int_0^t S(t - s)f'(s)ds + S(t)f(0) = \int_0^t S(s)f'(t - s)ds + S(t)f(0) . \end{aligned} \tag{6.37}$$

Interchanging the order of integration,

$$\begin{aligned} \int_0^t v(s)ds &= \int_0^t S(r)\left(\int_r^t f'(s - r)ds\right) dr + \\ &+ \int_0^t S(s)f(0)ds = \int_0^t S(t - s)f(s)\, ds . \end{aligned} \tag{6.38}$$

To prove (6.36) for $f \in \mathcal{H}_\alpha$ we select a sequence $\{f_n\}$ of continuously differentiable functions such that

$$\|f - f_n\|_{\mathcal{H}_\alpha} \to 0$$

as $n \to \infty$. Once this is done write (6.38) for f_n and take limits: that this is justified follows from (6.32) applied to $f - f_n$.

§VI.7 The inhomogeneous equation: convergence of $u(t;\varepsilon)$ and $u'(t;\varepsilon)$.

We examine in this section estimates and convergence results for the generalized solution $u(t;\varepsilon)$ of the nonhomogeneous initial value problem

$$\varepsilon^2 u''(t;\varepsilon) + u'(t;\varepsilon) = Au(t;\varepsilon) + f(t;\varepsilon) \quad (t \geq 0),$$
$$u(0;\varepsilon) = 0, \quad u_1(\varepsilon) = 0 . \tag{7.1}$$

The limit will be the solution of

$$u'(t) = Au(t) + f(t) \quad (t \geq 0), \qquad u(0) = 0 . \tag{7.2}$$

Roughly speaking, the estimates are "dynamic" (that is, ε-dependent) counterparts of the results in §VI.6 ; the main tools will be the singular integral methods there and the estimates on $\mathfrak{S}'$ and $\mathfrak{S}''$ obtained in §VI.4. The first results (Theorems 7.2 and 7.4) are actually independent of singular integrals and are valid in a general Banach space E; the others (Theorem 7.6) are restricted to Hilbert space-valued functions.

Throughout this section $u(\hat{t};\varepsilon)$ is the generalized solution of (7.1) and $u(\hat{t})$ is the generalized solution of (7.2): we have

$$u(t;\varepsilon) = \int_0^t \mathfrak{S}(t-s;\varepsilon)f(s;\varepsilon)\,ds, \quad u(t) = \int_0^t S(t-s)f(s)\,ds. \tag{7.3}$$

The first result (which will not be used in the sequel) is on conditions on $f(t;\varepsilon)$ that guarantee that $u(t;\varepsilon)$ is a genuine solution of (7.1).

LEMMA 7.1. *Assume that one of the following two conditions is satisfied:*

(a) $f(t) \in D(A)$ *and* $f(t)$, $Af(t)$ *are continuous in* $0 \leq t \leq T$,

or

(b) $f(t)$ *is continuously differentiable in* $0 \leq t \leq T$.

Then $u(t;\varepsilon)$ *is a genuine solution of* (7.1).

This result can be reduced to Lemma II.4.1 through the transformations used to transmute the solution of (2.1) into the solution of (2.3) and viceversa.

THEOREM 7.2. Assume that $f(\hat{t};\varepsilon)$, $f(\hat{t}) \in L^1(0,T;E)$ with $0 < T < \infty$. Then (a) $u(\hat{t};\varepsilon)$ (resp. $u(\hat{t})$) is continuously differentiable (resp. continuous) in $0 \le t \le T$. (b)

$$\|u(t;\varepsilon)\| \le C_0 e^{\omega^2 T}\|f(\hat{t};\varepsilon)\|_1, \quad \|u(t)\| \le C_0 e^{\omega^2 T}\|f(\hat{t})\|_1 \qquad (0 \le t \le T) . \tag{7.4}$$

(c) If $f(\hat{t};\varepsilon) \to f(\hat{t})$ in $L^1(0,T;E)$ then $u(\hat{t};\varepsilon) \to u(t)$ uniformly in $0 \le t \le T$.

Proof: Inequality (7.4) for $u(\hat{t})$ was already shown in Lemma I.5.2; the corresponding estimate for $u(t;\varepsilon)$ follows in the same way from (7.3) and the estimate (3.4) for $\mathfrak{S}(\hat{t};\varepsilon)$,

$$\|\mathfrak{S}(t;\varepsilon)\| \le C_0 e^{\omega^2 t} \qquad (t \ge 0) . \tag{7.5}$$

Continuity of $u(\hat{t})$ was shown as well in Lemma I.5.2, but we sketch the proof anew. If $0 \le t < t' \le T$ we have

$$\|u(t') - u(t)\| \le \int_t^{t'} \|S(t' - s)\|ds + \int_0^t \|S(t' - s) - S(t - s)\|\|f(s)\|ds . \tag{7.6}$$

The first integral is shown to converge to zero on account of the estimate

$$\|S(t)\| \le C_0 e^{\omega^2 t} \qquad (t \ge 0) . \tag{7.7}$$

The integrand of the second tends to zero due to continuity of $S(\hat{t})$ and is bounded in norm by a constant times $\|f(s)\|$, thus we show the integral tends to zero using the dominated convergence theorem. To prove that $u(t;\varepsilon)$ is continuously differentiable we note that it follows from (2.14) that $\mathfrak{S}(\hat{t};\varepsilon)$ is continuously differentiable in the norm of (E) in $t \ge 0$; this is easily seen to imply that $u'(t;\varepsilon)$ exists for all t and

$$u'(t,\varepsilon) = \int_0^t \mathfrak{S}'(t - s;\varepsilon)f(s)ds. \tag{7.8}$$

Continuity of $u'(\hat{t};\varepsilon)$ is dealt with in the same way as continuity of $u(\hat{t})$. To show (c) we note that

$$u(t;\varepsilon) - u(t) = \int_0^t \mathfrak{S}(t - s;\varepsilon)(f(s;\varepsilon) - f(s))ds + \int_0^t (\mathfrak{S}(t - s;\varepsilon) - S(t - s))f(s)ds. \tag{7.9}$$

If uniform convergence does not prevail then there exists a sequence

$\{t_n\}$ in the interval $0 \le t \le T$ such that $\|u(t_n;\varepsilon_n) - u(t)\|$ is bounded away from zero for some sequence $\{\varepsilon_n\}$ such that $\varepsilon_n \to 0$. We may assume that $t_n \to t$ $(0 \le t \le T)$. If $t = 0$ a contradiction is obtained with the help of (7.5) and (7.7); if $t \ne 0$ we deduce from Theorem 3.6 that

$$\mathfrak{S}(t_n - s;\varepsilon)f(s) \to S(t - s;\varepsilon)f(s)$$

as $n \to \infty$. This time the contradiction results from the dominated convergence theorem.

REMARK 7.3. When $\omega = 0$ we can take $T = \infty$ in the statement of Theorem 7.2; inequalities (7.4) become

$$\|u(t;\varepsilon)\| \le C_0\|f(\hat{t};\varepsilon)\|_1, \ \|u(t)\| \le C_0\|f(\hat{t})\|_1 \quad (0 \le t \le T). \quad (7.10)$$

Moreover, we obtain uniform convergence of $u(t;\varepsilon)$ in $t \ge 0$ (this is a consequence of (7.9)).

The next result establishes uniform boundedness and convergence of the derivative $u'(t;\varepsilon)$ under stronger assumptions on the functions $f(\hat{t};\varepsilon)$, $f(\hat{t})$.

THEOREM 7.4. <u>Assume that</u> $f(t;\varepsilon), f(t) \in \mathfrak{H}_\alpha(0,T;E)$, $0 < \alpha \le 1$. <u>Then</u> (a) $u(\hat{t};\varepsilon)$ <u>and</u> $u(\hat{t})$ <u>are continuously differentiable in</u> $0 \le t \le T$. (b) <u>There exists</u> C_α <u>such that</u>

$$\|u'(t;\varepsilon)\| \le C_\alpha|f(\hat{t};\varepsilon)|_\alpha, \quad \|u'(t)\| \le C_\alpha|f(\hat{t})|_\alpha \quad (0 \le t \le T). \quad (7.11)$$

(c) <u>If</u> $f(\hat{t};\varepsilon) \to f(\hat{t})$ <u>in</u> $\mathfrak{H}_\alpha(0,T;E)$ <u>then</u> $u(t;\varepsilon) \to u(t)$, $u'(t;\varepsilon) - \mathfrak{S}(t,\varepsilon)f(0) \to u'(t) - S(t)f(0)$ <u>uniformly in</u> $0 \le t \le T$.

<u>Proof</u>: The fact that $u(\hat{t})$ is a genuine solution of (7.2) in $0 \le t \le T$ has already been proved in Theorem 6.4 together with the second estimate (7.11): continuous differentiability of $u'(t;\varepsilon)$ follows from Theorem 7.2, under the only assumption that $f(\hat{t})$ is continuous. To show the first inequality (7.11) we write

$$u'(t;\varepsilon) = \int_0^t \mathfrak{S}'(t - s;\varepsilon)(f(s;\varepsilon) - f(t;\varepsilon)) + \mathfrak{S}(t;\varepsilon)f(t;\varepsilon). \quad (7.12)$$

and use the estimate (4.20) for $\mathfrak{S}'$, which implies that

$$\|\mathfrak{S}'(t;\varepsilon)\| \le C't^{-1} \quad (0 < t \le T). \quad (7.13)$$

Accordingly, it follows from (7.12) that

$$\|u'(t;\varepsilon)\| \le C'\|f(\hat{t};\varepsilon)\|_\alpha \int_0^t (t-s)^{\alpha-1}ds + C_0 e^{\omega^2 t}\|f(\hat{t};\varepsilon)\|_\alpha ,$$

which yields (7.11).

We prove finally the convergence statement (c). Using formulas (7.8) for $u'(t;\varepsilon)$ and (6.9) for $u'(t)$ we easily assemble the following expression:

$$u'(t;\varepsilon) - u'(t) = \int_0^t \mathfrak{S}'(t-s;\varepsilon)(f(s;\varepsilon) - f(t;\varepsilon)) + \mathfrak{S}(t;\varepsilon)f(t;\varepsilon)$$

$$- \int_0^t S'(t-s)(f(s) - f(t))ds + S(t)f(t)$$

$$= \int_0^t \mathfrak{S}'(t-s;\varepsilon)((f(s;\varepsilon) - f(s)) - (f(t;\varepsilon) - f(t)))ds + \mathfrak{S}(t;\varepsilon)f(t;\varepsilon)$$

$$+ \int_0^t (\mathfrak{S}'(t-s;\varepsilon) - S'(t-s))(f(s) - f(t))\, ds - S(t)f(t). \quad (7.14)$$

The first integral can be estimated in norm by an expression of the form

$$C\|f(\hat{t};\varepsilon) - f(\hat{t})\|_\alpha \int_0^t (t-s)^{\alpha-1}\, ds . \quad (7.15)$$

The integrand in the second integral tends to zero in $0 \le s < t$ by Theorem 4.2; moreover, we have, using (6.19) and (7.13),

$$\|(\mathfrak{S}'(t-s;\varepsilon) - S'(t-s))(f(s) - f(t))\| \le C(t-s)^{\alpha-1}\|f(\hat{t})\|_\alpha .$$

Finally, we have

$$\|(\mathfrak{S}(t;\varepsilon)f(t;\varepsilon) - S(t)f(t)) - (\mathfrak{S}(t;\varepsilon)f(0) - S(t)f(0))\|$$

$$\le \|\mathfrak{S}(t;\varepsilon)\|\|(f(t;\varepsilon) - f(t))\|$$

$$+ \|(\mathfrak{S}(t;\varepsilon) - S(t))(f(t) - f(0))\|$$

$$\le C_0 e^{\omega^2 t}\|f(\hat{t};\varepsilon) - f(\hat{t})\|_\alpha$$

$$+ C\|t^\alpha(\mathfrak{S}(t;\varepsilon) - S(t))\|\|f(\hat{t})\|_\alpha . \quad (7.16)$$

The first term obviously converges uniformly to zero in $0 \le t \le T$. The same is true of the second due to Theorem 3.6.

REMARK 7.5. We see easily that the correction terms $\mathfrak{S}(t;\varepsilon)f(0)$ and $S(t)f(0)$ that produce uniform convergence of the derivatives cannot be dropped; this is due to the fact there cannot be uniform convergence of $u'(t;\varepsilon)$ to $u'(t)$ (since

$$u'(0;\varepsilon) = 0,\ u'(0) = Au(0) + f(0) = f(0))$$

unless $f(0) = 0$.

In the rest of the chapter we shall assume that $E = H$ is a Hilbert space; the estimations and convergence results will be based on Theorem 6.1 on singular integrals.

THEOREM 7.6. _Assume_ H _is a Hilbert space, and let_ $f(\hat{t};\varepsilon)$, $f(\hat{t})$ _belong to_ $L^p(0,T;H)$ $(0 < T < \infty,\ 1 < p < \infty)$. _Then_ (a) $u(\hat{t};\varepsilon)$ _is continuously differentiable and_ $u(t)$ _has a derivative in_ $L^p(0,T;H)$ _in the sense that there exists_ $v \in L^p(0,T;H)$ _with_ $\int_0^t v(s)ds = u(t)$ (b) $u(t) \in D(A)$ _and_ $u'(t) = Au(t) + f(t)$ a.e. _in_ $0 \le t \le T$. (c) _There exists a constant_ C_p _such that_

$$\|u'(\hat{t};\varepsilon)\|_p \le C_p\|f(\hat{t};\varepsilon)\|_p,\quad \|u'(\hat{t})\|_p \le C_p\|f(\hat{t})\|_p\ . \tag{7.17}$$

(d) _If_ $f(\hat{t};\varepsilon) \to f(\hat{t})$ _in_ $L^p(0,T;H)$ _then_ $u(\hat{t};\varepsilon) \to u(\hat{t})$ _uniformly in_ $0 \le t \le T$ _and_ $u'(\hat{t};\varepsilon) \to u'(\hat{t})$ _in_ $L^p(0,T;H)$;

Proof: The statements concerning the solution $u(\hat{t})$ of (7.2), including the second inequality (7.17), were proved in Theorem 6.2. The basis of the treatment of $u'(t;\varepsilon)$ will be formula (7.8) expressing it as an integral with kernel $\mathfrak{S}'(t;\varepsilon)$ to which Theorem 6.1 can be applied in the way used to show Theorem 6.2. Since $\mathfrak{S}'(t;\varepsilon)$, unlike $S'(t)$, is not singular at the origin, the argument is actually simpler and the smoothing operator $\mu R(\mu;A)$ need not be used.

Let $\rho > \omega^2$ (ω the constant in (4.20) and (4.26)) and define

$$K_\rho(t;\varepsilon) = e^{-\rho t}\mathfrak{S}'(t;\varepsilon) \tag{7.18}$$

for $t \ge 0$, $K_\rho(t;\varepsilon) = 0$ for $t < 0$.

LEMMA 7.7. _There exist constants_ ε_0, $B > 0$ _independent of_ ε _and_ t _such that_

$$\int_{|s|\geq 2|t|} \|K_\rho(s - t;\varepsilon) - K_\rho(s;\varepsilon)\| \, ds \leq B \tag{7.19}$$

<u>for</u> $-\infty < t < \infty$, $0 \leq \varepsilon \leq \varepsilon_0$.

<u>Proof</u>: Write

$$\mathfrak{S}''(s;\varepsilon)u = \mathfrak{T}_0(s;\varepsilon)u + \mathfrak{T}_1(s;\varepsilon)u \tag{7.20}$$

for $u \in D(A)$, where $\mathfrak{T}_0$ is the first term on the right hand side of (4.27) and $\mathfrak{T}_1$ is the sum of the rest. Using Theorem 4.4 (or, rather, Theorem 4.7) we deduce

$$\|\mathfrak{T}_1(s;\varepsilon)\| \leq C(\omega^2 + \omega^2 s^{-1} + s^{-2})e^{\omega^2 s} \leq C' s^{-2} e^{\rho s} \quad (s > 0) . \tag{7.21}$$

Let $t > 0$, $u \in D(A)$. We have

$$K_\rho(s;\varepsilon)u - K_\rho(s - t;\varepsilon)u = \int_{s-t}^{s} K_\rho'(\sigma;\varepsilon)u \, d\sigma$$

$$= \int_{s-t}^{s} (e^{-\rho\sigma}\mathfrak{T}_1(\sigma;\varepsilon) - \rho e^{-\rho\sigma}\mathfrak{S}'(\sigma;\varepsilon))u \, d\sigma + \int_{s-t}^{s} e^{-\rho\sigma}\mathfrak{T}_0(\sigma;\varepsilon)u \, d\sigma. \tag{7.22}$$

By Theorem 4.1 we have

$$\|\mathfrak{S}'(s;\varepsilon)\| \leq C(\omega^2 + s^{-1})e^{\omega^2 s} \leq C' s^{-1} e^{\omega^2 s} \leq C'' s^{-2} e^{\rho s} \quad (s > 0) . \tag{7.23}$$

Putting together (7.21) and (7.23) we can estimate the integrand of the first integral in (7.22) by $C\sigma^{-2}$, thus the integral itself is bounded by a constant times

$$\frac{1}{s - t} - \frac{1}{s} . \tag{7.24}$$

The second integral in (7.22), after integration by parts, becomes

$$e^{-\rho s} \frac{e^{-s/2\varepsilon^2}}{\varepsilon^2} C(s/\varepsilon)u - e^{-\rho(s-t)} \frac{e^{-(s-t)/2\varepsilon^2}}{\varepsilon^2} C((s - t)/\varepsilon)u$$

$$- \int_{s-t}^{s} \left(e^{-\rho\sigma} \frac{e^{-\sigma/2\varepsilon^2}}{\varepsilon^2} \right)' C(\sigma/\varepsilon)u \, d\sigma . \tag{7.25}$$

A look at the integrand in (7.25) makes plain that it can be estimated by a constant times

$$\frac{1}{\sigma} \|u\| \left(\frac{\sigma}{\varepsilon^2}\right) \exp\left\{ - \frac{\sigma}{\varepsilon^2}\left(\frac{1}{2} - \omega\varepsilon + \rho\varepsilon^2\right)\right\}$$

$$+ \frac{1}{\sigma^2}\left(\frac{\sigma}{\varepsilon^2}\right)^2 \exp\left\{ - \frac{\sigma}{\varepsilon^2}\left(\frac{1}{2} - \omega\varepsilon + \rho\varepsilon^2\right)\right\} \leq C\sigma^{-2} \quad (\sigma > 0), \tag{7.26}$$

thus the integral contributes another serving of (7.24). Putting together all estimations and taking advantage of the fact that $D(A)$ is dense in E we deduce that

$$\|K(s-t)-K(s)\| \leq C\left(\frac{1}{s-t}-\frac{1}{s}\right)+\frac{C}{\varepsilon^2}e^{-\rho s}e^{-s/2\varepsilon^2}e^{\omega s/\varepsilon}$$

$$+\frac{C}{\varepsilon^2}e^{-\rho(s-t)}e^{-(s-t)/2\varepsilon^2}e^{\omega(s-t)/\varepsilon}, \tag{7.27}$$

the last two summands originating from estimation of the boundary terms in (7.25). On this basis, we proceed to estimate the integral

$$\int_{2t}^{\infty}\|K(s-t)-K(s)\|\,ds . \tag{7.28}$$

The integral of the first term in (7.27) is computed as in Lemma 6.3. The integral of the second term in (7.27) is

$$\frac{C}{\varepsilon^2}\left(\rho+\frac{1}{2\varepsilon^2}-\frac{\omega}{\varepsilon}\right)^{-1}\exp\left\{-2t\left(\rho+\frac{1}{2\varepsilon^2}+\frac{\omega}{\varepsilon}\right)\right\}$$

$$=C\left(\frac{1}{2}-\omega\varepsilon+\rho\varepsilon^2\right)^{-1}\exp\left\{-\frac{2t}{\varepsilon^2}\left(\frac{1}{2}-\omega\varepsilon+\rho\varepsilon^2\right)\right\} . \tag{7.29}$$

To compute the integral of the last term we make the change of variables $s-t=\sigma$; the domain of integration is then $s \geq t$ and we obtain (7.29) again. Estimation of the part of (7.28) with domain $-\infty < s \leq -2t$ is of course unnecessary since $K_\rho(t;\varepsilon)=0$ for $t<0$. The bounding of (7.28) for $t<0$ runs along the same lines and is omitted.

LEMMA 7.8. Let $\tilde{K}_\rho(\hat{\sigma};\varepsilon)$ be the Fourier transform of $K_\rho(t;\varepsilon)$. Then there exists B independent of t and ε such that

$$\|\tilde{K}_\rho(\sigma;\varepsilon)\| \leq B \quad (-\infty<\sigma<\infty,\ 0<\varepsilon\leq\varepsilon_0) . \tag{7.30}$$

Proof: We use equality (5.10):

$$(2\pi)^{-1/2}\tilde{K}_\rho(\sigma;\varepsilon)=\int_{-\infty}^{\infty}e^{i\sigma t}K_\rho(t;\varepsilon)dt=\int_0^{\infty}e^{-(\rho-i\sigma)t}\mathfrak{S}'(t;\varepsilon)dt$$

$$=(\rho-i\sigma)R(\varepsilon^2(\rho-i\sigma)^2+(\rho-i\sigma);A) . \tag{7.31}$$

We make then use of the first inequality in the sequence (II.2.22) characterizing generators of cosine functions; setting $\lambda=\rho-i\sigma$,

$$(2\pi)^{-1/2}\|\tilde{K}_\rho(\sigma;\varepsilon)\| \le r_\rho(\sigma;\varepsilon) = \frac{c_0|\lambda|}{(\mathrm{Re}(\varepsilon^2\lambda^2+\lambda)^{1/2}-\omega)|\varepsilon^2\lambda^2+\lambda|^{1/2}}$$

$$= \frac{c_0|\rho - i\sigma|}{(\mathrm{Re}(\varepsilon^2(\rho-i\sigma)^2+(\rho-i\sigma))^{1/2}-\omega)|\varepsilon^2(\rho-i\sigma)^2+(\rho-i\sigma)|^{1/2}} \tag{7.32}$$

Multiplying numerator and denominator by $|\lambda|^{-1}$, setting $\mu = \varepsilon^2\lambda$ and noting that $\lambda/|\lambda| = \mu/|\mu|$ we see that it is enough to show that

$$\left\{\mathrm{Re}((\mu+1)^{1/2}(\mu/|\mu|)^{1/2}) - \omega|\lambda|^{-1/2}\right\}|\mu+1|^{1/2} \tag{7.33}$$

is bounded away from zero in the strip $0 < \mathrm{Re}\,\mu \le \varepsilon_0^2\rho$ with ε_0 sufficiently small, where λ is the unique multiple of μ on the line $|\lambda| = \rho$. We check easily that (7.33) never vanishes, thus we only have to show that it is bounded away from zero for $|\mu| \to \infty$. Note that, for $|\mu| = r$ $\mathrm{Re}((\mu+1)^{1/2}(\mu/|\mu|)^{1/2})$ attains its minimum at $\mu = \pm ir$, thus

$$\mathrm{Re}((\mu+1)^{1/2}(\mu/|\mu|)^{1/2}) \ge \frac{1}{2}|\mu|^{-1} + O(|\mu|^{-2}) \tag{7.34}$$

On the other hand, $|\lambda| > \varepsilon_0^{-2}|\mu|$ so that $|\lambda|^{-1/2}|\mu+1|^{1/2} \le$ $\le \varepsilon_0|\mu|^{-1/2}|\mu+1|^{1/2}$, thus our claim holds for $\varepsilon_0 < 1, 1/2\omega$.

Proof of Theorem 7.6. That the kernel $K_\rho(t;\varepsilon)$ satisfies (a) in Theorem 6.1 is obvious from its definition. The estimate (6.4) with B independent of ε was shown in Lemma 7.7, while (6.3), with B likewise independent of ε was the subject of Lemma 7.8. Accordingly, the operator

$$f(\hat{t}) \to \int_0^t K_\rho(t-s;\varepsilon)f(s)\,ds \tag{7.35}$$

is bounded in $L^p(0,T;H)$. Using (6.22) we deduce that (7.6) defines as well a bounded operator in $L^p(0,T;H)$. This yields the first estimate (7.15).

We prove finally (d). The statement on convergence of $u(\hat{t};\varepsilon)$ is a consequence of Theorem 6.1. To show convergence of $u'(\hat{t};\varepsilon)$ in L^p we take f, say, in $\mathfrak{H}_1(0,T;E)$ and write the differentiated version of (7.9) as follows:

$$u'(t;\varepsilon) - u'(t) = \int_0^t \mathfrak{S}'(t-s;\varepsilon)(f(s;\varepsilon)-f(s))\,ds$$
$$+ \int_0^t (\mathfrak{S}'(t-s;\varepsilon) - S'(t-s))(f(s)-f(t))\,ds$$
$$+ (\mathfrak{S}(t;\varepsilon) - S(t))f(t). \tag{7.36}$$

Apply (7.11) to the first integral, Theorem 7.4 to the second and Theorem 3.6 to the last term: the conclusion is $u'(\hat{t};\varepsilon) \to u'(\hat{t})$ in $L^p(0,T;H)$. To show convergence for arbitrary $f(\hat{t}) \in L^p(0,T;H)$, let $g(\hat{t};\varepsilon), g(\hat{t}) \in \mathcal{H}_1(0,T;E)$ and $v(t;\varepsilon)$, $v(t)$ the respective solutions of (7.1), (7.2). We have

$$\|u(\hat{t};\varepsilon) - u(\hat{t})\|_p \leq \|u(\hat{t};\varepsilon) - v(\hat{t};\varepsilon)\|_p + $$
$$+ \|v(\hat{t};\varepsilon) - v(\hat{t})\|_p + \|v(\hat{t}) - u(\hat{t})\|_p . \tag{7.37}$$

The first and last terms on the right hand side of (7.33) can be made small using the fact that $\mathcal{H}_1(0,T;E)$ is dense in $L^p(0,T;H)$ and both inequalities (7.17); for the second term we use (7.32) and following comments. This ends the proof of Theorem 7.6.

§VI.8. Correctors at the initial layer. Asymptotic series.

We work in this section with the homogeneous initial value problem

$$\begin{aligned} &\varepsilon^2 u''(t;\varepsilon) + u'(t;\varepsilon) = Au(t;\varepsilon) \qquad (t \geq 0), \\ &u_0(0;\varepsilon) = u_0(\varepsilon), \quad u'(0;\varepsilon) = u_1(\varepsilon) , \end{aligned} \tag{8.1}$$

and the equation

$$u'(t) = Au(t) \quad (t \geq 0), \tag{8.2}$$

with initial condition to be fixed below. As pointed out before (see Remark 5.11), in the general conditions of Theorem 3.6 (where there may be crossover of initial conditions), uniform convergence of $u(t;\varepsilon)$ to $u(t)$ near $t = 0$ cannot be expected since in general $u_0(\varepsilon) \not\to u_0$. However, uniform convergence can be attained through addition of <u>correctors</u> (solutions of a different approximating equation) at the boundary. This method can be applied equally well to the case where the initial conditions $u_0(\varepsilon)$, $u_1(\varepsilon)$ have asymptotic expansions in powers of ε, as made clear below.

We assume that $u_0(\varepsilon)$ and $u_1(\varepsilon)$ have asymptotic developments of the form

$$u_0(\varepsilon) = u_0 + \varepsilon u_1 + \varepsilon^2 u_2 + \varepsilon^2 u_3 + \cdots + \varepsilon^N u_N + O(\varepsilon^{N+1}),$$
$$u_1(\varepsilon) = \varepsilon^{-2} v_0 + \varepsilon^{-1} v_1 + v_2 + \varepsilon v_3 + \cdots + \varepsilon^{N-2} v_N + O(\varepsilon^{N-1})$$
$$(\varepsilon \to 0). \tag{8.3}$$

The objective is to show that $u(t;\varepsilon)$ possesses a similar asymptotic development, uniformly on compacts of $t \geq 0$; to produce convergence near $t = 0$ we shall need to introduce correction terms at each step. We examine first the cases $N = 0,1$ where the method is unencumbered by details. For $N = 0$ the central idea is to approximate $u(\hat{t})$ near $t = 0$ not by $u(t;\varepsilon)$ but by $u(t;\varepsilon) - v_0(t;\varepsilon)$, where $v_0(t;\varepsilon)$ is the solution of

$$\varepsilon^2 v_0''(t;\varepsilon) + v_0'(t;\varepsilon) = 0, \quad v_0'(0;\varepsilon) = \varepsilon^{-2} v_0 , \qquad v_0(t;\varepsilon) \to 0 \text{ as } \varepsilon \to \infty . \tag{8.4}$$

We refer the reader to KEVORKIAN-COLE [1981:1] for a thorough discussion of the choice of $v_0(\hat{t})$ in the one dimensional case only pointing out that the initial condition is to eliminate the contribution of $u_1(\varepsilon)$ to u_0 (see Remark 5.11). In fact, it follows from (8.4) that

$$u'(0;\varepsilon) + v'(0;\varepsilon) = 0 .$$

On the other hand, since

$$v_0(t;\varepsilon) = -e^{-t/\varepsilon^2} v_0 , \tag{8.5}$$

$v_0(t;\varepsilon)$ will not disturb convergence in the region $t \geq t(\varepsilon)$ outside of the boundary layer (here $t(\varepsilon)$ satisfies (3.20)). The price to pay, of course, is that $u(t;\varepsilon) - v_0(t;\varepsilon)$ will not be a solution of the homogeneous equation (8.1), thus all the results below will use the theory of the nonhomogeneous equation (only to the extent of Theorem 7.2). Throughout this section, $u(t;\varepsilon)$ denotes the solution of the initial value problem (8.1) with $u_0(\varepsilon)$ and $u_1(\varepsilon)$ having asymptotic developments of the form (8.3). The term $0(\varepsilon^k)$ in the asymptotic expansions (8.3) is an element of E whose norm is $\leq C\varepsilon^k$ for some constant C as $\varepsilon \to 0$. Solutions of the equation (8.2), with initial conditions specified in the following results will be usually written $u_0(t)$, $u_1(t)$; the functions $u_2(t)$, $u_3(t), \ldots$ etc. are solutions of a different equation (see (8.21)).

THEOREM 8.1. <u>Assume that</u> (8.3) <u>holds for</u> $N = 0$, <u>that is</u>

$$u_0(\varepsilon) = u_0 + 0(\varepsilon), \; u_1(\varepsilon) = \varepsilon^{-2} v_0 + 0(\varepsilon^{-1}) \quad (\varepsilon \to 0) , \tag{8.6}$$

<u>and that</u> $u_0, v_0 \in D(A)$. <u>Then</u>

$$u(t;\varepsilon) = u_0(t) + v_0(t;\varepsilon) + O(\varepsilon) \tag{8.7}$$

uniformly on compacts of $t \geq 0$, where $u_0(\hat{t})$ is the solution of (8.2) with

$$u_0(0) = u_0 + v_0. \tag{8.8}$$

If $\omega = 0$, (8.7) holds uniformly in $t \geq 0$.

Proof. The function

$$w(\hat{t};\varepsilon) = u(\hat{t};\varepsilon) - v_0(\hat{t};\varepsilon) \tag{8.9}$$

is a solution of the initial value problem

$$\begin{aligned} &\varepsilon^2 w''(t;\varepsilon) + w'(t;\varepsilon) = Aw(t;\varepsilon) - e^{-t/\varepsilon^2} Av_0, \\ &w(0;\varepsilon) = u_0(\varepsilon) + v_0, \quad w'(0;\varepsilon) = u_1(\varepsilon) - \varepsilon^{-2} v_0. \end{aligned} \tag{8.10}$$

Write $w = w^1 + w^2$ where w^1 is the solution of the homogeneous equation with the assigned initial conditions and w^2 is the solution of the inhomogeneous equation with zero initial conditions and

$$f(t;\varepsilon) = -e^{-t/\varepsilon^2} Av_0. \tag{8.11}$$

We apply to w_1 Theorem 5.5 (with the simplified estimate (5.30)), while w^2 is handled by means of Theorem 7.2 (specifically, the first inequality (7.4)). The final estimate is

$$\begin{aligned} \|u(t;\varepsilon) - v_0(t;\varepsilon) - u_0(t)\| &\leq \|w^1(t;\varepsilon) - u_0(t)\| + \|w^2(t;\varepsilon)\| \\ &\leq C_0\varepsilon^2(1 + \omega^2 t)e^{\omega^2 t}\|A(u_0 + v_0)\| \\ &\quad + C_0 e^{\omega^2 t}\|u_0(\varepsilon) - u_0\| \\ &\quad + C_0 e^{\omega^2 t}\|\varepsilon^2 u_1(\varepsilon) - v_0\| \\ &\quad + C_0 \varepsilon^2 e^{\omega^2 T}\|Av_0\| , \end{aligned} \tag{8.12}$$

with the obvious modification in the last term if $\omega = 0$. This ends the proof.

For $N = 1$ an additional corrector must be used, namely

$$v_1(t;\varepsilon) = -e^{-t/\varepsilon^2} v_1.$$

THEOREM 8.2. Assume that (8.3) holds for $N = 1$, that is

$$u_0(\varepsilon) = u_0 + \varepsilon u_1 + o(\varepsilon^2),$$

$$u_1(\varepsilon) = \varepsilon^{-2} v_0 + \varepsilon^{-1} v_1 + o(1) \qquad (8.13)$$

and that $u_0, u_1, v_0, v_1 \in D(A^2)$. *Then*

$$u(t;\varepsilon) = u_0(t) + v_0(t;\varepsilon) + \varepsilon(u_1(t) + v_1(t;\varepsilon)) + o(\varepsilon^2) \qquad (8.14)$$

uniformly on compacts of $t \geq 0$, *where* $u_0(\hat{t})$ (*resp.* $u_1(\hat{t})$) *is the solution of* (8.2) *with*

$$u_0(0) = u_0 + v_0 \quad (\textit{resp.}\ u_1(0) = u_1 + u_1). \qquad (8.15)$$

If $\omega = 0$, (8.14) *holds uniformly in* $t \geq 0$.

Proof. We consider this time the function

$$w(\hat{t};\varepsilon) = u(\hat{t};\varepsilon) - v_0(\hat{t};\varepsilon) - \varepsilon v_1(\hat{t};\varepsilon) \qquad (8.16)$$

that solves the initial value problem

$$\varepsilon^2 w''(t;\varepsilon) + w'(t;\varepsilon) = Aw(t;\varepsilon) - e^{-t/\varepsilon^2} A v_0 - \varepsilon e^{-t/\varepsilon^2} A v_1, \qquad (8.17)$$

$$w(0,\varepsilon) = u_0(\varepsilon) + v_0 + \varepsilon v_1, \ w'(0,\varepsilon) = u_1(\varepsilon) - \varepsilon^{-2} v_0 - \varepsilon^{-1} v_1. \qquad (8.18)$$

As in Theorem 8.1, we write w as the sum of a solution w_1 of the homogeneous equation taking the assigned initial conditions and a solution w_2 of the nonhomogeneous equation with zero initial conditions. We apply again the second inequality (7.4) to w_2 and (5.30) to w_1, obtaining

$$\begin{aligned} &\|u(t;\varepsilon) + v_0(t;\varepsilon) + \varepsilon v_1(t;\varepsilon) - u_0(t) - \varepsilon u_1(t)\| \\ &\quad \leq C_0 \varepsilon^2 (1 + \omega^2 t) e^{\omega^2 t} \|A(u_0 + v_0) + \varepsilon A(u_1 + v_1)\| \\ &\qquad + C_0 e^{\omega^2 t} \|u_0(\varepsilon) - u_0 - \varepsilon u_1\| + C_0 e^{\omega^2 t} \|\varepsilon^2 u_1(\varepsilon) - v_0 - \varepsilon v_1\| \\ &\qquad + C_0 \varepsilon^2 e^{\omega^2 T} (\|A v_0\| + \varepsilon \|A v_1\|) . \end{aligned} \qquad (8.19)$$

Obviously, a different tack must be adopted for $N \geq 2$, since the first term on the right hand sides of (8.12) and (8.19) cannot be squeezed smaller than $O(\varepsilon^2)$. We proceed at first on a purely formal level. The approximating function will be of the form

$$u_N(t;\varepsilon) = u_0(t) + \varepsilon u_1(t) + \cdots + \varepsilon^N u_N(t) , \qquad (8.20)$$

where $u_0(\hat{t})$, $u_1(\hat{t})$ are defined as before and the $u_n(\hat{t})$, $n \geq 2$, satisfy the differential equations

$$u_n'(t) = Au_n(t) - u_{n-2}''(t) \qquad (t \geq 0) . \tag{8.21}$$

Noting that the correctors v_0, v_1 used in the cases $N = 0,1$ are of the form $v_0(t;\varepsilon) = v_0(t/\varepsilon^2)$, $v_1(t;\varepsilon) = v_1(t/\varepsilon^2)$, we shall use a combination of correctors of the form

$$\mathfrak{v}_N(t;\varepsilon) = v_0(t/\varepsilon^2) + \varepsilon v_1(t/\varepsilon^2) + \cdots + \varepsilon^N v_N(t/\varepsilon^2). \tag{8.22}$$

The v_n, $n \geq 2$ will satisfy the differential equations

$$v_n''(t) + v_n'(t) = Av_{n-2}(t) \qquad (t \geq 0) , \tag{8.23}$$

and the decay condition

$$v_n(t) \to 0 \quad \text{as} \quad t \to \infty . \tag{8.24}$$

Note that the equation satisfied by $u_n(\hat{t})$, $n = 1,2$ is

$$u_n'(t) = Au_n(t), \tag{8.25}$$

and v_n, $n = 1,2$ satisfies

$$v_n''(t) + v_n'(t) = 0 . \tag{8.26}$$

Consider now the function

$$\mathfrak{w}_N(t;\varepsilon) = u_N(t;\varepsilon) + \mathfrak{v}_N(t;\varepsilon) . \tag{8.27}$$

We have

$$\begin{aligned}
\varepsilon^2\mathfrak{w}_N''(t) + \mathfrak{w}_N'(t) &= \varepsilon^2 u_N''(t) + u_N'(t) + \varepsilon^2\mathfrak{v}_N''(t) + \mathfrak{v}_N'(t) \\
&= \varepsilon^2 \sum_{n=0}^{N} \varepsilon^n u_n''(t) + \sum_{n=0}^{N} \varepsilon^n u_n'(t) \\
&\quad + \varepsilon^{-2} \sum_{n=0}^{N} \varepsilon^n v_n''(t/\varepsilon^2) + \varepsilon^{-2} \sum_{n=0}^{N} \varepsilon^n v_n'(t/\varepsilon^2) \\
&= (\varepsilon^2 u_0''(t) + u_0'(t)) + \varepsilon(\varepsilon^2 u_1''(t) + u_1'(t)) \\
&\quad + \sum_{n=2}^{N} \varepsilon^n(\varepsilon^2 u_n''(t) + u_n'(t)) \\
&\quad + \varepsilon^{-2}(v_0''(t/\varepsilon^2) + v_0'(t/\varepsilon^2)) + \varepsilon^{-2}(\varepsilon(v_1''(t/\varepsilon^2) + v_1'(t/\varepsilon^2))) \\
&\quad + \varepsilon^{-2} \sum_{n=2}^{N} \varepsilon^n(v_n''(t/\varepsilon^2) + v_n'(t/\varepsilon^2))
\end{aligned}$$

$$= (\varepsilon^2 u_0''(t) + Au_0(t)) + \varepsilon(\varepsilon^2 u_1''(t) + Au_1(t))$$

$$+ \sum_{n=2}^{N} \varepsilon^n(\varepsilon^2 u_n''(t) + Au_n(t)) - \varepsilon^2 \sum_{n=0}^{N-2} \varepsilon^n u_n''(t)$$

$$+ \sum_{n=0}^{N-2} \varepsilon^n Av_n(t/\varepsilon^2)$$

$$= \sum_{n=0}^{N} \varepsilon^n A(u_n(t) + v_n(t/\varepsilon^2))$$

$$+ \varepsilon^{N+1} u_{N-1}''(t) + \varepsilon^{N+2} u_N''(t)$$

$$- \varepsilon^{N-1} Av_{N-1}(t/\varepsilon^2) - \varepsilon^N Av_N(t/\varepsilon^2)$$

$$= A\mathfrak{w}_N(t;\varepsilon) + \varepsilon^{N+1} u_{N-1}''(t) + \varepsilon^{N+2} u_N''(t)$$

$$- \varepsilon^{N-1} Av_{N-1}(t/\varepsilon^2) - \varepsilon^N Av_N(t/\varepsilon^2). \tag{8.28}$$

The initial conditions on u_n, $n = 0,1$ are those in Theorem 8.1:

$$u_0(0) = u_0 + v_0,\ u_1(0) = u_1 + v_1 . \tag{8.29}$$

On the other hand, the initial conditions on v_n, $n = 0, 1$ must be those that insure that $v_0(t;\varepsilon) = v_0(t/\varepsilon^2)$, $v_1(t;\varepsilon) = v_1(t/\varepsilon^2)$, where $v_0(t;\varepsilon)$, $v_1(t;\varepsilon)$ are the correctors used in Theorem 8.2. Accordingly,

$$v_0'(0) = v_0,\ v_1'(0) = v_1, \tag{8.30}$$

hence, taking (8.26) and (8.24) into account,

$$v_0(t) = -e^{-t}v_0,\ v_1(t) = -e^{-t}v_1 . \tag{8.31}$$

For $n \geq 2$, the initial conditions for $u_n(\hat{t})$ and $v_n(\hat{t})$ are, respectively

$$u_n(0) = u_n - v_n(0) \tag{8.32}$$

$$v_n'(0) = v_n - u_{n-2}'(0), \tag{8.33}$$

thus $u_n(\hat{t})$ must be constructed after $v_n(\hat{t})$. The initial conditions for $\mathfrak{w}(t;\varepsilon)$ are obtained from (8.29) (8.30) (8.32) and (8.33):

$$\mathfrak{w}_N(0;\varepsilon) = \sum_{n=0}^{N} \varepsilon^n u_n(0) + \sum_{n=0}^{N} \varepsilon^n v_n(0) =$$

$$= u_0 + v_0 + \varepsilon(u_1 + v_1) + \sum_{n=2}^{N} \varepsilon^n (u_n - v_n(0))$$

$$- v_0 - \varepsilon v_1 + \sum_{n=2}^{N} \varepsilon^n v_n(0) = \sum_{n=0}^{N} \varepsilon^n u_n \tag{8.34}$$

$$\mathfrak{w}_N'(0;\varepsilon) = \sum_{n=0}^{N} \varepsilon^n u_n'(0) + \varepsilon^{-2} \sum_{n=0}^{N} \varepsilon^n v_n'(0)$$

$$= \sum_{n=0}^{N} \varepsilon^n u_n'(0) + \sum_{n=0}^{N} \varepsilon^{n-2} v_n$$

$$- \sum_{n=0}^{N-2} \varepsilon^n u_n'(0) = \sum_{n=0}^{N} \varepsilon^{n-2} v_n + \varepsilon^{N-1} u_{N-1}'(0)$$

$$+ \varepsilon^N u_N'(0) \tag{8.35}$$

Hence, in view of (8.3),

$$\|u(0;\varepsilon) - \mathfrak{w}_N(0;\varepsilon)\| = 0(\varepsilon^{N+1}) \tag{8.36}$$

and

$$\|u'(0;\varepsilon) - \mathfrak{w}_N'(0;\varepsilon)\| = 0(\varepsilon^{N-1}) . \tag{8.37}$$

We face now the problem of making all these computations valid. Roughly speaking, this amounts to:

(a) showing that every derivative written (as in (8.17), (8.21), (8.24), etc.) actually exists.

(b) showing that every time we write Au (as in (8.17), (8.21), etc.), u actually belongs to the domain of A.

This will be done by requiring "smoothness" conditions of varying degree on the coefficients u_n, v_n in (8.3). We begin with

$$u_0(t) = S(t)(u_0 + v_0), \; u_1(t) = S(t)(u_1 + v_1) \tag{8.38}$$

while $v_0(t)$, $v_1(t)$ are made explicit in (8.31). To construct $v_2(t)$, $v_3(t)$ we solve (8.23) with the initial condition (8.33) at $t = 0$ and the decay condition (8.24) as $t \to \infty$:

$$v_2''(t) + v_2'(t) = -e^{-t}Av_0 \quad (t \geq 0),$$
$$v_2'(0) = v_2 - u_0'(0),\ v_2(t) \to 0 \quad \text{as} \quad t \to \infty, \tag{8.39}$$

$$v_3''(t) + v_3'(t) = -e^{-t}Av_1 \quad (t \geq 0),$$
$$v_3'(0) = v_3 - u_1'(0),\ v_3(t) \to 0 \quad \text{as} \quad t \to \infty. \tag{8.40}$$

Solving explicitly these equations,

$$v_2(t) = te^{-t}Av_0 - e^{-t}(v_2 - Au_0 - 2Av_0), \tag{8.41}$$

$$v_3(t) = te^{-t}Av_1 - e^{-t}(v_3 - Au_1 - 2Av_0), \tag{8.42}$$

where we have used the fact that $u_0'(0) = A(u_0 + v_0)$, $u_1'(0) = A(u_1 + v_1)$. We compute next $u_2(\hat{t})$, $u_3(\hat{t})$ using the equation (8.21) and the initial condition (8.32):

$$u_2''(t) = Au_2(t) - S(t)A^2(u_0 + v_0) \quad (t \geq 0),$$
$$u_2(0) = u_2 + v_0, \tag{8.43}$$

$$u_3''(t) = Au_3(t) - S(t)A^2(u_1 + v_1) \quad (t \geq 0),$$
$$u_3(0) = u_3 + v_1, \tag{8.44}$$

where we have used the facts that $u_0''(t) = S(t)A^2(u_0 + v_0)$, $u_1''(t) = S(t)A^2(u_1 + v_1)$, $v_0'(0) = v_0$, $v_1'(0) = v_1$ (see (8.29) and (8.37)). Hence

$$u_2(t) = S(t)(u_2 + v_0) - \int_0^t S(t-s)S(s)A^2(u_0 + v_0)\,ds$$
$$= S(t)(u_2 + v_0) - tS(t)A^2(u_0 + v_0), \tag{8.45}$$

$$u_3(t) = S(t)(u_3 + v_1) - tS(t)A^2(u_1 + v_1). \tag{8.46}$$

With $u_2(\hat{t})$, $u_3(\hat{t})$, $v_2(\hat{t})$, $v_3(\hat{t})$ already manufactured, we can easily see that $v_4(\hat{t})$, $v_5(\hat{t})$ will have the form

$$v_4(t) = e^{-t}P_4(t),\ v_5(t) = e^{-t}P_5(t), \tag{8.47}$$

where $P_4(\hat{t})$ is a polynomial of degree 2 whose coefficients are linear combinations of $A^j u_0$, $A^j v_0$ $(j \leq 3)$, Au_2 and Av_2 and $P_5(t)$ is the same polynomial with u_0, v_0, u_2, v_2 replaced by u_1, v_1, u_3, v_3 respectively. On the other hand, we have

$$u_4(t) = S(t)(u_4 - P_4(0)) + tS(t)A^2(u_2 + v_0) - 2tS(t)A^3(u_0 + v_0) - \frac{t^2}{2} S(t)A^4(u_0 + v_0), \tag{8.48}$$

$$u_5(t) = S(t)(u_5 - P_5(0)) + tS(t)A^2(u_3 + v_1) - 2tS(t)A^3(u_1 + v_1) - \frac{t^2}{2} S(t)A^4(u_1 + v_1), \tag{8.49}$$

thus $u_4(\hat{t})$ (resp. $u_5(\hat{t})$) can be constructed if u_0, $v_0 \in D(A^4)$, $u_2 \in D(A^2)$ (resp. $u_1, v_1 \in D(A^4)$, $u_3 \in D(A^2)$. However, if we wish (8.47) to be a genuine solutions of (8.23) we actually need that u_0, $v_0 \in D(A^5)$ and $u_2 \in D(A^3)$, u_4, $P_4(0) \in D(A)$; in view of our previous comments about P_4, it is sufficient for this that u_0, $v_0 \in D(A^4)$, $u_2 \in D(A^3)$ $v_2 \in D(A^2)$ and $u_4 \in D(A)$. Likewise, if we wish (8.48) to be a genuine solution of (8.21) we must ask that u_1, $v_1 \in D(A^5)$, $u_3 \in D(A^3)$ $v_3 \in D(A^2)$, $u_5 \in D(A)$. It will be of interest later to ascertain that $u_4(t)$, $u_5(t)$ are twice continuously differentiable. This will be the case if u_0, v_0, u_1, $v_1 \in D(A^6)$, u_2, $u_3 \in D(A^4)$, v_2, $v_3 \in D(A^3)$ and $u_4, u_5 \in D(A)$.

From these observations we surmise the following rules, valid for arbitrary $m \geq 1$. In the first place, we have

$$v_{2m}(t) = e^{-t}P_{2m}(t), v_{2m+1}(t) = e^{-t}P_{2m+1}(t) , \tag{8.50}$$

where $P_{2m}(\hat{t})$ (resp. $P_{2m+1}(\hat{t})$) is a polynomial whose coefficients are linear combinations of $A^j u_0, A^j v_0$ $(j \leq 2m - 1)$, $A^j u_2, A^j v_2$ $(j \leq 2m - 3), \ldots$ $A^j u_{2m-4}, A^j u_{2m-4}$ $(j \leq 3)$, Au_{2m-2}, Av_{2m-2} (resp. $A^j u_1$, $A^j v_1$ $(j \leq 2m - 1)$, $A^j u_3$, $A^j v_3 (j \leq 2m - 3), \ldots$ $A^j u_{2m-3}$, $A^j v_{2m-3}$ $(j \leq 3)$, Au_{2m-1}, Av_{2m-1}. Accordingly:

(a) $v_{2m}(\hat{t})$ <u>and</u> $v_{2m+1}(\hat{t})$ <u>can be constructed if</u>

$$u_0, v_0, u_1, v_1 \in D(A^{2m-1}),\ u_2, v_2, u_3, v_3 \in D(A^{2m-3}), \ldots$$
$$u_{2m-4}, v_{2m-4}, u_{2m-3}, v_{2m-3} \in D(A^3),$$
$$u_{2m-2}, v_{2m-2}, u_{2m-1}, v_{2m-1} \in D(A); \tag{8.51}$$

$v_{2m}(\hat{t})$ *and* $v_{2m+1}(\hat{t})$ *are always genuine solutions of* (8.23), (8.33). *Moreover,* $v_{2m}(t), v_{2m+1}(t) \in D(A)$ *with* $Av_{2m}(t)$, Av_{2m+1} *continuous if*

$$u_0, v_0, u_1, v_1 \in D(A^{2m}), u_2, v_2, u_3, v_3 \in D(A^{2m-2}), \ldots$$
$$u_{2m-4}, v_{2m-4}, u_{2m-3}, v_{2m-3} \in D(A^4),$$
$$u_{2m-2}, v_{2m-2}, u_{2m-1}, v_{2m-1} \in D(A^2) . \tag{8.52}$$

The statement about genuine solutions is obvious since (8.23) is essentially a scalar equation.

(b) $u_{2m}(\hat{t})$ *and* $u_{2m+1}(\hat{t})$ *can be constructed if* (8.51) *holds*. *These functions are genuine solutions of* (8.21) *if*

$$u_0, v_0, u_1, v_1 \in D(A^{2m+1}), u_2, v_2, u_3, v_3 \in D(A^{2m-1}), \ldots$$
$$u_{2m-2}, v_{2m-2}, u_{2m-1}, v_{2m-1} \in D(A^3),$$
$$u_{2m}, v_{2m}, u_{2m+1}, v_{2m+1} \in D(A) ; \tag{8.53}$$

$u_{2m}(t), u_{2m+1}(t)$ *are twice continuously differentiable if*

$$u_0, v_0, u_1, v_1 \in D(A^{2m+2}),\ u_2, v_2, u_3, v_3 \in D(A^{2m}),$$
$$u_{2m-2}, v_{2m-2}, u_{2m-1}, v_{2m-1} \in D(A^4),$$
$$u_{2m}, v_{2m}, u_{2m+1}, v_{2m+1} \in D(A^2). \tag{8.54}$$

Obviously, conditions (8.52), (8.53) and (8.54) are excessive for the various purposes at hand; we have "equalized indices" among the u_n and the v_n to simplify.

THEOREM 8.3. *Assume that* (8.3) *holds in the norm of* E *for* N *odd* ≥ 3 *and that* $u_N, u_{N-1}, v_N, v_{N-1} \in D(A^2)$, $u_{N-2}, u_{N-3}, v_{N-2}, v_{N-3} \in D(A^4), \ldots, u_1, u_0, v_1, v_0 \in D(A^{N+1})$. *Then*

$$u(t;\varepsilon) = \sum_{n=0}^{N} \varepsilon^n u_n(t) + \sum_{n=0}^{N} \varepsilon^n v_n(t/\varepsilon^2) + o(\varepsilon^{N+1}) \tag{8.55}$$

uniformly on compacts of $t \geq 0$ (*uniformly in* $t \geq 0$ if $\omega = 0$). *If*

$N \geq 2$ <u>is even the same result obtains under the assumption that</u> $u_N, v_N \in D(A^2)$, $u_{N-1}, u_{N-2}, v_{N-1}, v_{N-2} \in D(A^4), \ldots, u_1, u_0, v_1, v_0 \in D(A^{N+2})$.

<u>Proof</u>. We consider first the case N odd ≥ 3; we set here $N = 2m+1$ and apply rule (a). Since conditions (8.51) are satisfied (with something to spare) we deduce that $v_0(t), \ldots, v_N(t) \in D(A)$ with $Av_0(t), \ldots, Av_N(t)$ continuous. Taking (8.50) into account we deduce that

$$\int_0^\infty \|Av_n(t/\varepsilon^2)\| \, dt \leq C_n \varepsilon^2 \qquad (n = 0,1,\ldots,N). \tag{8.56}$$

This will be used to estimate the last two terms on the right hand side of (8.28): for the first two terms we simply use the fact, assured by (b), that $u_{N-1}(\hat{t})$ and $u_N(t)$ are twice continuously differentiable. Using the first inequality (7.10) in (8.28) we obtain

$$\begin{aligned}
&\|u(t;\varepsilon) - \sum_{n=1}^{N} \varepsilon^n u_n(t) + \sum_{n=1}^{N} \varepsilon^n v_n(t/\varepsilon^2)\| \\
&\leq C_0 e^{\omega^2 t}\Big(\|u_0(\varepsilon) - \sum_{n=1}^{N} \varepsilon^n u_n\| + \|\varepsilon^2 u_1(\varepsilon) - \sum_{n=1}^{N} \varepsilon^n v_n - \varepsilon^{N+1} u'_{N-1}(0) - \varepsilon^{N+2} u'_N(0)\|\Big) \\
&\quad + C_0 e^{\omega^2 T}\Big(\varepsilon^{N+1} \int_0^T \|u''_{N-1}(t)\| \, dt + \varepsilon^{N+2} \int_0^T \|u''_N(t)\| \, dt\Big) \\
&\quad + C_0 e^{\omega^2 T}\Big(\varepsilon^{N-1} \int_0^T \|Av_{N-1}(t/\varepsilon^2)\| \, dt + \varepsilon^N \int_0^T \|Av_N(t/\varepsilon^2)\| \, dt\Big)
\end{aligned} \tag{8.57}$$

where, in view of (8.56), the contribution of the last two terms is $O(\varepsilon^{N+1})$. This ends the proof. The case N even ≥ 2 is handled much in the same way and we omit the details.

§VI.9 Elliptic differential equations.

We apply the theory in the last eight sections to the differential operator

$$Au = \sum_{j=1}^{m} \sum_{k=1}^{m} D^j(a_{jk}(x) D^j u) + \sum_{j=1}^{m} b_j(x) D^j u + c(x) u \tag{9.1}$$

in a bounded domain Ω of m-dimensional space with boundary Γ; here

$A(\beta)$ denotes the restriction of A obtained by means of the Dirichlet boundary condition

$$u(x) = 0 \qquad (x \in \Gamma), \tag{9.2}$$

or by means of the variational boundary condition

$$D^{\tilde{\nu}}u(x) = \gamma(x)u(x) \qquad (x \in \Gamma). \tag{9.3}$$

The construction of $A(\beta)$ was carried out in Chapter IV in considerable detail; in particular, it was shown in §IV.3 (for the Dirichlet boundary condition) and in §IV.6 (for the boundary condition (9.3)) that $A(\beta)$ generates a strongly continuous cosine function, thus all the results in this chapter apply automatically. Of special interest are those in §VI.5, for instance the "intermediate" estimations in Theorems 5.12 and 5.13. Combining Theorem 5.12 with Lemma 5.15 we deduce that if $u_0 \in D((b^2I - A(\beta))^{\alpha})$ (b large enough) then

$$\|u(t;\varepsilon) - u(t)\| = O(\varepsilon^{2\alpha}) \quad \text{as} \quad \varepsilon \to 0 \tag{9.4}$$

uniformly on compacts of $t \geq 0$. The most interesting case is $\alpha = 1/2$, where $D((b^2I - A(\beta))^{\alpha})$ can be identified. In fact, we shall show that

$$D((b^2I - A(\beta))^{1/2} = H_0^1(\Omega) \tag{9.5}$$

if β is the Dirichlet boundary condition and $b_1,\dots,b_m$ belong to $H^1(\Omega)$. To show (9.5) we note that it has already been proved that $D((b^2I - A_0(\beta))^{1/2}) = H_0^1(\Omega)$ (see (IV.2.4)) and recall Theorem IV.2.2, especially (IV.2.6)). Thus, we only have to show that

$$D((b^2I - A_0(\beta))^{1/2}) = D((b^2I - A(\beta))^{1/2}) \ . \tag{9.6}$$

We sketch the proof of (9.6). Let $\mathcal{C}_0(t) = \cosh t\, A_0(\beta)^{1/2}$ be the cosine function generated by $A_0(\beta)$. It follows from the perturbation used to construct $\mathfrak{S}(t)$ from $\mathfrak{S}_0(t)$ (or directly) that $\mathcal{C}(t)$, the cosine function generated by $A(\beta)^{1/2}$ can be expressed by means of the perturbation series

$$\mathcal{C}(t)u = \mathcal{C}_0(t)u + \overline{\mathcal{S}_0(t)P}_*\mathcal{C}_0(t)u + \overline{\mathcal{S}_0(t)P}_*\overline{\mathcal{S}_0(t)P}_*\mathcal{C}_0(t)u + \cdots , \tag{9.7}$$

where $\overline{\mathcal{S}_0(t)P}$ denotes the (only) bounded extension of $\mathcal{S}_0(t)P$ (with domain $H^1(\Omega)$) to all of $L^2(\Omega)$; that this extension exists follows from the fact that $\mathcal{S}_0(t)P$ is bounded (in the norm of $L^2(\Omega)$) in $H_0^1(\Omega)$, since

$$\mathcal{S}_0(t)P = (\sinh t\, A_0(\beta)^{1/2})A_0(\beta)^{-1/2}P,$$

and

$$\int_\Omega (A_0(\beta)^{-1/2}Pu)v\,dx = \int_\Omega Pu(A_0(\beta)^{-1/2}v)\,dx =$$

$$= -\sum_{j=1}^{m}\int_\Omega uD^j(b_j(x)A_0(\beta)^{-1/2}v)\,dx +$$

$$+\int_\Omega cuA_0(\beta)^{-1/2}v\,dx$$

Using (1.5) and the "reciprocal" series

$$C_0(t)u = C(t)u - \overline{\mathcal{S}(t)P}_*C(t)u + \overline{\mathcal{S}(t)P}_*\overline{\mathcal{S}(t)P}_*C(t)u + \cdots \tag{9.8}$$

we show that $C(t)u$ is continuously differentiable if and only if $C_0(t)$ is continuously differentiable, thus (9.6) follows from Theorem III.6.4.

However, in the present situation, estimates on rates of convergence like (1.4) can be obtained under weaker assumptions by more elementary methods. We sketch below this theory in a suitably "abstract" version.

Let $E = H$ be a Hilbert space and A_0 a self adjoint operator such that

$$(A_0u, u) \leq -\kappa\|u\|^2 \qquad (u \in D(A_0)) \tag{9.9}$$

with $\kappa > 0$. We consider the operator

$$A = A_0 + P, \tag{9.10}$$

where P is such that

$$PB^{-1} \tag{9.11}$$

is bounded, where $B = (-A_0)^{1/2}$ defined as in §IV.3. Using essentially the same methods in §IV.4 we show that A generates a strongly continuous cosine function, thus all results in this chapter apply, in particular those in §VI.5. We exploit these below.

Using the functional calculus for self adjoint operators we can define fractional powers $(-A_0)^\zeta$ of $-A_0$ where $\zeta = \alpha + i\tau$ is an arbitrary complex number: we set

$$(-A_0)^{\zeta}u = \int_0^{\infty} \mu^{\zeta} P(-d\mu)u \ , \qquad (9.12)$$

where $P(d\mu)$ is the resolution of the identity for A_0 (so that $P(-d\mu)$ is the resolution of the identity for $-A_0$). Due to well known properties of the functional calculus (see for instance DUNFORD-SCHWARTZ [1963:1]) we have

$$\|(-A_0)^{\zeta}u\|^2 = \|(-A_0)^{\alpha+i\tau}u\| = \int_0^{\infty} \mu^{2\alpha}\|P(-d\mu)u\|^2 = \|(-A_0)^{\alpha}u\|^2 . \qquad (9.13)$$

Let $Q : H \to H$ be a linear operator such that

$$\begin{aligned} \|Qu\| &\leq K_0\|u\| \qquad (u \in H), \\ \|Qu\| &\leq K_1\|A_0u\| \qquad (u \in D(A)) . \end{aligned} \qquad (9.14)$$

To obtain estimates for Q in the intermediate spaces $D((-A_0)^{\alpha})$ we apply the <u>three lines theorem</u> below to the H-valued holomorphic function

$$\varphi(\zeta) = \varphi(\alpha + i\tau) = Q(-A_0)^{-\alpha-i\tau}u \qquad (0 \leq \alpha \leq 1). \qquad (9.15)$$

THREE LINES THEOREM 9.1. <u>Let</u> $f(\zeta)$ <u>be a Banach space valued analytic function defined and bounded in a strip</u> $a \leq \text{Re}\ \zeta \leq b$, $-\infty < \text{Im}\ \zeta < \infty$. <u>Then</u> $\log M(\alpha)$ <u>is a convex function of</u> α, <u>where</u>

$$M(\alpha) = \sup_{-\infty<\tau<\infty} \|f(\alpha + i\tau)\|.$$

The proof can be found in DUNFORD-SCHWARTZ [1958:1, p. 520] for a complex-valued function $f(\zeta)$: the proof of the general case is the same.

Using (9.13) we deduce from (9.14) that the function $\varphi(\zeta)$ defined in (9.15) satisfies

$$\|\varphi(i\tau)\| = \|Qu\| \leq K_0\|u\| \qquad (-\infty < \tau < \infty) \ , \qquad (9.16)$$

$$\|\varphi(1 + i\tau)\| = \|QA_0^{-1}u\| \leq K_1\|u\| \qquad (-\infty < \tau < \infty) \ . \qquad (9.17)$$

Since $\|\varphi(\alpha + i\tau)\| = \|Q(-A_0)^{-\alpha}u\|$ we deduce from Theorem 9.1 that

$$\|\varphi(\alpha + i\tau)\| \le K_0^{1-\alpha} K_1^{\alpha} \|u\| \qquad (0 \le \alpha \le 1),$$

hence

$$\|Qu\| \le K_0^{1-\alpha} K_1^{\alpha} \|(-A_0)^{\alpha} u\| \qquad (u \in D((-A_0)^{\alpha}),\ 0 \le \alpha \le 1). \quad (9.18)$$

We use (9.18) for the operator $\mathfrak{C}(t;\varepsilon) - S(t)$. Using (3.4) and (2.18) we obtain

$$\|\mathfrak{C}(t;\varepsilon) - S(t)\| \le 2C_0 e^{\omega^2 t} \qquad (t \ge 0). \quad (9.19)$$

On the other hand, (5.7) implies that

$$\|(\mathfrak{C}(t;\varepsilon) - S(t))u\| \le C_0 \varepsilon^2 (1 + \omega^2 t) e^{\omega^2 t} \|Au\|. \quad (9.20)$$

Since the norms $\|A_0 u\|$ and $\|Au\|$ are equivalent,

$$\|(\mathfrak{C}(t;\varepsilon) - S(t))u\| \le 2^{1-\alpha} C_1 \varepsilon^{2\alpha} (1 + \omega^2 t)^{\alpha} e^{\omega^2 t} \|(-A_0)^{\alpha} u\| \quad (u \in D((-A_0)^{\alpha}),\ t \ge 0), \quad (9.21)$$

and we deduce that if $u(\hat{t};\varepsilon)$ is the generalized solution of (2.1) and $u(\hat{t})$ is the generalized solution of (2.2) with $u_0 \in D((-A)^{\alpha})$ $(0 \le \alpha \le 1)$ we shall have

$$\begin{aligned} \|u(t;\varepsilon) - u(t)\| \le {} & 2^{1-\alpha} C_1 \varepsilon^{2\alpha} (1 + \omega^2 t)^{\alpha} e^{\omega^2 t} \|(-A_0)^{\alpha} u\| + \\ & + C_1 e^{\omega^2 t} \|u_0(\varepsilon) - u_0\| + \\ & + C_1 \varepsilon^{2\alpha} e^{\omega^2 t} (\varepsilon^{2(1-\alpha)} \|u_1(\varepsilon)\|) \quad (t \ge 0). \end{aligned} \quad (9.22)$$

If $u_0 \in D((-A_0)^{1+\alpha})$ and $u_0(\varepsilon) \in D(A)$ we can estimate the rate of convergence of the derivative $u'(t;\varepsilon)$:

$$\begin{aligned} \|u'(t;\varepsilon) - u'(t)\| \le {} & 2^{1-\alpha} C_1 \varepsilon^{2\alpha} (1 + \omega^2 t)^{\alpha} e^{\omega^2 t} \|(-A_0)^{1+\alpha} u\| + \\ & + C_1 e^{\omega^2 t} (\|u_1(\varepsilon) - Au_0\| + \|u_1(\varepsilon) - Au_0(\varepsilon)\|) \\ & (t \ge 0). \end{aligned} \quad (9.23)$$

The preceding two estimates are a notable improvement over (5.55) and (5.56) in that no ad-hoc assumptions on the b_j have to be made and in that the constants on the right-hand sides are far better identified.

We show below that the rates of convergence provided by the estimate (9.22) (or by (5.55)) are best possible.

EXAMPLE 9.2. We consider the operator

$$A = D^2 , \tag{9.24}$$

in $\Omega = (0,\pi) \in \mathbb{R}^1$ with the Dirichlet boundary condition β. Fourier analysis shows that $L^2(\Omega)$ is isomorphic to ℓ^2 and that $A(\beta)$ is just the operator in Example 3.10. We have

$$\mathfrak{S}(t;\varepsilon)\{u_n\} = \{\Phi_{in}(t;\varepsilon)u_n\} , \tag{9.25}$$

where $\Phi_\omega(t;\varepsilon)$ has been defined in (3.3). Using Taylor series we show that, for n fixed and $t > 0$

$$\Phi_{in}(t;\varepsilon) = e^{-n^2t} + \frac{1}{4}\varepsilon^2 n^4 te^{-n^2t} - 2\varepsilon^2 n^2 e^{-n^2t} + O(\varepsilon) \quad (\varepsilon \to 0) \tag{9.26}$$

Assume $\{u_n\} \in \ell^2$ is such that

$$\|\mathfrak{S}(t;\varepsilon)\{u_n\} - S(t)\{u_n\}\| \leq C\varepsilon^2$$

for small ε and t. Then

$$\sum \varepsilon^{-4}|\Phi_{in}(t;\varepsilon) - e^{-n^2t}|^2|u_n|^2 \leq C ,$$

thus, by (9.26),

$$\sum |n^4te^{-n^2t} - 8n^2e^{-n^2t}|^2|u_n|^2 < \infty$$

for small t. This implies that

$$\sum n^4|u_n|^2 < \infty ,$$

so that $\{u_n\} \in D(A)$.

§VI.10 Miscellaneous comments.

The singular perturbation problem for the system (2.1) was considered by KISYŃSKI [1963:1] in the case where A is a self adjoint, positive definite operator in Hilbert space. Similar assumptions are used by SMOLLER [1965:1], [1965:2], although in [1965:2] the operator A is not necessarily positive definite and approximations of (2.2) by equations of order ≥ 3 are studied as well. For other results in this setup see LATIL [1968:1].

The next step in the direction of increasing generality was taken by KISYŃSKI [1970:1], SOVA [1970:2], [1972:1] and SCHOENE [1970:1]. The first two authors work under the assumptions in this chapter, that is, require A to be the infinitesimal generator of a strongly continuous cosine function; SCHOENE assumes that $A = B^2$, where B is the infinitesimal generator of a strongly continuous group. Every such operator is the infinitesimal generator of a strongly continuous cosine function (see III.6.24 and comments there), but the converse is not true (Example III.6.5) thus KISYŃSKI's and SOVA's results have a wider range of applicability (on this matter see also NAGY [1976:1]). The most precise results are those of KISYŃSKI [1970:1] who proved Theorem 5.1, Theorem 5.5 and Theorem 5.6 for $u_0(\varepsilon)$, $u_1(\varepsilon)$ fixed; the crucial step is the estimate (5.7) using the theory of alternating functions. SOVA's results in [1970:2] and [1972:1] are somewhat less precise and obtained in a different way; instead of writing the solutions of (2.1) explicitly, SOVA estimates their Laplace transforms and uses the Post-Widder formula (I.3.14). A treatment of the singular perturbation problem in another vein was given by GRIEGO-HERSH [1971:1].

Convergence of the solution of (2.1) using explicit formulas of the type of (2.14) was studied by DETTMAN [1973:1]. The results in the first four sections in this chapter are in the author [1983:4]. The main novelty here is that they are uniform with respect to the initial conditions $u_0(\varepsilon)$, $u_1(\varepsilon)$ as long as $u_0(\varepsilon)$, $\varepsilon^2 u_1(\varepsilon)$ remain bounded, although there is an initial layer $0 \leq t \leq t(\varepsilon)$ where convergence is lost. Section 5 is also taken from [1983:4]; the results are slight generalizations of those in KISYŃSKI [1970:1] mentioned above in that the initial conditions are allowed to depend on ε. The "intermediate" estimates in Theorems 5.12 and 5.13 are in [1983:4].

The treatment of the nonhomogeneous equation in Sections 5 and 6 is also in the author [1983:4]. Theorem 6.2 is due to DE SIMON [1964:1].

Section 8 on asymptotic development of the solution of (2.1) is also in the author [1983:4]. The asymptotic series is an abstract version of a well known series used in diverse singular perturbation problems for partial differential equations (see for instance GEEL [1978:1]). The particular form used here is modelled on HSIAO-WEINACHT [1979:1], who consider the heat equation in dimension 1 in a space of continuous functions.

The "numerical perturbation" formula (2.8) (due to SOVA [1970:4]) is an abstract version of a well known formula for the solution of the telegraphist's equation (or of the Klein-Gordon equation) in terms of the solution of the wave equation. Other expressions of the same type are in SOVA [1970:4]. They are particular cases of transmutation formulas, which can be roughly described as transforming the solution of a differential equation into the solution of another differential equation. See the author [1983:1] for additional information and references.

An important application of (2.8) is that of estimating the growth of the solutions of $u''(t) = (A + b^2I)u(t)$ for any complex b in terms of the solutions of $u''(t) = Au(t)$. On this subject, we include some results (Exercises 3 to 8) taken from the author [1985:2].

EXERCISE 1. Let $a > 0$ be arbitrary,

$$m(\lambda) = (a\lambda^2 + \lambda)^{1/2} \qquad (\lambda \geq 0) \tag{10.1}$$

Show that $m'(\lambda)$ is alternating.

EXERCISE 2. Let $a > 0$, ζ an arbitrary complex number A an operator in $\mathfrak{C}^2$. Show that the Cauchy problem for

$$au''(t) + \zeta u'(t) = Au(t) \tag{10.2}$$

is well posed in $-\infty < t < \infty$. Write an explicit formula for the solution of (10.2) in terms of $C(\hat{t})$, the cosine function generated by A.

In Exercises 3 to 8, $C(t;A)$ is a strongly continuous cosine function with infinitesimal generator A and b is a complex number in the right half plane $\operatorname{Re} b \geq 0$.

EXERCISE 3. Let A be a normal operator in a Hilbert space H such that A generates a cosine function $C(t;A)$ with

$$\|C(t;A)\| \leq \cosh \omega t \qquad (-\infty < t < \infty). \tag{10.3}$$

Then $C(t;A + b^2I)$, the cosine function generated by $A + b^2I$ satisfies

$$\|C(t;A + b^2I)\| \leq \cosh(\omega^2 + |\operatorname{Re} b|^2)^{1/2}t \qquad (-\infty < t < \infty). \tag{10.4}$$

(Hint: use Exercise II.5).

EXERCISE 4. Let A be the infinitesimal generator of a strongly continuous cosine function $C(t;A)$ satisfying

$$\|C(t;A)\| \le C_0 \cosh \omega t \qquad (-\infty < t < \infty) . \tag{10.5}$$

Show that $C(t;A + b^2 I)$ satisfies

$$\|C(t;A + b^2 I)\| \le C_0 \cosh \omega t + C_0 K_\omega(b;t) \qquad (-\infty < t < \infty) , \tag{10.6}$$

where $K_\omega(b;t) = K_\omega(b;|t|)$ and

$$K_\omega(b;t) = |b| t \int_0^t \frac{|I_1(b(t^2 - s^2)^{1/2})|}{(t^2 - s^2)^{1/2}} \cosh \omega s \, ds \tag{10.7}$$

for $t \ge 0$.

EXERCISE 5. Show that there exists a strongly continuous cosine function $C(t;A)$ satisfying

$$C(t;A) = \cosh \omega t \qquad (-\infty < t < \infty) , \tag{10.8}$$

and

$$\|C(t;A + b^2 I)\| = \cosh \omega t + K_\omega(b;t) \qquad (-\infty < t < \infty) . \tag{10.9}$$

(Hint: use the cosine function (II.6.10) in the space E_ω of all continuous functions $u(x)$ defined in $-\infty < x < \infty$ and such that $|u(x)|e^{\omega x} \to 0$ as $|x| \to \infty$, endowed with the norm $\|u\|_\omega = \max|u(x)|e^{\omega x}$).

EXERCISE 6. Using the asymptotic development of the Bessel function $I_1(x)$ (GRADSTEIN-RYDZYK [1963:1, p. 975] or WATSON [1948:1, p. 199]) show that there exist a constant C such that, if $\eta \ge 0$

$$K_0(i\eta;t) \le C\eta^{1/2}|t|^{1/2} \qquad (-\infty < t < \infty) , \tag{10.10}$$

$$K_\omega(i\eta;t) \le C\eta^{1/2}|t|^{1/4}e^{\omega|t|} \qquad (-\infty < t < \infty) . \tag{10.11}$$

Show that, given $\delta > 0$ there exists $c > 0$ such that

$$K_0(i\eta;t) \ge c\eta^{1/2}|t|^{1/2} \qquad (\eta, \ |t| \ge \delta) , \tag{10.12}$$

$$K_\omega(i\eta;t) \ge c\eta^{1/2}|t|^{1/4}e^{\omega|t|} \qquad (\eta, \ |t| \ge \delta) . \tag{10.13}$$

EXERCISE 7. Using the fact that $I_1(x) \ge 0$, show: if $b \ge 0$

and $C(t;A)$ is a cosine function satisfying (10.5), then

$$\|C(t;A + b^2 I)\| \le C_0 \cosh(\omega^2 + b^2)^{1/2} t \qquad (-\infty < t < \infty) . \quad (10.14)$$

(Hint: apply (2.8) to the scalar cosine function $\cosh \omega t$). Using (10.6), (10.10), (10.11) and (10.14) plus the Phragmén-Lindelöf theorem obtain upper bounds for $K_0(b;t)$, $K_\omega(b;t)$ for b complex.

EXERCISE 8. Let $C(t;A)$ be a cosine function in a Hilbert space H satisfying

$$\|C(t;A)\| \le C_0 \qquad (-\infty < t < \infty) . \quad (10.15)$$

Show that there exists a constant C depending only on C_0 such that

$$\|C(t;A + b^2 I)\| \le Ce^{|\mathrm{Re}\ b||t|} \qquad (-\infty < t < \infty) . \quad (10.16)$$

(Hint: use the theory in Chapter V, specifically Corollary V.6.5).

FOOTNOTES TO CHAPTER VI

(1) We note an inconsistence of notation with §III.4, where we use $-b^2$ instead of b^2.

(2) Strictly speaking, we should write $C(\sigma)f(\varepsilon s;\varepsilon)d\sigma$ in the integrand, since $C(\hat{\sigma})$ is only strongly continuous. We shall ignore this here and in other places.

(3) The letter $\mathfrak{S}$ is used with a different meaning in Chapters III and VIII.

(4) Recall that, due to the first inequality (II.2.22) we have

$$\|\varepsilon^{-1}R(\varepsilon^{-1};A)\| \le C.$$

(5) This is the only place where the fact that $\omega > 0$ is significant (see Remark 4.5 on the case $\omega = 0$).

CHAPTER VII

OTHER SINGULAR PERTURBATION PROBLEMS

§ VII.1 A singular perturbation problem in quantum mechanics.

In relativistic quantum mechanics (see SCHOENE [1970:1]) one considers functions of the form

$$u(x,t) = v(x,t)\exp(imc^2 t/h)$$

where v is a solution of the Klein-Gordon equation for a free particle

$$-h^2 v_{tt} = -h^2 c^2 \Delta v + m^2 c^4 v$$

(here m is the mass of the particle, h is Planck's constant and c is the speed of light). If follows that $u(x,t)$ satisfies the equation

$$\frac{h}{2mc^2} u_{tt} - iu_t = \frac{h}{2m} \Delta u \; ;$$

since $h/2mc^2 \ll 1$ we expect u to differ by little from the solution of the Schrödinger equation

$$- ihu_t = \frac{h^2}{2m} \Delta u \; .$$

It is then natural to consider the equation

$$\varepsilon^2 u_{tt}(x,t;\varepsilon) - iu_t(x,t;\varepsilon) = Au(x,t;\varepsilon) \; , \qquad (1.1)$$

where $u(x,t;\varepsilon) = u(x_1,\ldots,x_m,t;\varepsilon)$ and A is a differential operator in the "space variables" $x_1,\ldots,x_m$ like

$$Au = \sum_{j=1}^{m} \sum_{k=1}^{m} a_{jk}(x)D^j D^k u + \sum_{j=1}^{m} b_j(x)D^j u + c(x)u, \qquad (1.2)$$

or a suitable higher order version; the problem at hand is that of determining in which way (if any)

$$u(x,t;\varepsilon) \to u(x,t) \qquad (1.3)$$

as $\varepsilon \to 0$, where $u(x,t)$ is the solution of the unperturbed equation

$$u_t(x,t) = iAu(x,t). \tag{1.4}$$

This will be done in the rest of the chapter for an abstract version of (1.1), (1.4).

§VII.2 The Schrödinger singular perturbation problem.

Throughout this chapter A will be the infinitesimal generator of a strongly continuous cosine function $C(t)$ in the Banach space E (that is, $A \in \mathfrak{C}^2$). We consider the Cauchy problems

$$\begin{gathered} \varepsilon^2 u''(t;\varepsilon) - iu'(t;\varepsilon) = Au(t;\varepsilon) + f(t;\varepsilon) \quad (-\infty < t < \infty), \\ u(0;\varepsilon) = u_0(\varepsilon), \quad u'(0;\varepsilon) = u_1(\varepsilon), \end{gathered} \tag{2.1}$$

and

$$u'(t) = iAu(t) + if(t), \quad u(0) = u_0 \quad (-\infty < t < \infty). \tag{2.2}$$

The Schrödinger singular perturbation problem is that of showing that

$$u(t;\varepsilon) \to u(t) \tag{2.3}$$

as $\varepsilon \to 0$, where convergence in (2.3) can be understood in various senses. A superficial kinship of (2.1) and (VI.2.1) is obvious, although the factor $-i$ before $u'(t)$ causes radical differences between these initial value problems, as put in evidence all throughout this chapter. An explicit solution of (2.1) can be constructed by the methods of §VI.2 (or it can be formally obtained from (VI.2.14) considering $u(it;i\varepsilon)$, where $u(t;\varepsilon)$ is the function in (2.14), and keeping in mind that $I_0(-x) = I_0(ix) = J_0(x)$ and that $I_1(-x) = -I_1(ix) = -iJ_1(x)$). The end result is an explicit expression for the solution of (2.1):

$$\begin{aligned} u(t;\varepsilon) &= e^{it/2\varepsilon^2} C(t/\varepsilon)u_0(\varepsilon) \\ &- \frac{te^{it/2\varepsilon^2}}{\varepsilon^2} \int_0^{t/\varepsilon} \frac{J_1(((t/\varepsilon)^2 - s^2)^{1/2}/2\varepsilon)}{((t/\varepsilon)^2 - s^2)^{1/2}} C(s)(\tfrac{1}{2} u_0(\varepsilon))\, ds \\ &- \frac{ie^{it/2\varepsilon^2}}{\varepsilon} \int_0^{t/\varepsilon} J_0(((t/\varepsilon)^2 - s^2)^{1/2}/2\varepsilon)C(s)(\tfrac{1}{2} u_0(\varepsilon) + i\varepsilon^2 u_1(\varepsilon))\, ds \\ &+ e^{it/2\varepsilon^2} \int_0^{t/\varepsilon} \left(\int_0^{t/\varepsilon - s} J_0(((t/\varepsilon - s)^2 - \sigma^2)^{1/2}/2\varepsilon)C(\sigma)\, d\sigma \right) e^{-is/2\varepsilon} f(\varepsilon s;\varepsilon)\, ds = \end{aligned}$$

$$= e^{it/2\varepsilon^2} C(t/\varepsilon)u_0(\varepsilon) + \mathfrak{R}_i(t;\varepsilon)(\tfrac{1}{2} u_0(\varepsilon)) - i\mathfrak{C}_i(t;\varepsilon)(\tfrac{1}{2}u_0(\varepsilon) + i\varepsilon^2 u_1(\varepsilon))$$

$$+ \int_0^t \mathfrak{S}_i(t - s;\varepsilon)f(s;\varepsilon)\, ds$$

$$= \mathfrak{C}_i(t;\varepsilon)u_0(\varepsilon) + \mathfrak{S}_i(t;\varepsilon)(\varepsilon^2 u_1(\varepsilon)) + \int_0^t \mathfrak{S}_i(t - s;\varepsilon)f(s;\varepsilon)\, ds. \qquad (2.4)$$

The obvious difference between the representation (2.4) for (generalized) solutions of the initial value problem (2.1) and the similar representation (VI.2.14) for the initial value problem (VI.2.1) lies in the different asymptotic behavior of the integrands and the exponential factors. In (VI.2.14) we can combine the asymptotic estimate

$$|I_\nu(x)| = O(x^{-1/2}e^x) \quad \text{as} \quad x \to \infty \qquad (2.5)$$

(see §VI.2) with the rapidly decreasing factor $e^{-t/2\varepsilon^2}$ and the bound (VI.2.17) for $\|C(\cdot)\|$ to show that we can take limits directly as $\varepsilon \to 0$ using the dominated convergence theorem: the results are uniform convergence results of the type of Theorems VI.3.6 and VI.3.8. In the same way we obtain uniform bounds for derivatives (Theorem VI.4.1). In contrast, the asymptotic estimates

$$|J_\nu(x)| = O(x^{-1/2}) \qquad (2.6)$$

combined with the indifferent factor e^{it/ε^2} in (2.4) do not allow direct passage to the limit; in fact, doing so formally, one can only hope for a relation similar to (VI.3.40) for $\mathfrak{C}_1$, $\mathfrak{S}_i$ with limit

$$S_i(t)u = S(it)u = \frac{1}{(\pi i t)^{1/2}} \int_0^\infty e^{is^2/4t} C(s)u\, ds \qquad (2.7)$$

which does not make sense even in the most favorable case where $\|C(t)\|$ is uniformly bounded in $-\infty < t < \infty$. Hence, computation of the limit (2.3) will have to be carried out by indirect means. In particular (and in contrast with the parabolic singular perturbation problem) the existence of the group $S_i(t) = \exp(iAt) = S(it)$ is not assured by the existence-uniqueness assumption for (2.1) above but will follow from the stronger assumptions in §VII.2. (we note in passing that $S_i(t)$ is the "boundary value" of the analytic semigroup constructed in §VI.2). As it may be expected, the convergence results for the operators $\mathfrak{C}_i(t;\varepsilon)$ and $\mathfrak{S}_i(t;\varepsilon)$ (which will be called in what follows the <u>propagators</u> or <u>solution operators</u> of (1.1)-(1.2)) are different in character from those for $\mathfrak{S}(t;\varepsilon)$, $\mathfrak{S}(t;\varepsilon)$ in Chapter VI, especially those in §VI.2 to §VI.4. The main differences between the results in this chapter and those in Chapter VI can be summarized as follows:

(a) *There are no uniform bounds (that is, bounds in the norm of (E)) for* $\mathfrak{S}_i'(t;\varepsilon)$, $\mathfrak{S}_i''(t;\varepsilon),\ldots$ *of the type of* (VI.4.20), (VI.4.54). (Example 4.9).

(b) *There are no convergence results in the norm of* (E) *for* $\mathfrak{S}_i(t;\varepsilon)$, $\mathfrak{S}_i(t;\varepsilon)$, $\mathfrak{S}_i'(t;\varepsilon)$, $\mathfrak{S}_i''(t;\varepsilon),\ldots$ *of the type of Theorem* VI.3.11, *Theorem* VI.4.6 or *Theorem* VI.4.7: *in fact* $\mathfrak{S}_i(t;\varepsilon)$ *is not even strongly convergent* (Example 4.6).

The nonexistence of uniform bounds for $\mathfrak{S}_i'(t;\varepsilon)$ and $\mathfrak{S}_i''(t;\varepsilon)$, as well as the lack of convergence of $\mathfrak{S}_i(t;\varepsilon)$ has consequences for the inhomogeneous equation (2.1): in fact,

(c) *There are no analogues of Theorem* VI.7.4 *for Hölder continuous functions or of Theorem* VI.7.6 *for functions in* L^p.

Among the results on the positive side, we have

(d) *There exists an analogue of Theorem* VI.7.2 *for a restricted class of operators* A (*Theorem* 7.1).

(e) *Most of the results in* §VI.5 *on rates of convergence can be proved in the present situation* (see §VII.4, especially Theorem 4.1).

We mention finally that the theory in §VI.8 concerning asymptotic series does not seem to extend to the Schrodinger singular perturbation problem: although the asymptotic series (VI.8.20) can be formally defined in the present situation, the L^1 norm of $v_n(t/\varepsilon^2)$ over any finite interval cannot be conveniently bounded.

§VII.3 Assumptions on the initial value problem.

As pointed out in §VII.2, the fact that $A \in \mathfrak{S}^2$ allows us to show that the Cauchy problem for (2.1) is well posed. The key condition on (2.1) that will allow us to prove (2.3) is

ASSUMPTION 3.1. *Let* $\mathfrak{S}_i(t;\varepsilon)$, $\mathfrak{S}_i(t;\varepsilon)$ *be the propagators of* (2.1)

<u>Then there exist constants</u> C_0, C_1, ω <u>independent of</u> t <u>and</u> ε <u>such that</u>

$$\|\mathfrak{C}_i(t;\varepsilon)\| \le C_0 e^{\omega|t|}, \quad \|\mathfrak{S}_i(t;\varepsilon)\| \le C_1 e^{\omega|t|}$$

$$(-\infty < t < \infty,\ 0 \le \varepsilon \le \varepsilon_0). \tag{3.1}$$

Assumption 3.1 can be given a considerably simpler form as follows. Let $u(t;\varepsilon)$ be a solution of (2.1). Set $v(t) = e^{-it/2\varepsilon}u(\varepsilon t)$ or, equivalently, $u(t) = e^{it/2\varepsilon^2}v(t/\varepsilon)$. Then $v(t)$ satisfies

$$\begin{aligned} v''(t) &= \Big(A - \frac{1}{4\varepsilon^2} I\Big)v(t) \qquad (-\infty < t < \infty), \\ v(0) &= u_0(\varepsilon),\ v'(0) = -\frac{i}{2\varepsilon}u_0(\varepsilon) + \varepsilon u_1(\varepsilon). \end{aligned} \tag{3.2}$$

Accordingly, $u(t)$ can be explicitly written in terms of the propagators $C(t;A - (2\varepsilon)^{-2}I)$, $\mathcal{S}(t;A - (2\varepsilon)^{-2}I)$ of the initial value problem (3.2):

$$\begin{aligned} u(t) &= e^{it/2\varepsilon^2}C(t/\varepsilon;A - (2\varepsilon)^{-2}I)u_0(\varepsilon) \\ &+ e^{it/2\varepsilon^2}\mathcal{S}(t/\varepsilon;A - (2\varepsilon)^{-2}I)\Big(-\frac{i}{2\varepsilon}u_0(\varepsilon) + \varepsilon u_1(\varepsilon)\Big). \end{aligned} \tag{3.3}$$

(We note, incidentally, that formula (2.4) can be obtained from (3.3) by Lemma VI.2.1, although this will play no role in what follows). Using (3.3) we deduce that

$$\begin{aligned} \mathfrak{C}_i(t;\varepsilon) &= e^{it/2\varepsilon^2}C(t/\varepsilon;A - (2\varepsilon)^{-2}I) \\ &- i(2\varepsilon)^{-1}e^{it/2\varepsilon^2}\mathcal{S}(t/\varepsilon;A - (2\varepsilon)^{-2}I), \end{aligned} \tag{3.4}$$

$$\mathfrak{S}_i(t;\varepsilon) = \varepsilon^{-1}e^{it/2\varepsilon^2}\mathcal{S}(t/\varepsilon;A - (2\varepsilon)^{-2}I), \tag{3.5}$$

$$C(t;A - (2\varepsilon)^{-2}I) = e^{-it/2\varepsilon}\mathfrak{C}_i(\varepsilon t;\varepsilon) + (i/2)e^{-it/2\varepsilon}\mathfrak{S}_i(\varepsilon t;\varepsilon), \tag{3.6}$$

$$\mathcal{S}(t;A - (2\varepsilon)^{-2}I) = \varepsilon e^{-it/2\varepsilon}\mathfrak{S}_i(\varepsilon t;\varepsilon). \tag{3.7}$$

Accordingly, Assumption 3.1 will hold if and only if

$$\|C(t;A - (2\varepsilon)^{-2}I)\| \le C_0' e^{\omega\varepsilon|t|}, \quad \|\mathcal{S}(t;A - (2\varepsilon)^{-2}I)\| \le C_1 \varepsilon e^{\omega\varepsilon|t|}$$

$$(-\infty < t < \infty,\ 0 < \varepsilon \le \varepsilon_0). \tag{3.8}$$

We shall examine in §VII.5 a class of operators satisfying Assumption 3.1: as seen in §VII.6, these operators include the partial differential operator (1.2) (with boundary conditions), the coefficients suitably restricted.

We point out below a consequence of Assumption 3.1.

THEOREM 3.2. *Let the operator* A *satisfy Assumption* 3.1. *Then* iA *generates a strongly continuous group* $S_i(\hat{t})$ *such that*

$$\|S_i(t)\| \leq C_0 e^{\omega|t|} \qquad (-\infty < t < \infty), \tag{3.9}$$

C_0 *and* ω *the constants in* (3.1).

Proof: If $u \in D(A^2)$ then it follows from formula (2.4) (or from the considerations at the beginning of this section) that $\mathfrak{C}_i(t;\varepsilon)u$ is four times continuously differentiable, $\mathfrak{C}_i''(t;\varepsilon)u \in D(A)$ and

$$\varepsilon^2\mathfrak{C}_i''''(t;\varepsilon)u - i\mathfrak{C}_i''(t;\varepsilon)u = A\mathfrak{C}_i''(t;\varepsilon)u \qquad (-\infty < t < \infty). \tag{3.10}$$

On the other hand, we have

$$\mathfrak{C}_i''(0;\varepsilon)u = \varepsilon^{-2}(A\mathfrak{C}_i(0;\varepsilon)u + i\mathfrak{C}_i'(0;\varepsilon)u) = \varepsilon^{-2}Au, \tag{3.11}$$

$$\mathfrak{C}_i'''(0;\varepsilon)u = \varepsilon^{-2}(A\mathfrak{C}_i'(0;\varepsilon)u + i\mathfrak{C}_i''(0;\varepsilon)u) = i\varepsilon^{-4}Au, \tag{3.12}$$

thus we obtain, applying (2.4),

$$\mathfrak{C}_i''(t;\varepsilon)u = \varepsilon^{-2}\mathfrak{C}_i(t;\varepsilon)Au + i\varepsilon^{-2}\mathfrak{S}_i(t;\varepsilon)Au.$$

Accordingly, it follows from Assumption 3.1 that

$$\|\mathfrak{C}_i''(t;\varepsilon)u\| \leq C\varepsilon^{-2}e^{\omega|t|}\|Au\| \qquad (-\infty < t < \infty,\ 0 < \varepsilon \leq \varepsilon_0). \tag{3.13}$$

We take now a sequence $\{\varepsilon_n\}$, $1 > \varepsilon_1 > \varepsilon_2 > \dots > 0$ to be specified later. For $u \in D(A)$ we have

$$\varepsilon_n^2(\mathfrak{C}_i''(t;\varepsilon_n)u - \mathfrak{C}_i''(t;\varepsilon_{n+1})u) - i(\mathfrak{C}_i'(t;\varepsilon_n)u - \mathfrak{C}_i'(t;\varepsilon_{n+1})u) =$$

$$= A(\mathfrak{C}_i(t;\varepsilon_n)u - \mathfrak{C}_i(t;\varepsilon_{n+1})u) + (\varepsilon_n^2 - \varepsilon_{n+1}^2)\mathfrak{C}_i''(t;\varepsilon_{n+1})u.$$

Using again (2.4),

$$\mathfrak{C}_i(t;\varepsilon_n)u - \mathfrak{C}_i(t;\varepsilon_{n+1})u = (\varepsilon_n^2 - \varepsilon_{n+1}^2)\int_0^t \mathfrak{S}_i(t - s;\varepsilon_n)\mathfrak{C}_i''(s;\varepsilon_{n+1})u\,ds.$$

In view of (2.12) and Assumption 3.1, if $u \in D(A^2)$ we have

$$\|\mathfrak{C}_i(t;\varepsilon_n)u - \mathfrak{C}_i(t;\varepsilon_{n+1})u\| \le C'|t|e^{\omega|t|}\left(1 - \frac{\varepsilon_{n+1}^2}{\varepsilon_n^2}\right)\|Au\|.$$

Selecting now a sequence $\{\varepsilon_n\}$ that tends to zero sufficiently slowly (say, $\varepsilon_n = n^{-1/2}$) we show that $\{\mathfrak{C}_i(t;\varepsilon_n)u\}$ is a Cauchy sequence, uniformly with respect to t on compact subsets of $-\infty < t < \infty$. Using the uniform bound (3.1) and the denseness of $D(A^2)$ we deduce that $\{\mathfrak{C}_i(t;\varepsilon_n)\}$ converges strongly, uniformly on compacts of $-\infty < t < \infty$ to a strongly continuous operator valued function $S_i(s)$ satisfying the estimate (3.9).

Denote by $\hat{\mathfrak{C}}_i(\lambda;\varepsilon)u$ the Laplace transform of $\mathfrak{C}_i(t;\varepsilon)u$. Writing equation (2.1) for $\mathfrak{C}_i(t;u)$ $(u \in D(A))$ and taking into account that $\mathfrak{C}_i(0;\varepsilon)u = u$, $\mathfrak{C}_i'(0;\varepsilon)u = 0$ we deduce, after integrating from 0 to t two times,

$$\varepsilon^2(\mathfrak{C}_i(t;\varepsilon)u - u) - i\int_0^t (\mathfrak{C}_i(s;\varepsilon)u - u)ds = \int_0^t (t - s)A\mathfrak{C}_i(s;\varepsilon)u\, ds .$$

Taking Laplace transforms, we obtain

$$\varepsilon^2(\lambda^2\hat{\mathfrak{C}}_i(\lambda;\varepsilon)u - \lambda u) - i(\lambda\hat{\mathfrak{C}}_i(\lambda;\varepsilon)u - u) = A\hat{\mathfrak{C}}_i(\lambda;\varepsilon)u ,$$

where the introduction of A under the integral sign is easily justified. It follows that

$$\hat{\mathfrak{C}}_i(\lambda;\varepsilon)u = (\varepsilon^2\lambda - i)R(\varepsilon^2\lambda^2 - i\lambda;A)u$$

for $\lambda > \omega$. Putting $\varepsilon = \varepsilon_n$ with $\{\varepsilon_n\}$ a sequence as above and taking limits, we obtain

$$\hat{S}_i(\lambda)u = -iR(-\lambda;A)u = R(\lambda;iA)u$$

for $u \in D(A)$ and thus, by denseness of $D(A)$, for all $u \in E$. It follows then from Theorem I.3.4 that $S_i(\hat{t})$ is a strongly continuous group with infinitesimal generator A. This ends the proof of Theorem 3.2.

We note that our proof includes a result on convergence of $\mathfrak{C}_i(t;\varepsilon)$ to $S_i(t)$. However, it's not worth formulating since it will be considerably improved in next section.

§VII.4 The homogeneous equation: convergence results.

Let u be an element of $D(A)$. The function

$$u(t;\varepsilon) = \mathfrak{C}_i'(t;\varepsilon)u$$

is a generalized solution of the homogeneous equation (2.1) with

$$u(0;\varepsilon) = 0,\ u'(0;\varepsilon) = \mathfrak{C}_i''(0;\varepsilon)u$$

$$= \varepsilon^{-2}A\mathfrak{C}_i(0;\varepsilon)u + i\varepsilon^{-2}\mathfrak{C}_i'(0;\varepsilon)u = \varepsilon^{-2}Au.$$

It follows that

$$\mathfrak{C}_i'(t;\varepsilon)u = \mathfrak{S}_i(t;\varepsilon)Au\ . \tag{4.1}$$

On the other hand,

$$v(t;\varepsilon) = \mathfrak{S}_i'(t;\varepsilon)u$$

is also a generalized solution of (2.1) with

$$v(0;\varepsilon) = \mathfrak{S}_i'(0;\varepsilon)u = \varepsilon^{-2}u,\ v'(0;\varepsilon)$$

$$= \mathfrak{S}_i''(0;\varepsilon)u = \varepsilon^{-2}A\mathfrak{S}_i(0;\varepsilon)u + i\varepsilon^{-2}\mathfrak{S}_i'(0;\varepsilon)u = i\varepsilon^{-4}u,$$

so that

$$\mathfrak{S}_i'(t;\varepsilon)u = \varepsilon^{-2}\mathfrak{C}_i(t;\varepsilon)u + i\varepsilon^{-2}\mathfrak{S}_i(t;\varepsilon)u\cdot \tag{4.2}$$

Since all operators in (4.2) are bounded, the equality can be extended to all $u \in E$. We combine (4.1) and (4.2), the latter inequality written for an element of the form Au. The result is

$$\mathfrak{C}_i'(t;\varepsilon)u = iA\mathfrak{C}_i(t;\varepsilon)u - i\varepsilon^2\mathfrak{S}_i'(t;\varepsilon)Au. \tag{4.3}$$

Since iA is the infinitesimal generator of the strongly continuous semigroup $S_i(\hat{t})$, we obtain applying formula (I.5.3) that

$$\mathfrak{C}_i(t;\varepsilon)u - S_i(t)u = -i\varepsilon^2\int_0^t S_i(t - s)\mathfrak{S}_i'(s;\varepsilon)Au\ ds$$

$$= -i\varepsilon^2\int_0^t \mathfrak{S}_i'(t - s;\varepsilon)S_i(s)Au\ ds. \tag{4.4}$$

If $u \in D(A^2)$ we can integrate by parts:

$$\mathfrak{C}_i(t;\varepsilon)u - S_i(t)u = -\varepsilon^2\mathfrak{S}_i(t;\varepsilon)Au - i\varepsilon^2\int_0^t \mathfrak{S}_i(t - s;\varepsilon)S_i(s)A^2u\ ds. \tag{4.5}$$

Estimating the integral in (4.5) in an obvious way, we obtain

THEOREM 4.1. *Let* A *be an operator satisfying Assumption* 3.1 *and let* $u(t;\varepsilon)$ *be a solution of the homogeneous problem* (2.1), $u(t)$ *a solution of the homogeneous problem* (2.2) *with* $u_0 \in D(A^2)$. *Then we have*

$$\|u(t;\varepsilon) - u(t)\| \le C_1\varepsilon^2 e^{\omega|t|}(\|Au_0\| + C_0|t|\,\|A^2u_0\|)$$
$$+ C_0e^{\omega|t|}\|u_0(\varepsilon) - u_0\| + C_1\varepsilon^2e^{\omega|t|}\|u_1(\varepsilon)\| \quad (-\infty < t < \infty)\,. \tag{4.6}$$

The proof of (4.6) is essentially the same as that of (VI.5.27): we only have to note that

$$u(t;\varepsilon) - u(t) = \mathfrak{C}_i(t;\varepsilon)u_0 - S_i(t)u_0$$
$$+ \mathfrak{C}_i(t;\varepsilon)(u_0(\varepsilon) - u_0) + \mathfrak{S}_i(t;\varepsilon)(\varepsilon^2u_1(\varepsilon))\,. \tag{4.7}$$

Theorem 4.1 implies that when $u_0 \in D(A^2)$ we have

$$\|u(t;\varepsilon) - u(t)\| = O(\varepsilon^2) \tag{4.8}$$

uniformly on compacts of $-\infty < t < \infty$ if

$$\|u_0(\varepsilon) - u_0\| = O(\varepsilon^2),\ \|u_1(\varepsilon)\| = O(1) \quad \text{as} \quad \varepsilon \to 0. \tag{4.9}$$

Estimates of the same sort can be easily obtained for the derivative $u'(t;\varepsilon)$ if $u_0 \in D(A^3)$ and $u_0(\varepsilon) \in D(A)$. In fact, $v(t;\varepsilon) = u'(t;\varepsilon)$ is the generalized solution of the homogeneous equation (2.1) with

$$v(0;\varepsilon) = u'(0;\varepsilon) = u_1(\varepsilon),\ v'(0;\varepsilon) = u''(0;\varepsilon) = \varepsilon^{-2}(Au_0(\varepsilon) + iu_1(\varepsilon)). \tag{4.10}$$

On the other hand, $v(t) = u'(t)$ is the solution of the homogeneous equation (2.2) with

$$v(0) = u'(0) = iAu_0. \tag{4.11}$$

Applying Theorem 4.1 to $v(t;\varepsilon)$, $v(t)$ we obtain

THEOREM 4.2. *Let* A *be as in Theorem* 4.1 *and let* $u(t;\varepsilon)$ *be a solution of the homogeneous problem* (2.1) *with* $u_0(\varepsilon) \in D(A)$, $u(t)$ *a solution of the homogeneous problem* (2.2) *with* $u_0 \in D(A^3)$. *Then we have*

$$\|u'(t;\varepsilon) - u'(t)\| \le C_1\varepsilon^2e^{\omega|t|}(\|A^2u_0\| + C_0|t|\,\|A^3u_0\|)$$
$$+ e^{\omega|t|}(C_0\|u_1(\varepsilon) - iAu_0\| + C_1\|u_1(\varepsilon) - iAu_0(\varepsilon)\|)\,. \tag{4.12}$$

It follows from this result that if $u_0 \in D(A^3)$ and $u_0(\varepsilon) \in D(A)$ then

$$\|u'(t;\varepsilon) - u'(t)\| = O(\varepsilon^2) \tag{4.13}$$

uniformly on compacts of $-\infty < t < \infty$ if

$$\|u_1(\varepsilon) - iAu_0\| = O(\varepsilon^2) \text{ and } \|u_1(\varepsilon) - iAu_0(\varepsilon)\| = O(\varepsilon^2) \text{ as } \varepsilon \to 0 , \tag{4.14}$$

or, equivalently, if

$$\|u_1(\varepsilon) - iAu_0\| = O(\varepsilon^2) \text{ and } \|Au_0(\varepsilon) - Au_0\| = O(\varepsilon^2) \text{ as } \varepsilon \to 0 . \tag{4.15}$$

Theorems 4.1 and 4.2 allow us to deduce convergence results for arbitrary initial conditions.

THEOREM 4.3. _Let_ $u(t;\varepsilon)$ _be a generalized solution of the homogeneous problem_ (2.1) _with_ $u_0(\varepsilon), u_1(\varepsilon) \in E$, $u(t)$ _a generalized solution of the homogeneous problem_ (2.2) _with_ $u_0 \in E$. _Assume that_

$$u_0(\varepsilon) \to u_0, \ \varepsilon^2 u_1(\varepsilon) \to 0 \text{ as } \varepsilon \to 0. \tag{4.16}$$

Then

$$u(t;\varepsilon) \to u(t) \text{ as } \varepsilon \to 0 \tag{4.17}$$

uniformly on compacts of $-\infty < t < \infty$.

Proof: Pick $\delta > 0$ and choose $\bar{u} \in D(A^2)$ such that $\|\bar{u} - u_0\| \le \delta$. Let $\bar{u}(t)$ be the solution of the initial value problem (2.2) with $\bar{u}(0) = \bar{u}$. Applying Theorem 4.1 we deduce that

$$\|u(t;\varepsilon) - u(t)\| \le \|u(t;\varepsilon) - \bar{u}(t)\| + \|\bar{u}(t) - u(t)\|$$

$$\le C_1 \varepsilon^2 e^{\omega|t|}(\|A\bar{u}\| + C_0|t|\,\|A^2\bar{u}\|) + C_0 e^{\omega|t|}\|u_0(\varepsilon) - \bar{u}\|$$

$$+ C_1 e^{\omega|t|}\varepsilon^2\|u_1(\varepsilon)\| + C_0\delta e^{\omega|t|} . \tag{4.18}$$

Taking $\varepsilon > 0$ sufficiently small we can obviously make the right-hand side of (4.18) $\le 2C_0\delta e^{\omega a}$ in $|t| \le a > 0$. This ends the proof.

THEOREM 4.4. _Let_ $u(t;\varepsilon)$, $u(t)$ _be as in Theorem_ 4.3. _Assume that_ $u_0(\varepsilon)$, $u_0 \in D(A)$ _and that_

$$Au_0(\varepsilon) \to Au, \ u_1(\varepsilon) \to iAu_0 \text{ as } \varepsilon \to 0 . \tag{4.19}$$

Then

$$u'(t;\varepsilon) \to u'(t) \quad \text{as} \quad \varepsilon \to 0 \tag{4.20}$$

uniformly on compacts of $-\infty < t < \infty$.

Proof: Given $\delta > 0$ choose $\bar{u} \in D(A^3)$ such that $\|A\bar{u} - A\bar{u}_0\| \leq \delta$ so that $\|\bar{u}'(t) - u'(t)\| = \|S_i'(t)(\bar{u} - u_0)\| = \|S_i(t)(A\bar{u} - Au_0)\| \leq C_0\delta e^{\bar{\omega}|t|}$. We use this time (4.12) with $\bar{u}$ instead of u_0: details are omitted.

We have already pointed out that no analogues of the uniform bounds in Chapter VI or of the uniform convergence results that we established there exist in this case. To put this in evidence we use (as we did already in Chapter VI, for instance in Example VI.3.10) the Hilbert space $E = \ell^2$ of all complex valued sequences $\{u_n;\ 1 \leq n < \infty\}$ such that $\|u\| = (\sum |u_n|^2)^{1/2} < \infty$ and A is the (self adjoint) operator

$$A\{u_n\} = \{\mu_n u_n\}, \tag{4.21}$$

where $\{\mu_n\}$ is a sequence of real numbers bounded above (the domain of A consists of all $\{u_n\}$ such that the right hand side of (4.21) belongs to ℓ^2). We shall also use the space $E = \mathbb{C}^m$ with its ordinary Euclidean norm and an operator A of the form (4.21). We check easily that the solution operators of (2.1) corresponding to the operator A are

$$\mathfrak{S}_i(t;\varepsilon)\{u_n\} = \left\{\frac{\lambda_n^+(\varepsilon)e^{\lambda_n^-(\varepsilon)t} - \lambda_n^-(\varepsilon)e^{\lambda_n^+(\varepsilon)t}}{\lambda_n^+(\varepsilon) - \lambda_n^-(\varepsilon)}\, u_n\right\}, \tag{4.22}$$

$$\mathfrak{S}_i(t;\varepsilon)\{u_n\} = \left\{\frac{e^{\lambda_n^+(\varepsilon)t} - e^{\lambda_n^-(\varepsilon)t}}{\varepsilon^2(\lambda_n^+(\varepsilon) - \lambda_n^-(\varepsilon))}\, u_n\right\}, \tag{4.23}$$

where $\lambda_n^+(\varepsilon)$, $\lambda_n^-(\varepsilon)$ are the roots of the characteristic polynomial

$$\varepsilon^2\lambda^2 - i\lambda - \mu_n = 0:$$

$$\lambda_n^+(\varepsilon) = \frac{i}{2\varepsilon^2}\left(1 + (1 - 4\varepsilon^2\mu_n)^{1/2}\right), \quad \lambda_n^-(\varepsilon) = \frac{i}{2\varepsilon^2}\left(1 - (1 - 4\varepsilon^2\mu_n)^{1/2}\right)$$

(note that, since the sequence $\{\mu_n\}$ is bounded above, the roots $\lambda_n^+(\varepsilon)$, $\lambda_n^-(\varepsilon)$ will be different for ε sufficiently small).

EXAMPLE 4.5. Convergence in Theorem 4.3 is not uniform with respect to u_0 even if $\|u_0\|$ is bounded. We take $E = \ell^2$, $\mu_n = -n^2$,

$\varepsilon_m^2 = 3/4m^2$. Then

$$\lambda_m^+(\varepsilon_m) = 2m^2 i, \quad \lambda_m^-(\varepsilon_m) = -2m^2 i/3 \tag{4.24}$$

and, since

$$S_i(t)\{u_n\} = \{e^{-in^2 t} u_n\}, \tag{4.25}$$

we have, for all m,

$$\|\mathfrak{C}_i(t;\varepsilon_m)\{u_n\} - S_i(t)\{u_n\}\| \geq \left| \frac{3}{4} e^{-i(2m^2/3)t} + \frac{1}{4} e^{-i(2m^2)t} - e^{-im^2 t} \right| |u_m|. \tag{4.26}$$

In constrast, there is uniform convergence outside of an initial layer $0 \leq t \leq t(\varepsilon)$ in the parabolic case: see Theorem VI.3.11.

EXAMPLE 4.6. The condition $\varepsilon^2 u_1(\varepsilon) \to 0$ in Theorem 4.3 for convergence of $u(t;\varepsilon)$ cannot be weakened. We take $E = \ell^1$ and rewrite formula (4.23) as follows:

$$\mathfrak{S}_i(t;\varepsilon)\varepsilon^2 u_1(\varepsilon) = 2ie^{it/2\varepsilon^2} \frac{\sin((1 - 4\varepsilon^2\mu)^{1/2}/2\varepsilon^2)t}{(1 - 4\varepsilon^2\mu)^{1/2}} \varepsilon^2 u_1(\varepsilon), \tag{4.27}$$

which does not have a limit as $\varepsilon \to 0$ unless $\varepsilon^2 u_1(\varepsilon) \to 0$.

For the abstract parabolic case see Theorem VI.3.11: we note that there is again convergence outside of an initial layer.

EXAMPLE 4.7. The rate of convergence in Theorem 4.1 is best possible. We use the space $E = \ell^2$ and the operator A in (4.21). Write (4.22) in the form

$$\mathfrak{C}_i(t;\varepsilon)\{u_n\} = \{\Theta_n(t;\varepsilon)u_n\}. \tag{4.28}$$

After some computation with Taylor series we check that

$$\Theta_n(t;\varepsilon) = e^{i\mu_n t} + \frac{i\varepsilon^2}{4} \mu_n^2 t e^{i\mu_n t} + \varepsilon^2 \mu_n e^{i\mu_n t} - \varepsilon^2 \mu_n e^{ir_n(\varepsilon)t} + O(\varepsilon^4) \tag{4.29}$$

where $r_n(\varepsilon)$ is a real number. Assume that $\{u_n\} \in \ell^2$ is such that

$$\|\mathfrak{C}_i(t;\varepsilon)\{u_n\} - S_i(t)\{u_n\}\| < C\varepsilon^2$$

as $\varepsilon \to 0$. Rewrite this inequality as

$$\sum \varepsilon^{-4} |\Theta_n(t;\varepsilon) - e^{i\mu_n t}|^2 |u_n|^2 \leq C^2 .$$

Taking (4.29) into account, we obtain, assuming that $\mu_n \to \infty$ in (4.21),

$$\sum \mu_n^4 |u_n|^2 < \infty , \tag{4.30}$$

so that $u \in D(A^2)$. We recall that in the abstract parabolic case, convergence of order ε^2 can be obtained under the weaker assumption that $u \in D(A)$ (see Theorem VI.5.5).

EXAMPLE 4.8. Convergence in Theorem 4.3 is not uniform in $t > 0$, even when $\omega = 0$ in (3.1). We take here $E = C^1$, $Au = \mu u$ with $\mu < 0$, $u_0(\varepsilon) = u$, $u_1(\varepsilon) = 0$; the fact that $\mathfrak{S}_i(t;\varepsilon)u$ does not converge uniformly to $S_i(t) = e^{i\mu t}$ is an obvious consequence of the fact that $e^{\lambda^+(\varepsilon)t}$ does not converge uniformly to $e^{i\mu t}$ uniformly in $t \geq 0$.

EXAMPLE 4.9. There are no uniform bounds for $\mathfrak{S}_i'(t;\varepsilon)$. We obtain from (4.23) that, with the notation of (4.28),

$$\Theta_n'(t;\varepsilon) \to i\mu_n e^{\mu_n t}$$

as $\varepsilon \to 0$; if a bound of the type of (VI.4.20) were valid, $S'(t)$ would be a bounded operator, absurd if A is unbounded.

VII.5 Verification of the hypoteses.

We examine in this section operators that satisfy Assumption 3.1, beginning with the case where $E = H$ is a Hilbert space and A is a normal operator. It follows from Exercise II.5 that $A \in \mathfrak{C}^2(C,\omega)$ (that is, that A generates a strongly continuous cosine function $C(t)$ satisfying

$$\|C(t)\| \leq Ce^{\omega|t|} \qquad (-\infty < t < \infty), \tag{5.1}$$

if and only if $\sigma(A)$, the spectrum of A, is contained in a region of the form

$$\left\{\lambda : \mathrm{Re}\,\lambda \leq \omega^2 - \frac{(\mathrm{Im}\,\lambda)^2}{4\omega^2}\right\} \tag{5.2}$$

(that is, the region to the left of the parabola passing through $\omega^2, \pm 2i\omega^2$). We note in passing that $C(t)$ can be computed using the functional calculus for normal operators:

$$C(t)u = c(t;A)u = \int_{\sigma(A)} c(t;\mu)P(d\mu)u ,$$

where $P(d\mu)$ is the resolution of the identity for A and $c(t;\mu) = \cosh t\mu^{1/2} = 1 + t^2\mu/2! + t^4\mu^2/4! + \ldots$ Moreover, the constant C in (5.1) can be taken equal to 1. (In fact, the estimate can be improved to $\|C(t)\| \leq \cosh \omega t$).

THEOREM 5.1. _The normal operator_ A _satisfies Assumption_ 3.1 _if and only if_ $\sigma(A)$ _is contained in a half-strip_.

$$\mathrm{Re}\mu \leq a, \quad |\mathrm{Im}\,\mu| \leq b . \tag{5.3}$$

Proof: We have seen (Theorem 3.2) that Assumption 3.1 implies that iA generates a strongly continuous group. Since, on the other hand, A generates a cosine function it follows that $\sigma(A)$ is contained in the intersection of a horizontal strip with a region defined by (5.2) which intersection is itself contained in a half-strip of the form (5.3).

Conversely, let $\sigma(A)$ be contained in a half-strip defined by both inequalities (5.3). The half-strip corresponding to $a(\varepsilon) = a - (2\varepsilon)^{-2}$ is a subset of a set of the form (5.2) if and only if

$$\omega \geq \omega(\varepsilon), \tag{5.4}$$

where $\omega(\varepsilon)$ is a solution of the equation

$$a - \frac{1}{4\varepsilon^2} = \omega(\varepsilon)^2 - \frac{b^2}{4\omega(\varepsilon)^2} , \tag{5.5}$$

so that, if $\varepsilon \leq (4a)^{-1/2}$ we have

$$\begin{aligned}\omega(\varepsilon) &= \frac{1}{2^{1/2}}\left\{\left(\left(\frac{1}{4\varepsilon^2} - a\right)^2 + b^2\right)^{1/2} - \frac{1}{4\varepsilon^2} + a\right\}^{1/2} \\ &= \frac{1}{2^{1/2}}\left(\frac{1}{4\varepsilon^2} - a\right)^{1/2}\left\{\left(1 + b^2\left(\frac{1}{4\varepsilon^2} - a\right)^{-2}\right)^{1/2} - 1\right\}^{1/2} \\ &\leq \frac{1}{2^{1/2}}\left(\frac{1}{4\varepsilon^2} - a\right)^{1/2}\frac{b}{2^{1/2}}\left(\frac{1}{4\varepsilon^2} - a\right)^{-1} = \frac{b\varepsilon}{(1 - 4a\varepsilon^2)^{1/2}} ,\end{aligned} \tag{5.6}$$

and the first estimate (3.8) is verified, say for $0 \leq \varepsilon \leq (8a)^{-1/2}$ (it is actually satisfied for all $\varepsilon \geq 0$, as we see easily estimating (5.6) away from zero). We note next that

$$\mathcal{S}(t;A - (2\varepsilon)^{-2}I) = s(t;A - (2\varepsilon)^{-2}I), \tag{5.7}$$

where $s(t;\mu) = \mu^{-1/2}\sinh t\mu^{1/2} = t - t^3\mu^2/3! + t^5\mu^3/5! - \ldots$, thus the norm $\|\mathcal{S}(t;A - (2\varepsilon)^{-2}I)\|$ does not surpass the supremum of $|\mu^{-1/2}\sin t\mu^{1/2}|$ in the half strip defined by

$$\mathrm{Re}\mu \leq a - \frac{1}{4\varepsilon^2}, \quad |\mathrm{Im}\,\mu| \leq b. \tag{5.8}$$

If μ belongs to the region defined by (5.3) then

$$|\mu^{1/2}| > \left(\frac{1}{4\varepsilon^2} - a\right)^{1/2} = \frac{(1 - 4a\varepsilon^2)^{1/2}}{2\varepsilon}. \tag{5.9}$$

On the other hand

$$|\sinh t\mu^{1/2}| \leq \cosh \mathrm{Re}(t\mu^{1/2}) \leq e^{\mathrm{Re}(t\mu^{1/2})} \leq e^{\omega(\varepsilon)|t|}$$

for μ limited by (5.8) (recall that μ must be contained in the region defined by (5.2) with $\omega = \omega(\varepsilon)$). Hence

$$\|\mathcal{S}(t;A - (2\varepsilon)^{-2}I)\| \leq 2(1 - 4a\varepsilon^2)^{-1/2}\varepsilon\, e^{\omega(\varepsilon)|t|}$$

$$\leq 2\varepsilon(1 - 4a\varepsilon^2)^{-1/2}\exp((1 - 4a\varepsilon^2)^{-1/2}b\varepsilon|t|),$$

which is the second inequality (3.8). This ends the proof of Theorem 5.1.

We note the important particular case where A is self-adjoint with $-A \geq 0$, in which case we can take $\omega = 0$ in (3.8). Another case that can be reduced to this is covered by the following result

THEOREM 5.2. *Let* A *generate an uniformly bounded cosine function* $C(t)$,

$$\|C(t)\| \leq C \qquad (-\infty < t < \infty) \tag{5.10}$$

in a Hilbert space H. *Then* A *satisfies Assumption* 2.1 *with* $\omega = 0$.

Proof: We have shown in Theorem V.6.7 that if A is an operator that satisfies the assumptions in Theorem 5.2 then there exists a (self-adjoint) bounded, invertible operator Q and a self adjoint operator B with $B \leq 0$ such that

$$A = Q^{-1}BQ. \tag{5.11}$$

In obvious notation,

$$C(t;A) = Q^{-1}C(t;B)Q, \quad \mathcal{S}(t;A) = Q^{-1}\mathcal{S}(t;B)Q, \tag{5.12}$$

thus it follows from Theorem 5.1 that A satisfies Assumption 3.1 as claimed. To compute explicitly the constants C_0, C_1 in (3.1) or the constants C_0', C_1 in (3.8), explicit estimations for the norms $\|Q\|$, $\|Q^{-1}\|$ are needed. These are given in Theorem V.6.7 although it is not clear whether they are best possible.

Another class of operators satisfying Assumption 3.1 is identified below.

THEOREM 5.3. Let E be an arbitrary Banach space, A_0 an operator satisfying Assumption 3.1, B a bounded operator. Then $A = A_0 + B$ satisfies Assumption 3.1 as well.

The proof of Theorem 5.3 is based on the following result on perturbation of cosine functions, where E is again an arbitrary Banach space.

THEOREM 5.4. Let A_0 be the infinitesimal generator of a strongly continuous cosine function $C(\hat{t}) = C(t;A_0)$, B a bounded operator. Then $A = A_0 + B$ generates a strongly continuous cosine function $C(t;A) = C(t;A_0 + B)$.

Theorem 5.4 is an obvious kin of Theorem IV.2.3, where a similar "perturbation-by-bounded-operator" result is shown for generators of strongly continuous groups. Although Theorem 5.4 can be shown by means of estimations of the resolvent as in Theorem IV.2.3, we shall use here another method that yields an explicit formula for $C(t;A_0 + B)$.

Given two (E) - valued functions $F(\hat{t})$ and $G(\hat{t})$, defined and strongly continuous in $-\infty < t < \infty$ we define their convolution by

$$(F * G)(t)u = \int_0^t F(t - s)G(s)u \, ds. \tag{5.13}$$

We check easily that $(F * G)(\hat{t})$ is a strongly continuous (E) - valued function defined in $-\infty < t < \infty$. Note that, by the uniform boundedness theorem, any strongly continuous (E) - valued function must be bounded in the norm of (E) on compact subsets of $-\infty < t < \infty$; moreover, $\|F(\hat{t})\|$ is the supremum of the family $\{\|F(t)u\|; \|u\| \leq 1\}$ of continuous functions, thus is lower semicontinuous as a function of t. This gives sense to the following estimation:

$$\|(F * G)(t)\| \le \int_0^t \|F(t - s)\| \|G(s)\| \, ds \tag{5.14}$$

for $t \geq 0$, with an obvious counterpart for $t < 0$. We check easily that the convolution of (E) - valued strongly continuous functions enjoys all the privileges of its scalar counterpart (such as associativity) that make sense in the present context. Define a function $Q(t)$ as the sum of the series

$$C(t;A_0)u + C(t;A_0) * B\mathcal{S}(t;A_0)u + C(t;A_0) * B\mathcal{S}(t;A_0) * B\mathcal{S}(t;A_0)u + \ldots \tag{5.15}$$

To show convergence of (5.15) we pick constants $C_0, \omega > 0$ such that

$$\|C(t)\| \le C_0 e^{\omega|t|} \qquad (-\infty < t < \infty), \tag{5.16}$$

and make use of (5.14) in each of the terms. The final result is

$$\begin{aligned} &\|C(t;A_0) * B\mathcal{S}(t;A_0)^{*n}\| \\ &\le \omega^{-n} C_0^{n+1} \|B\|^n \frac{|t|^n}{n!} e^{\omega|t|} \qquad (-\infty < t < \infty,\ n = 1,2,\ldots). \end{aligned} \tag{5.17}$$

It follows that (5.15) converges uniformly on compact subsets of $-\infty < t < \infty$ to $Q(t)u$; since each term in (5.15) is a strongly continuous function, $Q(\hat{t})$ is a strongly continuous (E) - valued function satisfying

$$\|Q(t)\| \le C e^{(\omega + C_0\|B\|)|t|} \qquad (-\infty < t < \infty). \tag{5.18}$$

Using (5.17) and (5.18) we see that we can multiply (5.15) by $\exp(-\lambda t)$ if $\lambda > \omega + C_0\|B\|$ and then integrate term by term. By virtue of the convolution theorem for Laplace transforms in each term we obtain, using the first equality (II.2.11),

$$\begin{aligned} \hat{Q}(\lambda)u &= \int_0^\infty e^{-\lambda t} Q(t)u \, dt \\ &= \lambda R(\lambda^2;A_0)u + \lambda R(\lambda^2;A_0)BR(\lambda^2;A_0)u \\ &\quad + \lambda R(\lambda^2;A_0)BR(\lambda^2;A_0)BR(\lambda^2;A_0)u + \ldots \\ &= \lambda R(\lambda^2;A_0 + B) = \lambda R(\lambda^2;A)u \quad (\lambda > \omega + C_0\|B\|), \end{aligned} \tag{5.19}$$

where the justification of the last inequality follows the lines of the argument used in Theorem IV.2.3. We apply then Theorem II.2.3. This completes the proof of Theorem 5.4.

We note the following formula, which is obtained integrating (5.15) in $0 \leq s \leq t$:

$$\mathcal{S}(t;A)u = \mathcal{S}(t;A_0 + B)u = \mathcal{S}(t;A_0)u + \mathcal{S}(t;A_0) * B\mathcal{S}(t;A_0)u$$
$$+ \mathcal{S}(t;A_0) * B\mathcal{S}(t;A_0) * B\mathcal{S}(t;A_0)u + \dots \tag{5.20}$$

<u>Proof of Theorem</u> 5.3: We use the series (5.15) to express $C(t;A_0 + B - (2\varepsilon)^{-2}I) = C(t;A_0 - (2\varepsilon)^{-2}I + B)$ in terms of $C(t;A_0 - (2\varepsilon)^{-2}I)$ and $\mathcal{S}(t;A_0 - (2\varepsilon)^{-2}I)$ and estimate the convolutions in an obvious way, using as a basis both inequalities (3.8). We obtain

$$\|C(t;A_0 + B - (2\varepsilon)^{-2}I)\| \leq C_0' e^{\omega\varepsilon|t|} + C_0' C_1 \|B\| |t| \varepsilon\, e^{\omega\varepsilon|t|}$$
$$+ C_0' C_1^2 \|B\|^2 \frac{|t|^2}{2!} \varepsilon^2 e^{\omega\varepsilon|t|} + \dots = C_0' e^{(\omega + C_1\|B\|)\varepsilon|t|}$$
$$(-\infty < t < \infty, \quad 0 \leq \varepsilon \leq \varepsilon_0). \tag{5.21}$$

A totally similar estimation of $\mathcal{S}(t;A_0 + B - (2\varepsilon)^{-2}I)$ written in terms of $\mathcal{S}(t;A_0 - (2\varepsilon)^{-2}I)$ using (5.20) yields:

$$\|\mathcal{S}(t;A_0 + B - (2\varepsilon)^{-2}I)\| \leq C_1 \varepsilon e^{\omega\varepsilon|t|} + C_1^2 \|B\| \; |t| \varepsilon^2 e^{\omega\varepsilon|t|}$$
$$+ C_0^3 \|B\|^2 \frac{|t|^2}{2!} \varepsilon^3 e^{\omega\varepsilon|t|} + \dots = C_1 \varepsilon e^{(\omega + C_1\|B\|)\varepsilon|t|}$$
$$(-\infty < t < \infty, \quad 0 \leq \varepsilon \leq \varepsilon_0). \tag{5.22}$$

This ends the proof of Theorem 5.3.

We examine in the rest of the section a complement to Theorems 4.1 and 4.2 obtained by means of interpolation theory. We assume that $E = H$ is a Hilbert space and that

$$A = A_0 + B, \tag{5.23}$$

where B is bounded and A_0 is self-adjoint and bounded above (that A satisfies Assumption 2.1 has been proved in Theorem 5.3). To simplify, we also assume that $\sigma(A_0) \subseteq (-\infty, 0)$, which can always be achieved by an obvious decomposition of A_0. Finally, it is required that

$$BD(A_0) \subseteq D(A_0) . \tag{5.24}$$

We have already pointed out in §VI.8 how to defined fractional powers $(-A_0)^\zeta$, where $\zeta = \alpha + i\tau$ is an arbitrary complex number. The pertinent formula is

$$(-A_0)^\zeta u = \int_0^\infty \mu^\zeta P(-d\mu)u , \tag{5.25}$$

where $P(-d\mu)$ is the resolution of the identity for A_0. It was already proved in Section VI.8 that

$$\|(-A_0)^\zeta u\|^2 = \|(-A_0)^{\alpha+i\tau}u\|^2$$

$$= \int_0^\infty \mu^{2\alpha}\|P(-d\mu)u\|^2 = \|(-A_0)^\alpha u\|^2. \tag{5.26}$$

Let $Q : H \to H$ be a linear operator such that

$$\|Qu\| \le K_0\|u\| \quad (u \in H), \quad \|Qu\| \le K_2\|A_0^2 u\| \quad (u \in D(A_0^2)) . \tag{5.27}$$

Consider the H-valued holomorphic function

$$\varphi(\zeta) = \varphi(\alpha + i\tau) = Q(-A_0)^{-\alpha+i\tau}u \quad (0 \le \alpha \le 2),$$

for $u \in H$ fixed. Making use of both inequalities (5.27) and applying Hadamard's three-lines theorem to φ (see again §VI.8) we deduce that

$$\|\varphi(\alpha + i\tau)\| \le K_0^{(2-\alpha)/2}K_2^{\alpha/2}\|u\| \quad (0 \le \alpha \le 2),$$

hence

$$\|Qu\| \le K_0^{(2-\alpha)/2}K_2^{\alpha/2}\|(-A_0)^\alpha u\| \quad (u \in D((-A_0)^\alpha u, 0 \le \alpha \le 2). \tag{5.28}$$

We apply this argument to the operator $\mathfrak{C}_i(t;\varepsilon) - S_i(t)$. Using (3.1) and (3.9) we obtain

$$\|\mathfrak{C}_i(t;\varepsilon)u - S_i(t)u\| \le 2C_0e^{\omega|t|}\|u\| . \tag{5.29}$$

The second estimate is somewhat less trivial, since (4.5) provides bounds in terms of $\|Au\|$ and of $\|A^2u\|$, rather than $\|A_0^2u\|$ as needed. To perform the conversion we note that

$$A^2A_0^{-2} = (A_0 + B)^2A_0^{-2} = (A_0^2 + A_0B + BA_0 + B^2)A^{-2}$$

$$= I + A_0BA_0^{-2} + BA_0^{-1} + B^2A_0^{-2} , \tag{5.30}$$

where the first, third and fourth operators are trivially bounded; the second (in fact, even $A_0BA_0^{-1}$) is bounded because of (5.24) and the closed graph theorem. On the other hand, we have

$$AA_0^{-2} = (A_0 + B)A_0^{-2} = A_0^{-1} + BA_0^{-2}. \tag{5.31}$$

It follows from (5.30) and (5.31) that

$$\|A^2u\| \le C\|A_0^2u\|, \quad \|Au\| \le C\|A_0^2u\|. \tag{5.32}$$

We combine (5.32) with (4.5), obtaining

$$\|\mathfrak{C}_i(t;\varepsilon)u - S_i(t)u\| \le C'\varepsilon^2(1+|t|)e^{\omega|t|}\|A_0^2u\| \quad (u \in D(A^2) = D(A_0^2)). \tag{5.33}$$

Combining (5.29) and (5.33) with the preceding remarks (especially (5.28)) we deduce that if $u \in D((-A_0)^{\alpha})$ $0 < \alpha \le 2$, we have

$$\|\mathfrak{C}_i(t;\varepsilon)u - S_i(t)u\| \le C(\alpha)\varepsilon^{\alpha}(1+|t|)^{\alpha/2}e^{\omega|t|}\|(-A_0)^{\alpha}u\|$$

$$(u \in D((-A)^{\alpha}), \quad -\infty < t < \infty). \tag{5.34}$$

THEOREM 5.5. *Let* $E = H$ *be a Hilbert space,* $A = A_0 + B$ *with* B *bounded and* A_0 *self-adjoint and bounded above, and let* $u(t;\varepsilon)$ *be a solution of the homogeneous problem* (2.1), $u(t)$ *a solution of the homogeneous problem* (2.2) *with* $u_0 \in D((-A_0)^{\alpha})$, $0 < \alpha \le 2$. *Then, if* (5.24) *holds there exists a constant* $C(\alpha)$ *such that*

$$\|u(t;\varepsilon) - u(t)\| \le C(\alpha)\varepsilon^{\alpha}(1 + |t|)^{\alpha/2}e^{\omega|t|}\|(-A_0)^{\alpha}u_0\|$$

$$+ C_0e^{\omega|t|}\|u_0(\varepsilon) - u_0\| + C_1e^{\omega|t|}\varepsilon^2\|u_1(\varepsilon)\| \quad (-\infty < t < \infty). \tag{5.35}$$

The proof follows that of Theorem VI.8.2; we omit the details. Theorem 5.5 implies that

$$\|u(t;\varepsilon) - u(t)\| = O(\varepsilon^{\alpha}) \tag{5.36}$$

uniformly on compacts of $-\infty < t < \infty$ if

$$\|u_0(\varepsilon) - u_0\| = O(\varepsilon^{\alpha}), \quad \|u_1(\varepsilon)\| = O(\varepsilon^{\alpha-2}) \quad \text{as} \quad \varepsilon \to 0. \tag{5.37}$$

THEOREM 5.6. *Let* E, A *be as in Theorem* 5.5, *and let* $u(t;\varepsilon)$ *be a solution of the homogeneous problem* (2.1) *with* $u_0(\varepsilon) \in D(A)$, $u(t)$ *a solution of the homogeneous problem* (2.2) *with* $u_0 \in D((-A_0)^{1+\alpha})$. *Then, if* (5.24) *holds there exists a constant* $C(\alpha)$ *such that*

$$\|u'(t;\varepsilon) - u'(t)\| \le C(\alpha)\varepsilon^{\alpha}(1+|t|)^{\alpha/2}e^{\omega|t|}\|(-A_0)^{1+\alpha}u_0\|$$

$$+ e^{\omega|t|}(C_0\|u_1(\varepsilon) - iAu_0\| + C_1\|u_1(\varepsilon) - iAu_0(\varepsilon)\|). \tag{5.38}$$

The proof consists in applying Theorem 5.5 to $u'(t;\varepsilon)$, $u'(t)$ (see the proof of Theorem 4.2). As a consequence, we obtain easily that, if $u_0 \in D((-A_0)^{1+\alpha})$, $u_0(\varepsilon) \in D(A)$ then

$$\|u'(t;\varepsilon) - u'(t)\| = O(\varepsilon^{\alpha}) \tag{5.39}$$

uniformly on compacts of $-\infty < t < \infty$ if

$$\|u_1(\varepsilon) - iAu_0\| = O(\varepsilon^{\alpha}) \quad \text{and} \quad \|u_1(\varepsilon) - iAu_0(\varepsilon)\| = O(\varepsilon^{\alpha}) , \tag{5.40}$$

or, equivalently, if

$$\|u_1(\varepsilon) - iAu_0\| = O(\varepsilon^{\alpha}) \quad \text{and} \quad \|Au_0(\varepsilon) - Au_0\| = O(\varepsilon^{\alpha}) . \tag{5.41}$$

§VII.6 Elliptic differential operators.

We examine the operator (1.2) written in divergence or variational form,

$$A = \sum_{j=1}^{m} \sum_{k=1}^{m} D^j(a_{jk}(x)D^k u) + \sum_{j=1}^{m} \hat{b}_j(x)D^j u + c(x)u \tag{6.1}$$

in an arbitrary domain Ω of m-dimensional Euclidean space $\mathbb{R}^m$; here $D^j = \partial/\partial x_j$ and $x = (x_1,..,x_m)$. As pointed out in §IV.1, (6.1) is equivalent to (1.2) if the coefficients a_{jk} are differentiable. We assume again that $a_{jk} = a_{kj}$. The symbol $A(\beta)$ will denote the restriction of (6.1) obtained by imposition of a boundary condition β, either the Dirichlet boundary condition

$$u(x) = 0 \quad (x \in \Gamma), \tag{6.2}$$

or the variational boundary condition

$$D^{\tilde{\nu}}u(x) = \sum\sum a_{jk}(x)\nu_j D^k u(x) = \gamma(x)u(x) \quad (x \in \Gamma) , \tag{6.3}$$

where Γ is the boundary of Ω and $\nu = (\nu_1,\ldots,\nu_m)$ is the exterior normal (unit) vector at **x**; $D^{\tilde{\nu}}u$ is called the conormal derivative of u at x.

The operator $A(\beta)$ has already been constructed in Chapter IV under minimal assumptions on the coefficients. However, additional hypoteses will have to be placed on the first order coefficients $\hat{b}_j$ to force $A(\beta)$ to satisfy Assumption 3.1. We assume that the a_{jk} and c are measurable and bounded in Ω. Complex values for c are allowed; the a_{jk} are real and satisfy the uniform ellipticity condition

$$\sum\sum a_{jk}(x)\xi_j\xi_k \geq \kappa|\xi|^2 \quad (\xi \in \mathbb{R}^m) . \tag{6.4}$$

for some $\kappa > 0$. Finally, we assume that the first order coefficients $\hat{b}_j(x)$ are imaginary, that is

$$\hat{b}_j(x) = ib_j(x) \tag{6.5}$$

with b_j real (we shall see later that this requirement cannot be eliminated) and that each b_j belongs to $W^{1,\infty}(\Omega)$ (i.e. it has first partial derivatives in $L^\infty(\Omega)$).

As we see below, the requirements that the a_{jk} be real and the b_j be imaginary cannot be omitted.

EXAMPLE 6.1. Let $m = 1$, $\Omega = \mathbb{R}$, A the constant coefficient operator

$$Au = au'' + bu' + cu .$$

Using the Fourier-Plancherel transform we show that

$$\sigma(A) = \{-a\sigma^2 - ib\sigma + c;\ -\infty < \sigma < \infty\} ,$$

so that: (a) $\sigma(A)$ will not be contained in a region of the form (5.2) if a is not real (b) if b is not imaginary, $\sigma(A)$ will not be contained in a half-strip of the form (5.3).

The construction of the operator $A(\beta)$ has been carried out in Chapter IV without the special requirements on the b_j present here. In the case of the Dirichlet boundary condition, no restrictions on the domain Ω were necessary: for a boundary condition of type (6.3) it is enough to require that Ω be a bounded domain of class $C^{(1)}$ (or a domain of class $C^{(1)}$ with bounded boundary Γ) and that $\gamma \in L^\infty(\Gamma)$. However, our job is now to show that $A(\beta)$ satisfies Assumption 3.1 under the reinforced assumptions on the b_j, and we shall do so by proving that

$$A(\beta) = A_0(\beta) + B , \tag{6.6}$$

with $A_0(\beta)$ self-adjoint and B bounded. This will be achieved by slight modifications of the arguments in Chapter IV that we outline below, beginning with the Dirichlet boundary condition. For $u,v \in H_0^1(\Omega)$ define

$$[u,v]_\alpha = \int_\Omega \{\sum\sum a_{jk} D^j\bar{u}D^k v - \frac{i}{2}\sum b_j(D^j\bar{u}v - \bar{u}D^j v) + \alpha\bar{u}v\}dx , \tag{6.7}$$

where $\alpha > 0$ is a parameter to be fixed below. Obviously, $[u,v]_\alpha$ is linear in v, conjugate linear in u and we check easily that

$[v,u]_\alpha = \overline{[u,v]_\alpha}$. Using the uniform ellipticity assumption (6.4), the inequality $|D^j u)v| \le (\varepsilon/2)|D^j\bar{u}|^2 + (1/2\varepsilon)|v|^2$ and its counterpart for $|\bar{u}D^j v|$ we easily show that if α is large enough, the first inequality

$$c^2(u,u) \le [u,u]_\alpha \le C^2(u,u) \quad (u \in H_0^1(\Omega)) \tag{6.8}$$

holds for some $c > 0$, where (u,v) is the original scalar product of $H_0^1(\Omega)$; the second inequality (6.8) is a consequence of the assumptions on the coefficients. We shall from now on assume $H_0^1(\Omega)$ endowed with the scalar product (6.7) and its associated norm $\|u\|_\alpha = [u,u]_\alpha^{1/2}$.

The operator $A_0(\beta)$ is defined by

$$((\alpha I - A_0(\beta))u,w) = [u,w]_\alpha \quad (w \in H_0^1(\Omega)), \tag{6.9}$$

the domain of $A_0(\beta)$ consisting of all $u \in H_0^1(\Omega)$ which make the right-hand side of (6.9) continuous in the norm of $L^2(\Omega)$. The rest of the theory of $A_0(\beta)$ unfolds exactly as in Chapter IV, since it is based on the properties of the scalar product $[u,v]_\alpha$ which are the same $(u,v)_\alpha$ has: we check in the same way that $A_0(\beta)$ is symmetric and densely defined, that its construction does not depend on α, that

$$(\lambda I - A_0(\beta))D(A_0(\beta)) = E \quad (\lambda \ge \alpha), \tag{6.10}$$

and that $\lambda I - A_0(\beta)$ is one-to-one for $\lambda \ge \alpha$. We also obtain as a byproduct of (6.8) that $(\lambda I - A_0(\beta))^{-1}$ is <u>bounded</u>, so that $\lambda \in \rho(A_0(\beta))$ if $\lambda \ge \alpha$. This is known to imply that $A_0(\beta)$ is self adjoint (see Lemma IV.1.1).

The full operator $A(\beta)$ is constructed by perturbation. Let

$$Bu = \frac{i}{2}\sum (b_j D^j u - D^j(b_j u)) + cu = -\frac{i}{2}\sum (D^j b_j)u + cu . \tag{6.11}$$

The assumptions on the b_j and on c imply that B is a bounded operator. We define

$$A(\beta) = A_0(\beta) + B , \tag{6.12}$$

and it follows from Theorem 5.1 that $A(\beta)$ satisfies Assumption 3.1.

The case where the boundary condition (6.3) is used is handled again as in Chapter IV. This time, however, the functional is

$$[u,v]_\alpha' = [u,v]_\alpha - \int_\Gamma (\gamma + \frac{i}{2}\sum b_j)\bar{u}\, v\, d\sigma \tag{6.13}$$

for $u,v \in H^1(\Omega)$. The definition of the operator $A_0(\beta)$ is

$$((\alpha I - A_0(\beta))u,w) = [u,w]'_\alpha \quad (w \in H_0^1(\Omega)). \tag{6.14}$$

The operator $A_0(\beta)$ is again self adjoint: the full operator $A(\beta)$ is obtained by formula (6.12), where B is the bounded operator defined by (6.11). It follows again from Theorem 5.1 that $A(\beta)$ satisfies Assumption 3.1. Summarizing:

THEOREM 6.2. _Let_ Ω _be a domain in_ $\mathbb{R}^m$, A _the operator_ (6.1) _with_ a_{jk}, $c \in L^\infty(\Omega)$, $\hat{b}_j \in W^{1,\infty}(\Omega)$, _Assume, moreover that the_ a_{jk} _are real and satisfy the uniform ellipticity assumption_ (6.4) _and that the_ $\hat{b}_j$ _are purely imaginary. If_ β _is the Dirichlet boundary condition_ (6.2) _the operator_ $A(\beta)$ _defined by_ (6.7) _and_ (6.9) _satisfies Assumption_ 3.1. _If_ Ω _is bounded and of class_ $C^{(1)}$ _and_ β _is the boundary condition_ (6.3) _with_ γ _measurable and bounded in_ Γ _then the operator_ $A(\beta)$ _defined by_ (6.13) _and_ (6.9) _satisfies Assumption_ 3.1.

REMARK 6.3. Theorem 5.5 has an interesting application here. Although $D((-A_0(\beta))^\alpha)$ is not easily identifiable even for $\alpha = 1$ under the present smoothness assumptions, we have show in Theorem IV.2.2 and Theorem IV.5.1 that

$$D((-A_0(\beta))^{1/2}) = H_0^1(\Omega) \tag{6.15}$$

when β is the Dirichlet boundary condition (6.2), and

$$D((-A_0(\beta))^{1/2}) = H^1(\Omega) \tag{6.16}$$

when β is the variational boundary condition (6.3). Using Theorem 5.5 we deduce that if β is the Dirichlet boundary condition,

$$\|u(t;\varepsilon) - u(t)\| = O(\varepsilon^{1/2}) \tag{6.17}$$

if $u_0 \in H_0^1(\Omega)$ and

$$\|u_0(\varepsilon) - u_0\| = O(\varepsilon^{1/2}), \ \|u_1(\varepsilon)\| = O(\varepsilon^{-3/2}). \tag{6.18}$$

The same result holds for boundary conditions β of type (6.3) where we assume that $u_0 \in H^1(\Omega)$. However, we can only guarantee (6.17) if condition (5.24) holds. This is easily seen to be the case if

$$\sum D^j b_j,\ c \in W^{2,\infty}(\Omega).$$

§VII.7 The inhomogeneous equation.

As pointed out in §VII.2, the explicit (generalized) solution of the initial value problem (2.1) with null initial conditions $u_0(\varepsilon)$, $u_1(\varepsilon)$ is

$$u(t;\varepsilon) = \int_0^t \mathfrak{S}_i(t - s;\varepsilon)f(s;\varepsilon)ds \cdot \tag{7.1}$$

We have already noted (in Example 4.6) that $\mathfrak{S}_i(t;\varepsilon)$ is not even strongly convergent as $\varepsilon \to 0$. However (and somewhat surprisingly) (7.1) turns out to translate convergence of $f(t;\varepsilon)$ into convergence of $u(t;\varepsilon)$ at least for a class of operators containing the differential operators in §VII.6.

THEOREM 7.1. *Let* $E = H$ *be a Hilbert space,*

$$A = A_0 + B, \tag{7.2}$$

where A_0 *is a self adjoint operator bounded above,* B *a bounded operator. Let* $T > 0$, $\{f(s;\varepsilon);\ 0 < \varepsilon \le \varepsilon_0\}$ *a family of functions in* $L^1(-T,T;H)$ *such that*

$$f(s;\varepsilon) \to f(s) \quad \text{as} \quad \varepsilon \to 0 \tag{7.3}$$

in $L^1(-T,T;H)$. *Finally, let* $u(t;\varepsilon)$ *be the (weak) solution of the initial value problem*

$$\begin{aligned} \varepsilon^2 u''(t;\varepsilon) - iu'(t;\varepsilon) &= Au(t;\varepsilon) + f(t;\varepsilon) \quad (|t| \le T), \\ u(0;\varepsilon) = 0,\ u'(0;\varepsilon) &= 0 . \end{aligned} \tag{7.4}$$

Then

$$u(t;\varepsilon) \to u(t) \tag{7.5}$$

uniformly in $|t| \le T$, *where* $u(t;\varepsilon)$ *is the weak solution of*

$$\begin{aligned} u'(t) &= iAu(t) + if(t) \quad (|t| \le T), \\ u(0) &= 0 . \end{aligned} \tag{7.6}$$

Proof: We can obviously assume that $\sigma(A_0) \subseteq (0,\infty)$ (if not we incorporate into B the "part of A_0 with spectrum in $\mu \ge 0$"). We shall show first Theorem 7.1 for A_0 and then mix the perturbation B, considering first the case $f(t;\varepsilon) = f(t)$ independent of ε. Let $P(d\mu)$ be the resolution of the identity for A_0 and $\mathfrak{S}_i(t;\varepsilon;A_0)$ the (second) propagator of the equation (7.3) with $B = 0$. The same considerations leading to Examples 4.5 and 4.6 show that

$$\mathfrak{S}_i(t;\varepsilon;A_0)u = \int_{-\infty}^{0} \mathfrak{s}(t;\varepsilon;\mu)P(d\mu)u \tag{7.7}$$

for $u \in E$, where

$$\mathfrak{s}(t;\varepsilon;\mu) = \frac{e^{\lambda^+(\mu;\varepsilon)t} - e^{\lambda^-(\mu;\varepsilon)t}}{\varepsilon^2(\lambda^+(\mu;\varepsilon) - \lambda^-(\mu;\varepsilon))} \tag{7.8}$$

and

$$\lambda^+(\mu;\varepsilon) = i(2\varepsilon^2)^{-1}(1 + (1 - 4\varepsilon^2\mu)^{1/2}), \quad \lambda^-(\mu;\varepsilon) = i(2\varepsilon^2)^{-1}(1 - (1 - 4\varepsilon^2\mu)^{1/2})$$

are the roots of the characteristic polynomial

$$\varepsilon^2\lambda^2 - i\lambda - \mu = 0 \qquad (-\infty < \mu \leq 0). \tag{7.9}$$

Let $0 \leq t \leq T$. We can write

$$u(t;\varepsilon) = \int_0^t \mathfrak{S}_i(t - s;\varepsilon;A_0)f(s)\,ds = \int_{-\infty}^{0} P(d\mu) \int_0^t \mathfrak{s}(t - s;\varepsilon;\mu)f(s)\,ds \tag{7.10}$$

after an easily justified interchange in the order of integration. We note next that

$$|\mathfrak{s}(t - s;\varepsilon;\mu)| \leq 2, \tag{7.11}$$

hence

$$\left\| \int_0^t \mathfrak{s}(t - s;\varepsilon;\mu)f(s)\,ds \right\| \leq 2\|f\|_{L^1(-T,T;E)} \tag{7.12}$$

On the other hand,

$$\int_0^t \mathfrak{s}(t - s;\varepsilon;\mu)f(s)\,ds = \frac{e^{\lambda^+(\mu;\varepsilon)t}}{i(1 - 4\varepsilon^2\mu)^{1/2}} \int_0^t e^{-\lambda^+(\mu;\varepsilon)s} f(s)\,ds$$

$$- \frac{e^{\lambda^-(\mu;\varepsilon)t}}{i(1 - 4\varepsilon^2\mu)^{1/2}} \int_0^t e^{-\lambda^-(\mu;\varepsilon)s} f(s)\,ds$$

$$= I_1(t;\mu;\varepsilon) + I_2(t;\mu;\varepsilon) = I(t;\mu;\varepsilon) \tag{7.13}$$

Since

$$\lambda^-(\mu;\varepsilon) \to i\mu \quad \text{as} \quad \varepsilon \to 0$$

we deduce that, for μ fixed,

$$I_2(t;\mu;\varepsilon) \to ie^{i\mu t} \int_0^t e^{-\mu s} f(s)\,ds \tag{7.14}$$

uniformly in $0 \le t \le T$. To handle the first integral we note that

$$\lambda^+(\mu;\varepsilon) \to i\infty \quad \text{as} \quad \varepsilon \to 0$$

and use the following uniform version of the Riemann-Lebesgue lemma: <u>if</u> $g(t)$ <u>is a (scalar or vector-valued) function in</u> $L^1(0,T)$ <u>then</u>

$$\lim_{\alpha \to \infty} \int_0^t e^{i\alpha s} g(s)\, ds = 0 \tag{7.15}$$

<u>uniformly in</u> $0 \le t \le T$; the proof is achieved approximating g in the L^1 norm by smooth functions. Applying (7.15) to the first integral in (7.13) we obtain

$$I_1(t;\mu;\varepsilon) \to 0 \quad \text{as} \quad \varepsilon \to 0 \tag{7.16}$$

uniformly in $0 \le t \le T$.

Assume that $u(t;\varepsilon) \not\to u(t)$ uniformly in $0 \le t \le T$. Then there exists a sequence $\{t_n\}$, $0 \le t_n \le T$ and a sequence $\{\varepsilon_n\}$, $\varepsilon_n \to 0$ such that

$$\|u(t_n;\varepsilon_n) - u(t_n)\| \ge \delta > 0. \tag{7.17}$$

Making use of (7.14), (7.16) and a variant of the dominated convergence theorem we obtain a contradiction with (7.17). This shows that (7.5) holds uniformly in $0 \le t \le T$. An entirely similar argument takes care of the range $-T \le t \le 0$. The case where f depends on ε is handled writing

$$\begin{aligned} u(t;\varepsilon) &= \int_0^t \mathfrak{S}_i(t-s;\varepsilon;A_0) f(s;\varepsilon)\, ds \\ &= \int_0^t \mathfrak{S}_i(t-s;\varepsilon;A_0)(f(s;\varepsilon) - f(s))\, ds \\ &\quad + \int_0^t \mathfrak{S}_i(t-s;\varepsilon;A) f(s)\, ds \end{aligned} \tag{7.18}$$

and making use of the uniform bound (7.11).

We incorporate finally the perturbation B. It results from (3.7) and from the perturbation formula (5.20) (or directly) that we have

$$\begin{aligned} \mathfrak{S}_i(t;\varepsilon;A)u &= \mathfrak{S}_i(t;\varepsilon;A_0 + B)u = \mathfrak{S}_i(t;\varepsilon;A_0)u \\ &\quad + \mathfrak{S}_i(t;\varepsilon;A_0) * B\mathfrak{S}_i(t;\varepsilon;A_0)u \\ &\quad + \mathfrak{S}_i(t;\varepsilon;A_0) * B\mathfrak{S}_i(t;\varepsilon;A_0) * B\mathfrak{S}_i(t;\varepsilon;A_0)u + \dots \end{aligned} \tag{7.19}$$

hence

$$u(t;\varepsilon) = \mathfrak{S}_i(t;\varepsilon;A_0) * f(t;\varepsilon) + \mathfrak{S}_i(t;\varepsilon;A_0) * B\mathfrak{S}_i(t;\varepsilon;A_0) * f(t;\varepsilon)$$
$$+ \mathfrak{S}_i(t;\varepsilon;A_0) * B\mathfrak{S}_i(t;\varepsilon;A_0) * B\mathfrak{S}_i(t;\varepsilon;A_0) * f(t;\varepsilon) + \dots \qquad (7.20)$$

Now, using (3.1) we show that the n-th term of the series (7.20) is bounded in norm by

$$C_1^n \|B\|^{n-1} \frac{|t|^{n-1}}{(n-1)!} e^{\omega|t|} \|f(t;\varepsilon)\|_{L^1(-T,T;H)} . \qquad (7.21)$$

On the other hand, using repeatedly the previously proved result on convergence of $\mathfrak{S}_i(t;\varepsilon;A_0) * f(t;\varepsilon)$ in each term of (7.20) we deduce that $\mathfrak{S}_i(t;\varepsilon;A_0) * B\mathfrak{S}_i(t;\varepsilon;A_0) * f(t;\varepsilon)$, $\mathfrak{S}(t;\varepsilon;A_0) * B\mathfrak{S}_i(t;\varepsilon;A_0) * B\mathfrak{S}_i(t;\varepsilon;A_0) * f(t;\varepsilon), \dots$ all converge uniformly in $|t| \le T$; the limit of the n-th term of (7.20) is

$$iS(\hat{t};iA_0) * BiS(\hat{t};iA_0) * \dots * BiS(\hat{t};iA_0) * f(t)$$
$$= S(\hat{t};iA_0) * iBS(\hat{t};A_0) * \dots * iBS(\hat{t};A_0) * if(\hat{t}) ,$$

thus the sum of the series converges uniformly, as $\varepsilon \to 0$, to

$$S(\hat{t};iA_0) * if(\hat{t}) + S(\hat{t};iA_0) * iBS(\hat{t};iA_0) * if(\hat{t})$$
$$+ S(\hat{t};iA_0) * iBS(\hat{t};iA_0) * iBS(\hat{t};iA_0) * if(\hat{t}) + \dots$$
$$= S(\hat{t};i(A_0 + B)) * if(\hat{t}) ,$$

where $S(t;iA_0)$ (resp. $S(t;i(A_0 + B))$ is the group generated by iA_0(resp. by $i(A_0 + B)$). This completes the proof of Theorem 7.1.

§VII.8 Miscellaneous comments

The Schrödinger linear perturbation problem was discussed by SCHOENE [1970:1] in the case where E is a Hilbert space and A a self adjoint operator; in that particular setting, Theorem 4.1 was proved by Schoene with somewhat stronger assumptions on the initial conditions. The material in this chapter is entirely taken from the author [1985:1].

Problems of traffic flow (see WHITHAM [1974:1]) lead to equations whose linearized version is

$$\varepsilon^2\left(\frac{\partial^2 u}{\partial t^2} - a\frac{\partial^2 u}{\partial x^2}\right) + \left(\frac{\partial u}{\partial t} - b\frac{\partial u}{\partial x}\right) = 0 \tag{8.1}$$

where ε is a small parameter; the main problem about (8.1) is that of showing that the solution tends to the solution of the limit equation

$$\frac{\partial u}{\partial t} - b\frac{\partial u}{\partial x} = 0 \tag{8.2}$$

with due attention being paid, among other things, to the loss of one initial condition incurred in going from (8.1) to (8.2). For a classical treatment of the problem see WHITHAM [1974:1] or ZAUDERER [1983:1]. An attempt to treat this problem in an operator theoretic way was made in the author [1985:1]; some of the key results are given below in Exercises 3 to 12. The methods used resemble both those in the Schrödinger singular perturbation problem (such as the uniform bounds in Assumption 3.1) and those in the parabolic singular perturbation problem (such as the asymptotic developments in §VI.8).

EXERCISE 1. Write and interpret formula (2.4) for the operator

$$A = d^2/dx^2$$

in the spaces $L^p(-\infty,\infty)$ and $C_0(-\infty,\infty)$ (see Exercises II.3 and II.4).

EXERCISE 2. Show that the operators in Exercise 1 do not satisfy Assumption 3.1 (except in $L^2(-\infty,\infty)$).

EXERCISE 3. Let $E = H$ be a (complex) Hilbert space, A a self adjoint operator such that

$$(Au,u) \leq -\kappa(u,u) \qquad (u \in D(A)) \tag{8.3}$$

with $\kappa > 0$, Q the (only) positive self adjoint square root of $-A$, B a closed, densely defined operator such that

$$D(B) \supseteq D(Q). \tag{8.4}$$

Show that the Cauchy problem

$$\begin{aligned} &\varepsilon^2 u''(t) + u'(t) = (\varepsilon^2 A + B)u(t) \qquad (-\infty < t < \infty), \\ &u(0) = u_0,\ u'(0) = u_1, \end{aligned} \tag{8.5}$$

is well posed for any $\varepsilon > 0$ (Hint: this follows from a standard theorem on perturbation of generators of cosine functions. See the author [1971:1], SHIMIZU-MIYADERA [1978:1], TAKENAKA-OKAZAWA [1978:1] or TRAVIS-WEBB [1981:1].

EXERCISE 4. If $u(\hat{t})$ is a solution of (8.5) we have

$$\varepsilon^2(u(t),u'(t)) - \varepsilon^2(u_0,u_1) - \varepsilon^2\int_0^t \|u'(s)\|^2 ds + \frac{1}{2}\|u(t)\|^2 - \frac{1}{2}\|u_0\|^2$$
$$= \varepsilon^2\int_0^t (Au(s),u(s))ds + \int_0^t (Bu(s),u(s))ds \qquad (t \geq 0). \tag{8.6}$$

(Hint: multiply the equation (8.5) scalarly by $u(t)$ and integrate).

EXERCISE 5. If $u(\hat{t})$ is a solution of (8.5) we have

$$\frac{\varepsilon^2}{2}\|u'(t)\|^2 - \frac{\varepsilon^2}{2}\|u_1\|^2 + \int_0^t \|u'(s)\|^2 ds = \frac{\varepsilon^2}{2}(Au(t),u(t)) - \frac{\varepsilon^2}{2}(Au_0,u_0)$$
$$+ \int_0^t (Bu(s),u'(s))ds \qquad (t \geq 0). \tag{8.7}$$

(Hint: multiply the equation (8.5) scalarly by $u'(t)$ and integrate).

EXERCISE 6. The assumptions are the same in Exercises 3, 4, and 5; we require in addition that

$$\|Bu\| \leq \|Qu\| \qquad (u \in D(Q)). \tag{8.8}$$

Show that if $u(t)$ is a solution of (8.5) then

$$\varepsilon^4\|u'(t)\|^2 - \varepsilon^4(Au(t),u(t)) + \frac{1}{2}\|u(t)\|^2 + \varepsilon^2 \mathrm{Re}(u(t),u'(t)) \tag{8.9}$$
$$\leq \varepsilon^4\|u_1\|^2 - \varepsilon^4(Au_0,u_0) + \frac{1}{2}\|u_0\|^2 + \varepsilon^2\mathrm{Re}(u_0,u_1) + \varepsilon^2\mathrm{Re}\int_0^t (Bu(s),u(s))ds$$

(Hint: multiply (8.7) by $2\varepsilon^2$ and add to (8.6). Take real parts. Combine the first three integral terms into one and use the fact that

$$\|Qu\|^2 + \|v\|^2 - 2\mathrm{Re}(Bu,v) \geq 0 \tag{8.10}$$

for all $u \in D(Q)$, $v \in H$, consequence of (8.8)).

EXERCISE 7. Assume that

$$\operatorname{Re}(Bu,u) \le \omega\|u\|^2 \qquad (u \in D(B)) . \tag{8.11}$$

Using the inequalities $2|(u(t),u'(t))|^2 \le \alpha^2\|u'(t)\|^2 + \alpha^{-2}\|u(t)\|^2$, $2|(u_0,u_1)|^2 \le \beta^2\|u_1\|^2 + \beta^{-2}\|u_0\|^2$ and (8.9) show that

$$\begin{aligned}&\left(\frac{1}{2} - \frac{1}{2\alpha^2}\right)\|u(t)\|^2 + \varepsilon^4\|Qu(t)\|^2 + \left(1 - \frac{\alpha^2}{2}\right)\varepsilon^4\|u'(t)\|^2 \\ &\le \left(\frac{1}{2} + \frac{1}{2\beta^2}\right)\|u_0\|^2 + \varepsilon^4\|Qu_0\|^2 + \left(1 + \frac{\beta^2}{2}\right)\varepsilon^4\|u_1\|^2 + \omega\int_0^t \|u(s)\|^2 ds.\end{aligned} \tag{8.12}$$

EXERCISE 8. Setting $\alpha = \sqrt{2}$ in (8.12), letting $\beta \to \infty$ and using Gronwall's inequality show that, if $u_1 = 0$,

$$\|u(t)\|^2 + 4\varepsilon^4\|Qu(t)\|^2 \le 2(\|u_0\|^2 + 4\varepsilon^4\|Qu_0\|^2)e^{4\omega t} . \tag{8.13}$$

EXERCISE 9. Setting $\alpha = \sqrt{2}$ in (8.12) letting $\beta \to 0$ and using Gronwall's inequality show that, if $u_0 = 0$,

$$\|u(t)\|^2 + 4\varepsilon^4\|Qu(t)\|^2 \le 4\varepsilon^4\|u_1\|^2 e^{4\omega t}. \tag{8.14}$$

EXERCISE 10. Let $\mathfrak{C}_h(t;\varepsilon)$, $\mathfrak{S}_h(t;\varepsilon)$ be the solution operators of (8.5) (defined in the same way $\mathfrak{C}(t;\varepsilon)$, $\mathfrak{S}(t;\varepsilon)$ are defined for the equation (VI.2.1)). Using (8.13) and (8.14) show that

$$\|(I + 2\varepsilon^2 Q)\mathfrak{C}_h(t;\varepsilon)(I + 2\varepsilon^2 Q)^{-1}\| \le 2e^{2\omega t} \qquad (t \ge 0) , \tag{8.15}$$

$$\|\mathfrak{S}_h(t;\varepsilon)\| \le 2e^{2\omega t} \qquad (t \ge 0) . \tag{8.16}$$

EXERCISE 11. The assumptions are the same in Exercises 6, 7, 8, 9, and 10. We require in addition: if $u \in D(A)$ then $Bu \in D(Q)$ and

$$\|(QB - BQ)u\| \le C\|Qu\| \tag{8.17}$$

for some constant C. Show that (8.15) can be "rectified" to

$$\|\mathfrak{C}_h(t;\varepsilon)\| \le 2(1 + Ct)e^{2\omega t} \qquad (t \ge 0) . \tag{8.18}$$

EXERCISE 12. Show that (8.18) (in fact, the improved estimate

$$\| \mathfrak{C}_h(t;\varepsilon) \| \le 2e^{2\omega t} \qquad (t \ge 0))$$

can be shown without assuming (8.17); we require instead that B^* satisfy as well (8.4), (8.8) and (8.11).

EXERCISE 13. Assume that $D((\varepsilon^2 A + B)) \subseteq D(BA) \subseteq D(AB)$ for all $\varepsilon > 0$ and that for $u \in D(BA)$,

$$\|(AB - BA)u\| \le K\|Au\|$$

with the same assumptions on B^* in Exercise 12. Show that $A\mathfrak{C}_h(t;\varepsilon)A^{-1}$ is a strongly continuous (H)-valued function satisfying

$$\|A\, \mathfrak{C}_h(t;\varepsilon)A^{-1}\| \le 2e^{2(\omega + K)t} \qquad (t \ge 0) .$$

EXERCISE 14. The hypotheses are the same in Exercise 13; we assume in addition that $BD(A) \subseteq D(Q)$ and that $(\lambda I - B)D(A)$ is dense in H for $\operatorname{Re} \lambda > c$. Prove an analogue of Theorem 4.1 for the initial value problem (8.5).

CHAPTER VIII

THE COMPLETE SECOND ORDER EQUATION

§VIII.1 The Cauchy problem.

Let E be, as usual, a complex Banach space, A,B linear operators with domains $D(A)$ and $D(B)$ dense in E and range in E. Solutions in $t \geq 0$ of the abstract differential equation

$$u''(t) + Bu'(t) + Au(t) = 0^{(1)} \tag{1.1}$$

(which, by obvious reasons, is called the complete second order abstract differential equation) are assumed to satisfy the following conditions: $u(\hat{t})$ is twice continuously differentiable, $u(t) \in D(A)$, $u'(t) \in D(B)$, $Au(\hat{t})$ and $Bu'(\hat{t})$ are continuous and (1.1) is satisfied in $t \geq 0$. Similar definitions are used in intervals other than $[0,\infty)$.

The theory of (1.1) is considerably more complicated than that of its incomplete counterpart (II.1.1) and can hardly be said to be in definitive form. In fact, all we shall do in this chapter is to try to decide which is the correct notion of well posed Cauchy problem for (1.1). As we shall see in §VIII.2 use of (the obvious extension of) the definition in §II.1 for the incomplete equation leads to paradoxical situations entailing loss of exponential growth of solutions and nonexistence of phase spaces; moreover, the Cauchy problem may be well posed in an interval $0 \leq t \leq T$ without being well posed in $t \geq 0$ (Example 2.2). This motivates the introduction of an additional assumption in §3 (Assumption 3.1). In this setting a phase space generalizing that constructed in §III.1 for the incomplete equation (III.1.1) is assembled in §4.

Modifying slightly the definition in §II.1 we say that the Cauchy problem for (1.1) is well posed or properly posed in $t \geq 0$ if and only if

(a) There exist dense subspaces D_0, D_1 of E such that if $u_0 \in D_0$, $u_1 \in D_1$ then there is a solution $u(\hat{t})$ of (1.1) in $t \geq 0$ satisfying

$$u(0) = u_0, u'(0) = u_1 . \tag{1.2}$$

(b) *There exists a nonnegative, finite function* $C(\hat{t})$ *defined in* $t \geq 0$ *such that*

$$\|u(t)\| \leq C(t)(\|u(0)\| + \|u'(0)\|) \tag{1.3}$$

for any solution of (1.1) *in* $t \geq 0$.

If the function $C(t)$ in (1.3) can be chosen nondecreasing in $t \geq 0$ (or, more generally, bounded on compacts of $t \geq 0$) then we say that the Cauchy problem for (1.1) is *uniformly well posed* (or *uniformly properly posed*) in $t \geq 0$.

The *propagators* or *solution operators* of (1.1) are defined by

$$\mathcal{C}(t)u = u(t), \quad \mathcal{S}(t)u = v(t), \tag{1.4}$$

where $u(\hat{t})$ (resp. $v(\hat{t})$) is the solution of (1.1) with $u(0) = u$, $u'(0) = 0$ (resp. $v(0) = 0$, $v'(0) = u$). The definition of $\mathcal{C}(t)$ (resp. of $\mathcal{S}(t)$) makes sense for $u \in D_0$ (resp. for $u \in D_1$). Since both D_0 and D_1 are dense in E we can extend (using (1.3)) $\mathcal{C}(t)$ and $\mathcal{S}(t)$ to all of E as bounded operators; these operator-valued functions result strongly continuous in $t \geq 0$ and satisfy

$$\|\mathcal{C}(t)\| \leq C(t), \quad \|\mathcal{S}(t)\| \leq C(t) \qquad (t \geq 0). \tag{1.5}$$

Moreover, by definition, $\mathcal{C}(0) = I$, $\mathcal{S}(0) = 0$. Finally, we prove easily that if $u(\hat{t})$ is an arbitrary solution of (1.1) in $t \geq 0$ then

$$u(t) = \mathcal{C}(t)u(0) + \mathcal{S}(t)u'(0). \tag{1.6}$$

The proof is the same as that of (II.1.6).

We shall assume from now on that the operators A and B are *closed*.

§VIII.2 Growth of solutions and existence of phase spaces.

The definition of phase space is, except for small modifications, the same in §III.1. A *phase space* in $t \geq 0$ for the equation (1.1) is a product space $\mathfrak{E} = E_0 \times E_1$, equipped with any of its product norms, where E_0 and E_1 are Banach spaces satisfying the following assumptions:

(a) $E_0, E_1 \subseteq E$ *with bounded inclusion; moreover,* $D_0 \cap E_0$ (*resp.* $D_1 \cap E_1$) *is dense in* E_0 *in the topology of* E_0 (*resp. is dense in* E_1 *in the topology of* E_1).

(b) *There exists a strongly continuous semigroup* $\mathfrak{S}(t)$ *in*

$E = E_0 \times E_1$ *such that*

$$\mathfrak{S}(t)\begin{bmatrix} u(0) \\ \\ u'(0) \end{bmatrix} = \begin{bmatrix} u(t) \\ \\ u'(t) \end{bmatrix} \tag{2.1}$$

in $t \geq 0$ *for any solution* $u(\hat{t})$ *with* $u(0) \in E_0$, $u'(0) \in E_1$.

The comments after the definition of phase space in §III.1 apply here: we omit the details.

We examine in the rest of the section the relation of this notion with that of well posed Cauchy problem in the case where $E = \ell^2$ is the set of all sequences $u = \{u_n; n \geq 1\} = \{u_n\}$ of complex numbers with $\|\{u_n\}\|^2 = \sum |u_n|^2 < \infty$ and A, B are the operators

$$A\{u_n\} = \{a_n u_n\}, \quad B\{u_n\} = \{b_n u_n\}, \tag{2.2}$$

$\{a_n\}$ and $\{b_n\}$ sequences of complex numbers to be determined later: the domain of A consists of all $\{u_n\} \in E$ with $\{a_n u_n\} \in E$. The domain of B is similarly defined; observe that both A and B are normal operators commuting with each other. If $u(\hat{t}) = \{u_n(\hat{t})\}$ is a solution of (1.1) then each $u_n(\hat{t})$ satisfies the scalar equation $u_n''(t) + b_n u'(t) + a_n u_n(t) = 0$. On this basis, we deduce that the propagators $C(\hat{t})$, $\mathcal{S}(\hat{t})$ of (1.1) must be given by

$$C(t)\{u_n\} = \left\{\frac{\lambda_n^+ e^{\lambda_n^- t} - \lambda_n^- e^{\lambda_n^+ t}}{\lambda_n^+ - \lambda_n^-} u_n\right\}, \tag{2.3}$$

$$\mathcal{S}(t)\{u_n\} = \left\{\frac{e^{\lambda_n^+ t} - e^{\lambda_n^- t}}{\lambda_n^+ - \lambda_n^-} u_n\right\}, \tag{2.4}$$

where λ_n^+, λ_n^- are the roots of the characteristic equation

$$\lambda^2 + b_n \lambda + a_n = 0, \tag{2.5}$$

with the modifications de rigueur when $\lambda_n^+ = \lambda_n^-$ (a case that we will avoid here). Obviously, a necessary condition for the Cauchy problem for (1.1) to be well posed in $t \geq 0$ is that

$$\sigma(t) = \|C(t)\| = \sup_{n \geq 1} \left| \frac{\lambda_n^+ e^{\lambda_n^- t} - \lambda_n^- e^{\lambda_n^+ t}}{\lambda_n^+ - \lambda_n^-} \right| \tag{2.6}$$

and

$$\tau(t) = \|\mathcal{S}(t)\| = \sup_{n \geq 1} \left| \frac{e^{\lambda_n^+ t} - e^{\lambda_n^- t}}{\lambda_n^+ - \lambda_n^-} \right| \tag{2.7}$$

be bounded on compacts of $0 \leq t < \infty$. Conversely, the preceding conditions imply that the Cauchy problem for (1.1) is well posed: for, if (say) the Fourier coefficients of u_0, u_1 are all zero except for a finite number, then

$$u(t) = C(t)u_0 + \mathcal{S}(t)u_1$$

furnishes a solution of (1.1) in $t \geq 0$ with $u(0) = u_0$, $u_1'(0) = u_1$. Moreover we obtain taking coordinates that any solution $u(\hat{t})$ of (1.1) must be of the form

$$u(t) = C(t)u(0) + \mathcal{S}(t)u'(0) \qquad (t \geq 0) , \tag{2.8}$$

then

$$\|u(s)\| \leq (\sup_{0 \leq s \leq t} \sigma(s))\|u(0)\| + (\sup_{0 \leq s \leq t} \tau(s))\|u'(0)\| . \tag{2.9}$$

EXAMPLE 2.1. Let $\omega(\cdot)$ be an arbitrary function in $t \geq 0$, bounded on compact subsets. Then there exist A, B of the form (2.2) such that (a) The Cauchy problem for (1.1) is well posed in $t \geq 0$. (b)

$$\|C(t)\| \geq \omega(t), \quad \|\mathcal{S}(t)\| \geq \omega(t) \qquad (t \geq 1) . \tag{2.10}$$

In fact, let $\Omega = \{\omega_n\}$, $n \geq 1$ be a sequence of positive numbers such that

$$1 \leq \omega_1 \leq \omega_2^{1/2} \leq \ldots , \quad \lim_{n \to \infty} \omega_n^{1/n} = \infty \tag{2.11}$$

but otherwise arbitrary. Define

$$\alpha_n = \omega_n^{1/n} \exp\left(\frac{1}{n}\, \omega_n^{1/n}\right), \tag{2.12}$$

$$\beta_n = \exp \omega_n^{1/n} , \tag{2.13}$$

for $n \geq 1$, and let

$$m(t) = \sup_{n \geq 1} \frac{\alpha_n^t}{\beta_n} \quad (t \geq 0) . \tag{2.14}$$

Noting that

$$\frac{\alpha_n^t}{\beta_n} = \frac{\omega_n^{t/n} \exp\left(\frac{t}{n}\, \omega_n^{1/n}\right)}{\exp \omega_n^{1/n}} = \frac{\left(\omega_n^{1/n}\right)^t}{\exp\left(\left(1 - \frac{t}{n}\right)\omega_n^{1/n}\right)} \leq \frac{\left(\omega_n^{1/n}\right)^t}{\exp\left(\frac{1}{2}\, \omega_n^{1/n}\right)}$$

for $1 - t/n \geqq 1/2$ we see that $\alpha_n^t = o(\beta_n)$ as $n \to \infty$ for all t; then $m(t) < \infty$ for all $t \geq 0$. Moreover, for each t there exists an integer $n = n(t)$ such that

$$m(t) = \frac{\alpha_n^t}{\beta_n} . \tag{2.15}$$

Let now $t < t'$; since $\alpha_n > 1$ for all n,

$$m(t) = \frac{\alpha_{n(t)}^t}{\beta_{n(t)}} < \frac{\alpha_{n(t)}^{t'}}{\beta_{n(t)}} \leq m(t'); \tag{2.16}$$

accordingly the function $m(\hat{t})$ is increasing in $t \geq 0$, thus bounded on compacts. Also,

$$m(n) \geq \frac{\alpha_n^n}{\beta_n} = \omega_n \qquad (n \geq 1) . \tag{2.17}$$

Define

$$\gamma_n = \log \alpha_n = \log \omega_n^{1/n} + \frac{1}{n} \omega_n^{1/n}. \tag{2.18}$$

In view of the inequality $\log x + x \leq 2^{-1/2} e^x$, valid for $x > 0$, we have

$$\gamma_n \leq 2^{-1/2} \exp \omega_n^{1/n} = 2^{-1/2} \beta_n. \tag{2.19}$$

We select now a_n, b_n in (2.5) in such a way that

$$\lambda_n^+ = \gamma_n + i(\beta_n^2 - \gamma_n^2)^{1/2}, \ \lambda_n^- = 1 . \tag{2.20}$$

We have

$$|\lambda_n^+| = \beta_n \geq e . \tag{2.21}$$

On the other hand, in view of (2.19),

$$(\beta_n^2 - \gamma_n^2)^{1/2} \geq \gamma_n,$$

thus the sequence $\Lambda = \{\lambda_n^+\}$ is contained in the region

$$|\lambda| \geq e, \ 0 \leq \operatorname{Re} \lambda \leq \operatorname{Im} \lambda.$$

Accordingly there exist constants $\Theta > \theta > 0$ independent of Ω such that

$$\theta \leq \frac{|\lambda_n^+|}{|\lambda_n^+ - 1|} \leq \Theta . \tag{2.22}$$

We calculate now the functions $\sigma(\hat{t})$, $\tau(\hat{t})$ in (2.6), (2.7). We have

$$\frac{|\lambda_n^+|}{|\lambda_n^+ - 1|} \left(\frac{|e^{\lambda_n^+ t}|}{|\lambda_n^+|} - e^t \right) \leq \left| \frac{\lambda_n^+ e^{\lambda_n^- t} - \lambda_n^- e^{\lambda_n^+ t}}{\lambda_n^+ - \lambda_n^-} \right| \leq \frac{|\lambda_n^+|}{|\lambda_n^+ - 1|} \left(\frac{|e^{\lambda_n^+ t}|}{|\lambda_n^+|} + e^t \right), \tag{2.23}$$

$$\frac{|\lambda_n^+|}{|\lambda_n^+ - 1|}\left(\frac{e^{\lambda_n^+ t}}{|\lambda_n^+|} - \frac{e^t}{|\lambda_n^+|}\right) \leq \left|\frac{e^{\lambda_n^+ t} - e^{\lambda_n^- t}}{\lambda_n^+ - \lambda_n^-}\right| \leq \frac{|\lambda_n^+|}{|\lambda_n^+ - 1|}\left(\frac{|e^{\lambda_n^+ t}|}{|\lambda_n^+|} + \frac{e^t}{|\lambda_n^+|}\right). \quad (2.24)$$

In view of the definition (2.14) of $m(t)$ we obtain from (2.23) and (2.24) that

$$\theta(m(t) - e^t) \leq \sigma(t) \leq \Theta(m(t) + e^t), \quad (2.25)$$

$$\theta(m(t) - e^t) \leq \tau(t) \leq \Theta(m(t) + e^t), \quad (2.26)$$

in $t \geq 0$. The inequalities on the right-hand sides of (2.25), (2.26) imply that the Cauchy problem for (1.1) is well posed in $t \geq 0$. It only remains to choose the sequence Ω in such a way that the inequalities (2.10) are satisfied. To do this, we assume (as we obviously may) that $\omega(\hat{t})$ is nondecreasing. Keeping in mind that the constant in (2.22) is independent of the choice of Ω, we set

$$\omega_n = \left(\frac{\omega(n+1)}{\theta} + e^{n+1}\right)^n . \quad (2.27)$$

Both conditions (2.11) are obvious. On the other hand let $t \geq 1$, $n = [t]$, the greatest integer $\leq t$. Then, taking (2.16) and (2.27) into account, we obtain

$$\begin{aligned} \theta m(t) &\geq \theta m(n) \geq \theta\omega_n \geq \theta\omega_n^{1/n} \\ &\geq \omega(n+1) + \theta e^{n+1} \geq \omega(t) + \theta e^t \end{aligned} \quad (2.28)$$

whence the first inequality (2.10) results from (2.25); the second follows in a similar way from (2.26).

EXAMPLE 2.2. Let $a > 0$. Then there exist A, B of the form (2.2) such that the Cauchy problem for (1.1) is well posed in $0 \leq t \leq a$, but not well posed in any interval $0 \leq t \leq a'$, $a' > a$.

For this example we set $A = 0$ (so that $a_n = \lambda_n^+ = 0$); we pick the b_n in such a way that

$$\lambda_n^- = \frac{1}{a}\log n + \frac{i}{a}(n^2 - (\log n)^2)^{1/2} \qquad (n \geq 1) \quad (2.29)$$

(so that $b_n = -\lambda_n^-$). We check immediately that $C(t) = I$ for all t, thus we only have to check the boundedness of $\tau(\hat{t})$ in (2.7). We have

$$a(n^{(t-a)/a} - n^{-1}) \leq \left|\frac{e^{\lambda_n^+ t} - e^{\lambda_n^- t}}{\lambda_n^+ - \lambda_n^-}\right| \leq a(n^{(t-a)/a} + n^{-1}). \quad (2.30)$$

Consequently, $\tau(t) \leq 2a$ if $0 \leq t \leq a$, $\tau(t) = \infty$ if $t > a$.

Example 2.1 has important consequences in the theory of phase spaces for the equation (1.1). In fact, let $\mathfrak{E} = E_0 \times E_1$ be a phase space for (1.1), the choice of E_0 and E_1 entirely arbitrary.

The basic exponential growth relation (I.1.9) for the semigroup $\mathfrak{S}(\hat{t})$ in (2.1) implies that there exist constants C, ω such that

$$\|u(t)\|_{E_0} \leq Ce^{\omega t}(\|u(0)\|_{E_0} + \|u'(0)\|_{E_1}) \quad (t \geq 0) \tag{2.31}$$

for any solution $u(\hat{t})$ of (1.1) with $u(0) \in E_0$, $u'(0) = E_1$, where C and ω do not depend on the solution $u(\hat{t})$. If E_0 includes all finite sequences (that is, all sequences whose elements, except for a finite number vanish), we can take in (2.8), say, $u(0) = \{\delta_{mn}\}$, $u'(0) = 0$ (δ_{mn} the Kronecker delta). The fact that (2.31) cannot hold for all m is obvious since $\operatorname{Re} \lambda_m^+ \to \infty$ (see (2.3) and (2.4)). A similar observation applies to E_1. We have then shown:

THEOREM 2.3 <u>Let</u> A, B <u>be the operators employed in Example</u> 2.1. <u>Then the Cauchy problem is well posed in</u> $t \geq 0$ <u>for the equation</u> (1.1) <u>but</u> (1.1) <u>does not admit any phase space</u> $\mathfrak{E} = E_0 \times E_1$ <u>where</u> E_0 <u>or</u> E_1 <u>contain all finite sequences in</u> E.

Example 2.2 is even more intriguing; in fact, although we could easily define a notion of "phase space in $0 \leq t \leq a$" for (1.1), the Cauchy problem for a first order equation must be well posed in $t \geq 0$ if it is well posed in $0 \leq t \leq a$. It then follows that an equivalence between (1.1) and a first order equation is unlikely in any sense.

Examples 2.1 and 2.2 seem to indicate that "there is something missing" from the definition of well posed Cauchy problem we have introduced in §VIII.1. This holds in the sense that an additional assumption in next section will guarantee existence of state spaces.

§VIII.3 <u>Exponential growth of solutions and existence of phase spaces.</u>

We return to a general Banach space E. Assume $\mathfrak{E} = E_0 \times E_1$ is a phase space for the equation

$$u''(t) + Bu'(t) + Au(t) = 0 \tag{3.1}$$

such that $E_1 = E$ with its original norm (such was the case for all phase spaces constructed in Chapter III for the incomplete equation

(III.1.1). Then, by definition of phase space we must have

$$\mathfrak{S}(t) = \begin{bmatrix} C(t) & \mathcal{S}(t) \\ C'(t) & \mathcal{S}'(t) \end{bmatrix} .$$

It follows that $\mathcal{S}'(t)$ must be a bounded operator in E (we check easily that this is not the case in Examples 2.1 and 2.2). This justifies at least half of the following assumption:

ASSUMPTION 3.1. (a) $\mathcal{S}(\hat{t})u$ <u>is continuously differentiable in</u> $t \geq 0$ <u>for all</u> $u \in E$.

(b) $\mathcal{S}(t)E \subseteq D(B)$ <u>and</u> $B\mathcal{S}(t)u$ <u>is continuous in</u> $t \geq 0$ <u>for all</u> $u \in E$.

As we shall see below, Assumption 3.1 guarantees exponential growth of the solutions (3.1), as well as existence of a state space.

THEOREM 3.2. <u>Let The Cauchy problem for</u> (3.1) <u>be well posed in</u> $t \geq 0$, <u>and let Assumption</u> 3.1 <u>be satisfied. Then there exist constants</u> C, ω <u>such that</u>

$$\|C(t)\|, \ \|\mathcal{S}(t)\| \leq Ce^{\omega t} \quad (t \geq 0) . \tag{3.2}$$

The proof of Theorem 3.2 is lenghty and will be carried out in several steps. We examine first some immediate consequences of Assumption 3.1. Let $a > 0$ and let $C(0,a;E)$ be the Banach space of all continuous E-valued functions with its usual supremum norm. The operator $u \to \mathcal{S}'(t)u$ from E to $C(0,a;E)$ is easily seen to be closed; since it is everywhere defined, by the closed graph theorem it is as well bounded. But then $\mathcal{S}'(t)$ is a bounded operator for all t and $\|\mathcal{S}'(t)\|$ is bounded in $0 \leq t \leq a$. Taking into account that a is arbitrary we see that $\mathcal{S}'(\hat{t})$ is strongly continuous in $t \geq 0$ with $\|\mathcal{S}'(t)\|$ bounded on compacts therein. The same argument applied to $B\mathcal{S}(t)$ shows that $B\mathcal{S}(\hat{t})$ is a strongly continuous operator valued function such that $\|B\mathcal{S}(t)\|$ is bounded on compacts of $t \geq 0$.

We shall need later to solve the inhomogeneous equation

$$u''(t) + Bu'(t) + Au(t) = f(t) . \tag{3.3}$$

Solutions of (3.3) are defined in the same way as for the homogeneous equation (3.1).

LEMMA 3.3. <u>Let</u> $f(\hat{t})$ <u>be continuously differentiable in</u> $t \geq 0$. <u>Then</u> (a)

$$u(t) = (\mathcal{S} * f)(t) = \int_0^t \mathcal{S}(t-s)f(s)\,ds$$

$$= \int_0^t \mathcal{S}(s)f(t-s)\,ds \qquad (t \geq 0) \tag{3.4}$$

<u>is a solution of</u> (3.3) <u>with</u> $u(0) = u'(0) = 0$. (b) <u>If</u> $v(\hat{t})$ <u>is a solution of</u> (3.3) <u>then</u>

$$v(t) = C(t)u(0) + \mathcal{S}(t)u'(0) + u(t) \qquad (t \geq 0), \tag{3.5}$$

<u>where</u> $u(t)$ <u>is defined by</u> (3.4).

<u>Proof</u>: Integrating (3.4) by parts we obtain

$$u(t) = \int_0^t \mathcal{S}(s)f(0)\,ds + \int_0^t\int_0^{t-s} \mathcal{S}(r)f'(s)\,dr\,ds . \tag{3.6}$$

Differentiating,

$$u'(t) = \mathcal{S}(t)f(0) + \int_0^t \mathcal{S}(t-s)f'(s)\,ds, \tag{3.7}$$

$$u''(t) = \mathcal{S}'(t)f(0) + \int_0^t \mathcal{S}'(t-s)f'(s)\,ds. \tag{3.8}$$

Let now $u \in D_1$. We have

$$A\mathcal{S}(s)u = -B\mathcal{S}'(s)u - \mathcal{S}''(t)u.$$

Integrating in $0 \leq s \leq t$,

$$A\int_0^t \mathcal{S}(s)u\,ds = -B\mathcal{S}(t)u - \mathcal{S}'(t)u + u. \tag{3.9}$$

Since the right-hand side is a continuous operator of u, it follows from closedness of A and the fact that D_1 is dense in E that

$$\int_0^t \mathcal{S}(s)u\,ds \in D(A) \tag{3.10}$$

for all $u \in E$ and that (3.9) actually holds for every $u \in E$. These observations make clear (after the integration by parts (3.6)) that if $u(\hat{t})$ is the function defined by (3.4) then $u(t) \in D(A)$ and $Au(\hat{t})$ is continuous in $t \geq 0$. Moreover, we obtain from (3.7) that $u'(t) \in D(B)$ with $Bu(\hat{t})$ continuous in $t \geq 0$ and also that $u(t)$ is in fact a solution of (3.3) as claimed. To show (b) we note that if $v(\hat{t})$ is an arbitrary solution of (3.3) and $u(\hat{t})$ the solution provided by (3.4), $v(\hat{t}) - u(\hat{t})$ is a solution of the homogeneous equation (1.1) so that formula (1.6) applies.

LEMMA 3.4. (a) Let $u \in D(A)$. Then (a) $C(\hat{t})u$ is continuously differentiable and we have

$$C'(t)u = -\mathcal{S}(t)Au \qquad (t \geq 0) . \tag{3.11}$$

(b) Let $u \in D_0 \cap D(B)$. Then

$$\mathcal{S}'(t)u = C(t)u - \mathcal{S}(t)Bu \qquad (t \geq 0) . \tag{3.12}$$

Proof: (a) According to Lemma 3.3 the function

$$u(t) = -\int_0^t \mathcal{S}(s)Au \, ds$$

is a solution of the equation $u''(t) + Bu'(t) + Au(t) = -Au$ with $u(0) = u'(0) = 0$. Consequently, $v(t) = u(t) + u$ satisfies the homogeneous equation (1.1) with initial conditions $v(0) = u$, $v'(0) = 0$. By virtue of (3.5) we have $v(t) = C(t)u$, that is,

$$C(t)u - u = -\int_0^t \mathcal{S}(s)Au \, ds,$$

which is the integrated version of (3.11). To show (b), let

$$u(t) = -\int_0^t \mathcal{S}(s)Bu \, ds.$$

Applying again Lemma 3.3 we see that $u(\hat{t})$ satisfies $u''(t) + Bu'(t) + Au(t) = -Bu$. On the other hand, let

$$v(t) = \int_0^t C(s)u \, ds .$$

We have

$$v'(t) = C(t)u = \int_0^t C'(s)u \, ds + u,$$

$$v''(t) = C'(t)u = \int_0^t C''(s)u \, ds,$$

thus $v(\hat{t})$ satisfies $v''(t) + Bv'(t) + Av(t) = Bu$, with initial conditions $v(0) = 0$, $v'(0) = u$. Accordingly, $w(t) = v(t) - u(t)$

satisfies (1.1) with initial conditions $w(0) = 0$, $w'(0) = u$, so that $w(t) = \mathcal{S}(t)u$ or, equivalently,

$$\mathcal{S}(t)u = \int_0^t (C(s)u - \mathcal{S}(s)Bu)\, ds, \tag{3.13}$$

from which (3.12) is deduced by differentiation.

COROLLARY 3.5. (a) $D_0 = D(A)$. (b) $D = D(A) \cap D(B)$ <u>is dense in</u> E. (c) $D_1 \supseteq D(A) \cap D(B)$. (d) <u>If</u> $u \in D(A) \cap D(B)$,

$$\mathcal{S}''(t)u + \mathcal{S}'(t)Bu + \mathcal{S}(t)Au = 0 \quad (t \geq 0) . \tag{3.14}$$

<u>Proof</u>: As a byproduct of the proof of Lemma 3.4 (a) it was proved that $C(\hat{t})u$ is a solution of (3.1) for all $u \in D(A)$; since it is always true that $D_0 \subseteq D(A)$, (a) follows. Similarly, one of the steps in the proof of Lemma 3.4 (b) was to show that for any $u \in D_0 \cap D(B) = D(A) \cap D(B), \mathcal{S}(\hat{t})u$ is a solution of (3.1), thus (c) results (as we shall see below in Example 4.5 it is not in general true that $D_1 = D(A) \cap D(B)$). To show the denseness statement (b) we consider the subspace Ψ of E consisting of all elements of the form

$$\int \psi(s)\mathcal{S}'(s)u\, ds$$

where $\psi(\hat{s})$ is any infinitely differentiable (scalar) function with support contained in $t > 0$. Let $u \in E$ be arbitrary. Then $\mathcal{S}'(\hat{s})u$ is a continuous function with $\mathcal{S}'(0)u = u$. Accordingly, if $\{\psi_n\}$ is a sequence of nonnegative infinitely differentiable functions, each ψ_n with support in $0 \leq t \leq 1$ and such that $\int \psi_n ds = 1$, we have

$$\int \psi_n(s)\mathcal{S}'(s)u\, ds \to u$$

as $n \to \infty$. Hence, Ψ is dense in E. Observe next that, integrating by parts, any element of Ψ can be written in the form

$$-\int \psi'(s)\mathcal{S}(s)u\, ds = \int \psi''(s)\left(\int_0^s \mathcal{S}(r)u\, dr\right) ds . \tag{3.15}$$

We deduce from the first expression (3.15) that $\Psi \subseteq D(B)$; on the other hand, the second combined with (3.10) and the comments preceding it, implies that $\Psi \subseteq D(A)$, which completes the proof of (b). As for (d), it follows from differentiating (3.12) and then expressing $C'(t)u$ by means of (3.11). We have then completed the proof of

Lemma 3.5.

We point out that, as a consequence of equality (3.12), the operator $\mathcal{S}(t)B$ (with $D(A) \cap D(B)$ as domain) admits a bounded extension $\overline{\mathcal{S}(t)B}$ to all of E, namely

$$\overline{\mathcal{S}(t)B} = C(t) - \mathcal{S}'(t) . \tag{3.16}$$

Since $C(\hat{t})$, $\mathcal{S}'(\hat{t})$ are strongly continuous functions in $t \geq 0$, so is $\overline{\mathcal{S}(\hat{t})B}$.

We consider in what follows the <u>characteristic polynomial</u> of (3.1),

$$P(\lambda) = \lambda^2 I + \lambda B + A . \tag{3.17}$$

For each λ, $P(\lambda)$ is a linear operator with domain

$$D(P(\lambda)) = D = D(A) \cap D(B) .$$

LEMMA 3.6. (a) <u>$P(\lambda)$ is closable for all</u> λ. (b) <u>There exist constants</u> $\alpha, \beta \geq 0$ <u>such that</u> $P(\lambda)$ <u>is one-to-one for</u>

$$\operatorname{Re}\lambda \geq \alpha + \beta \log(1 + |\lambda|). \tag{3.18}$$

<u>Proof</u>: Assume (a) is false for some λ. Then there exists a sequence $\{u_n\} \subset D(P(\lambda))$ such that $u_n \to 0$, $v_n = P(\lambda)u_n \to v \neq 0$. Let $u_n(t) = e^{\lambda t}u_n$ $(t \geq 0)$. Clearly $u(\hat{t})$ satisfies the inhomogeneous equation (2.3) with $f(t) = e^{\lambda t}v_n$. We get as a consequence of Lemma 3.3 (b) that

$$e^{\lambda t}u_n = C(t)u_n + \lambda \mathcal{S}(t)u_n + \int_0^t \mathcal{S}(t-s)e^{\lambda(t-s)}v_n \, ds.$$

Letting $n \to \infty$ we obtain

$$\int_0^t e^{-\lambda s}\mathcal{S}(s)v \, ds = 0 \qquad (t \geq 0).$$

Differentiating twice,

$$e^{-\lambda t}\mathcal{S}'(t)v - \lambda e^{-\lambda t}\mathcal{S}(t)v = 0 \qquad (t \geq 0).$$

If we set $t = 0$ in this last expression we obtain $v = 0$, a contradiction.

Regarding (b), assume $P(\lambda)$ is not one-to-one for some λ. Then there exists $u \in D(P(\lambda))$, $u \neq 0$ such that $P(\lambda)u = 0$. Obviously $u(t) = e^{\lambda t}u$ is a solution of (3.1); making use of the estimate (1.3) for any fixed $t > 0$ we deduce that

$$e^{(\mathrm{Re}\,\lambda)t} \leq C(t)(1 + |\lambda|) .$$

Taking logarithms, we obtain the inequality opposite to (3.18) for $\alpha = t^{-1} \log C(t)$, $\beta = t^{-1}$.

<u>End of the proof of Theorem</u> 3.2: Let φ be a twice continuously differentiable scalar valued function with compact support and such that $\varphi(t) \equiv 1$ near zero. Consider the (plainly bounded) operator in the space E,

$$R(\lambda;\varphi)u = \int_0^\infty e^{-\lambda t}\varphi(t)\mathcal{S}(t)u\,dt, \tag{3.19}$$

defined for all complex λ. We have $R(\lambda;\varphi)E \subseteq D(B)$. Moreover, we can write, integrating by parts,

$$R(\lambda;\varphi)u = -\int_0^\infty (e^{-\lambda t}\varphi(t))'\left(\int_0^t \mathcal{S}(s)u\,ds\right)dt, \tag{3.20}$$

thus it follows from (3.10) that $R(\lambda;\varphi)E \subseteq D(A)$. Hence

$$R(\lambda;\varphi)E \subseteq D(A) \cap D(B) = D(P(\lambda)). \tag{3.21}$$

Assume $u \in D(A) \cap D(B) \subseteq D_1$. Using the fact that $\mathcal{S}(\hat{t})u$ is a solution of (3.1) (Corollary 3.5 (c)) we show, integrating by parts, that

$$\begin{aligned} P(\lambda)R(\lambda,\varphi)u &= u + \int_0^\infty e^{-\lambda t}((\varphi\mathcal{S})''(t)u + B(\varphi\mathcal{S})'(t)u + A(\varphi\mathcal{S})(t)u)\,dt \\ &= u + \int_0^\infty e^{-\lambda t}M(t,\varphi)u\,dt = u + \hat{M}(\lambda,\varphi)u , \end{aligned} \tag{3.22}$$

where $M(t,\varphi) = 2\varphi'(t)\mathcal{S}'(t) + \varphi''(t)\mathcal{S}(t) + \varphi'(t)B\mathcal{S}(t)$ is plainly a strongly continuous (E)-valued function in $t \geq 0$ with compact support contained in $(0,\infty)$. Let $\omega > 0$ be so large that

$$\int_0^\infty e^{-\omega t}\|M(t,\varphi)\|dt = \gamma < 1. \tag{3.23}$$

Then, indicating Laplace transforms with $\hat{}$,

$$\|\hat{M}(\lambda,\varphi)\| < \gamma \tag{3.24}$$

for $\mathrm{Re}\,\lambda \geq \omega$ and thus $I + \hat{M}(\lambda,\varphi)$ has a bounded inverse in this region.

Define a bounded operator $R(\lambda)$ by means of the formula

$$R(\lambda) = R(\lambda,\varphi)(I + \hat{M}(\lambda,\varphi))^{-1} = R(\lambda,\varphi)\sum_{n=0}^{\infty}(-1)^n\hat{M}(\lambda,\varphi)^n. \qquad (3.25)$$

We rewrite (3.22) in the form

$$\overline{P(\lambda)}R(\lambda,\varphi)u = u + \hat{M}(\lambda,\varphi)u \quad (u \in D), \qquad (3.26)$$

where $\overline{P(\lambda)}$ denotes the closure of $P(\lambda)$. It follows immediately from the facts that $\overline{P(\lambda)}$ is closed and $\hat{M}(\lambda,\varphi)$ is bounded that (3.26) must in fact hold for all $u \in E$. Then

$$\overline{P(\lambda)}R(\lambda)u = u \quad (u \in E). \qquad (3.27)$$

Observe now that, since $R(\lambda,\varphi)E \subseteq D(P(\lambda))$, $R(\lambda)E \subseteq D(P(\lambda))$; hence we may replace $\overline{P(\lambda)}$ by $P(\lambda)$ in (3.27). The equality thus obtained implies that

$$R(\lambda)E = D(P(\lambda)) = D = D(A) \cap D(B), \qquad (3.28)$$

at least for these values of λ for which $P(\lambda)$ is one-to-one. In fact, let $v \in D(P(\lambda))$ be such that $v \notin R(\lambda)E$, and let $u = P(\lambda)v$. Then $P(\lambda)(v - R(\lambda)u) = 0$, a contradiction.

We show next that $P(\lambda)$ is one-to-one in $\operatorname{Re}\lambda > \omega$. To do this, observe first that $R(\hat{\lambda};\varphi)$, $\hat{M}(\hat{\lambda},\varphi)$ are entire functions of λ (as Laplace transforms of functions with compact support). Note that, by virtue of the estimate (3.24), the series on the right-hand side of (3.25) converges uniformly in $\operatorname{Re}\lambda \geq \omega$, hence $R(\lambda)$ is analytic there. Let now $v \in D(P(\lambda)) = D(A) \cap D(B)$, $v \neq 0$ and let λ be, say, in the region defined by (3.18). By the preceding comments, $v = R(\lambda)u$ (u some element of E). Then

$$R(\lambda)P(\lambda)v = R(\lambda)P(\lambda)R(\lambda)u = R(\lambda)u = v. \qquad (3.29)$$

The left hand side of (3.28) is analytic in $\operatorname{Re}\lambda > \omega$; since it equals v in the region defined by (3.18), it must equal v as well in $\operatorname{Re}\lambda > \omega$, which shows that $P(\lambda)v \neq 0$ throughout $\operatorname{Re}\lambda > \omega$ as claimed.

Collecting all the observations made about $P(\lambda)$ and $R(\lambda)$ we can write

$$R(\lambda) = P(\lambda)^{-1} \quad (\operatorname{Re}\lambda > \omega). \qquad (3.30)$$

We obtain now some rough estimates for $R(\lambda)$, $BR(\lambda)$, $AR(\lambda)$ in the half plane $\mathrm{Re}\lambda \geq \omega$. Plainly $R(\lambda,\varphi)$, $BR(\lambda,\varphi)$ are bounded there: on the other hand, it follows from the expression (3.20) for $R(\lambda;A)$ that $\|AR(\lambda,\varphi)\| \leq C|\lambda|$. Finally, in view of (3.24),

$$\|(I + \hat{M}(\lambda,\varphi))^{-1}\| \leq \sum_{n=0}^{\infty} \gamma^n = (1-\gamma)^{-1}.$$

Accordingly,

$$\|R(\lambda)\|, \|BR(\lambda)\| \leq C, \ \|AR(\lambda)\| \leq C|\lambda| \tag{3.31}$$

in $\mathrm{Re}\lambda > \omega$ for some constant C. For $\bar{\omega} > \omega$ and $u \in E$ define

$$u(t) = \frac{1}{2\pi i}\int_{\bar{\omega}-i\infty}^{\bar{\omega}+i\infty} \frac{e^{\lambda t}}{\lambda^4} R(\lambda)u \, d\lambda . \tag{3.32}$$

In view of the first estimate (3.31), we can differentiate twice under the integration sign,

$$u^{(k)}(t) = \frac{1}{2\pi i}\int_{\bar{\omega}-i\infty}^{\bar{\omega}+i\infty} \frac{e^{\lambda t}}{\lambda^{4-k}} R(\lambda)u \, d\lambda \tag{3.33}$$

$(k = 0,1,2)$ with $u^{(k)}$ continuous in $t \geq 0$. Using the other estimates we deduce in addition that $Bu'(\hat{t})$, $Au(\hat{t})$ exist and are continuous in $t \geq 0$, introduction of A and B under the corresponding integrals being possible. Hence

$$\begin{aligned} u''(t) + Bu'(t) + Au(t) &= \frac{1}{2\pi i}\int_{\bar{\omega}-i\infty}^{\bar{\omega}+i\infty} \frac{e^{\lambda t}}{\lambda^4} P(\lambda)R(\lambda)u \, d\lambda \\ &= \left(\frac{1}{2\pi i}\int_{\bar{\omega}-i\infty}^{\bar{\omega}+i\infty} \frac{e^{\lambda t}}{\lambda^4} d\lambda\right) u = \left(\frac{1}{3!}\left(\frac{d}{d\lambda}\right)^3 e^{\lambda t}\right) u = \frac{t^3}{3!} u. \end{aligned} \tag{3.34}$$

Pushing the contour of integration in (3.33) for $k = 0,1$ to the right we show that

$$u(0) = u'(0) = 0 . \tag{3.35}$$

We express then the solution of the initial value problem (3.34)-(3.35) by means of formula (3.4), obtaining

$$\frac{1}{3!}\int_0^t (t-s)^3 \mathcal{S}(s)u \, ds = \frac{1}{2\pi i}\int_{\bar{\omega}-i\infty}^{\bar{\omega}+i\infty} \frac{e^{\lambda t}}{\lambda^4} R(\lambda)u \, d\lambda, \tag{3.36}$$

an equality that suggests - as will be proved later - that $R(\lambda)$ is the Laplace transform of $\mathcal{S}(\hat{t})$.

We now try to find a new representation for $R(\lambda)$. Let $u \in D$;

operating as in the lines leading to (3.22) and making use of Corollary 3.5 (d) we show that, if $u \in D(P(\lambda)) = D(A) \cap D(B)$,

$$R(\lambda,\varphi)P(\lambda)u = u + \int_0^\infty e^{-\lambda t}((\varphi\mathcal{S})''(t)u + (\varphi\mathcal{S})'(t)Bu + (\varphi\mathcal{S})(t)Au)dt$$

$$= u + \int_0^\infty e^{-\lambda t}N(t;\varphi)udt = u + \hat{N}(\lambda;\varphi)u \ , \qquad (3.37)$$

where now $N(t;\varphi) = 2\varphi'(t)\mathcal{S}'(t) + \varphi''(t)\mathcal{S}(t) + \varphi'(t)\,\overline{\mathcal{S}(t)B}$. If $\omega' \geq 0$ is such that

$$\int_0^\infty e^{-\omega' t}\|N(t;\varphi)\|dt = \gamma < 1, \qquad (3.38)$$

[2]

then $\|\hat{N}(\lambda;\varphi)\| \leq \gamma$ in $\mathrm{Re}\lambda \geq \omega'$ and $I + \hat{N}(\lambda;\varphi)$ is invertible there. Let $Q(\lambda) = (I + \hat{N}(\lambda;\varphi))^{-1}R(\lambda;\varphi)$. It follows from (3.37) that

$$Q(\lambda)P(\lambda)u = u \qquad (3.39)$$

for $u \in D$, which plainly shows that $Q(\lambda) = R(\lambda)$ in

$$\mathrm{Re}\lambda \geq \omega_1 = \max(\omega,\omega') \ . \qquad (3.40)$$

[3]

Accordingly,

$$R(\lambda) = (I + \hat{N}(\lambda;\varphi))^{-1}R(\lambda;\varphi) = \left(\sum_{n=0}^{\infty}(-1)^n\hat{N}(\lambda;\varphi)^n\right)R(\lambda;\varphi) . \qquad (3.41)$$

This formula suggests (by inversion of Laplace transforms) that

$$\mathcal{S}(t) = \left(\sum_{n=0}^{\infty}(-1)^n N(t;\varphi)^{*n}\right) * (\varphi\mathcal{S})(t) , \qquad (3.42)$$

where $*$ denotes convolution and the exponent $*n$ indicates the n-th convolution power. We attempt to justify (3.42) directly. By virtue of (3.38) and of Young's inequality,

$$\int_0^\infty e^{-\omega' t}\|N(t;\varphi)^{*2}\|dt \leq \int_0^\infty \|e^{-\omega' t}N(t;\varphi)\|^{*2}dt \leq \gamma^2 ,$$

and, in general,

$$\int_0^\infty e^{-\omega' t}\|N(t;\varphi)^{*n}\|dt \leq \gamma^n \quad (n \geq 1). \qquad (3.43)$$

If C is a constant such that

$$\|N(t;\varphi)\| \leq Ce^{\omega' t} \qquad (t \geq 0) , \qquad (3.44)$$

we can combine (3.44) and (3.43) as follows:

$$\|N(t;\varphi)^{*n}\| = \|N(t;\varphi)^{*(n-1)} * N(t;\varphi)\| \le C\gamma^{n-1}e^{\omega' t} \quad (t \ge 0,\ n = 1,\ldots) . \tag{3.45}$$

Consequently, the series

$$\sum_{n=1}^{\infty} (-1)^n N(t;\varphi)^{*n} \tag{3.46}$$

converges uniformly on compacts of $t \ge 0$ to a (E)-valued function $\mathfrak{n}(t;\varphi)$ such that

$$\|\mathfrak{n}(t;\varphi)\| \le C(1-\gamma)^{-1}e^{\omega' t} \qquad (t \ge 0). \tag{3.47}$$

Since the partial sums of the series (3.46) satisfy the estimate (3.47) as well and each of its terms is strongly continuous, its Laplace transform in the region $\operatorname{Re}\lambda \ge \omega'$ can be computed by term-by-term integration. Let $\tilde{\mathfrak{S}}(\hat{t})$ be the function defined by the right hand side of (3.42), that is

$$\begin{aligned}\tilde{\mathfrak{S}}(t)u &= (\delta \otimes I + \mathfrak{n}(t;\varphi)) * (\varphi\mathfrak{S})(t)u \\ &= (\varphi\mathfrak{S})(t)u + \int_0^t \mathfrak{n}(t-s;\varphi)(\varphi\mathfrak{S})(s)u\, ds .\end{aligned} \tag{3.48}$$

Plainly,

$$\|\tilde{\mathfrak{S}}(t)\| \le Ce^{\omega' t} \qquad (t \ge 0) \tag{3.49}$$

for some constant C. Computing the Laplace transform of $\mathfrak{S}(\hat{t})$ by application of the convolution theorem and likewise applying the convolution theorem to each of the terms in the series of $\mathfrak{n}(\hat{t};\varphi)$ we easily see that it equals

$$\Big(\sum_{n=0}^{\infty} (-1)^n \hat{N}(\lambda;\varphi)^n\Big) R(\lambda;\varphi) = R(\lambda) \tag{3.50}$$

by (3.41). But then, by a well known result on Laplace transforms of antiderivatives, we have, after inversion,

$$\frac{1}{3!}\int_0^t (t-s)^3 \tilde{\mathfrak{S}}(s)u\, ds = \frac{1}{2\pi i}\int_{\bar{\omega}-i\infty}^{\bar{\omega}+i\infty} \frac{e^{\lambda t}}{\lambda^4} R(\lambda)u\, d\lambda$$

for $\bar{\omega} > \omega_1$, $u \in E$. Compare this with (3.36) and differentiate three times the equality obtained from uniqueness of Laplace transforms. The result is

$$\tilde{\mathcal{S}}(t) = \mathcal{S}(t) \qquad (t \geq 0), \tag{3.51}$$

thus, in view of (3.49) we have

$$\|\mathcal{S}(t)\| \leq Ce^{\omega t}. \tag{3.52}$$

Apply (3.48) to an arbitrary $u \in E$ and differentiate: taking into account that $(\varphi\mathcal{S})(0) = 0$ we obtain

$$\mathcal{S}'(t)u = (\delta \otimes I + \mathfrak{n}(t;\varphi)) * (\varphi\mathcal{S})'(t)u. \tag{3.53}$$

Similarly, applying both sides of (3.48) to elements of the form Bu (with $u \in D(A) \cap D(B)$) we obtain

$$\overline{\mathcal{S}(t)B}u = (\delta \otimes I + \mathfrak{n}(t;\varphi)) * (\varphi\overline{\mathcal{S}B})(t)u; \tag{3.54}$$

it is obvious that the equality can be extended to all $u \in E$. We use now (3.16) to deduce that

$$\begin{aligned} C(t)u &= (\delta \otimes I + \mathfrak{n}(t;\varphi)) * ((\varphi\overline{\mathcal{S}B})(t)u + (\varphi\mathcal{S})'(t)u) \\ &= (\delta \otimes I + \mathfrak{n}(t;\varphi)) * ((\varphi C)(t)u + (\varphi'\mathcal{S})(t)u). \end{aligned} \tag{3.55}$$

We obtain from (3.55) and (3.53), in conjunction with (3.47), that

$$\|C(t)\| \leq Ce^{\omega' t}, \quad \|\mathcal{S}'(t)\| \leq Ce^{\omega' t} \qquad (t \geq 0) \tag{3.56}$$

for some constant C. The estimate (3.52) and the first inequality (3.56) comprise the claims in Theorem 3.2, whose proof is thus complete.

COROLLARY 3.7. Under the assumptions in Theorem 3.2 we have

$$\|\mathcal{S}'(t)\| \leq Ce^{\omega t}, \quad \|B\mathcal{S}(t)\| \leq Ce^{\omega t} \qquad (t \geq 0) \tag{3.57}$$

for some constants C, ω.

Proof: The first inequality (3.57) is the second inequality (3.56). To show the second we make use of the "left-handed" companion of (3.48), namely

$$\mathcal{S}(t)u = (\varphi\mathcal{S})(t) * (\delta \otimes I + \mathfrak{m}(t;\varphi))u , \tag{3.58}$$

where

$$\mathfrak{M}(t;\varphi) = \sum_{n=1}^{\infty} (-1)^n M(t;\varphi)^{*n}. \tag{3.59}$$

Formula (3.59) can be justified along the same lines as the "right-handed" formula was. On the basis of (3.23) we deduce that

$$\|M(t;\varphi)^{*n}\| \le C\gamma^{n-1}e^{\omega t} \quad (n \ge 1), \tag{3.60}$$

thus (3.59) converges uniformly on compacts of $t \ge 0$, its limit $\mathfrak{M}$ satisfying

$$\|\mathfrak{M}(t;\varphi)\| \le C(1-\gamma)^{-1}e^{\omega t} \quad (t \ge 0). \tag{3.61}$$

Equality (3.58) can be established, as in the case of (3.48) by taking the Laplace transform of both sides and using (3.25) and (3.33). Using the fact that $B\hat{\mathcal{S}}(t)$ is a (E)-valued strongly continuous function of t we deduce from (3.58) that

$$B\mathcal{S}(t)u = (\varphi B\mathcal{S}(t)) * (\delta \otimes I + \mathfrak{M}(t;\varphi))u\ . \tag{3.62}$$

We obtain the second inequality (3.57) from (3.58) and (3.61). This ends the proof of Corollary 3.7.

We note in closing that other equalities, like (3.53) and (3.55) have as well left-handed companions; these are

$$\mathcal{S}'(t)u = (\varphi\mathcal{S})'(t) * (\delta \otimes I + \mathfrak{M}(t;\varphi))u \tag{3.63}$$

and

$$C(t)u = ((\varphi C)(t) + (\varphi'\mathcal{S})(t)) * (\delta \otimes I + \mathfrak{M}(t;\varphi))u \tag{3.64}$$

respectively. Nothing much can be obtained from these identities that was not known previously on the basis of (3.53) and (3.55), except that the exponential bounds emanating from (3.63) and (3.64) employ the exponent ω in (3.23) instead of ω' in (3.38), which may be an improvement if $\omega < \omega'$ (this is not very exciting, however since both exponents are probably very far from optimal).

REMARK 3.8. Using techniques similar to those in the proof of Theorem 3.2 we can show that if the Cauchy problem for (3.1) is well posed in $0 \le t \le a$ then a suitable version of Theorem 3.2 holds. See Exercise 14 and the author [1970:2]) for details.

REMARK 3.9. Integrating by parts in (3.19) we obtain

$$R(\lambda;\varphi)u = \frac{1}{\lambda}\int_0^\infty e^{-\lambda t}(\varphi \mathcal{S})'(t)u\,dt, \tag{3.65}$$

thus

$$\|R(\lambda;\varphi)\| \le C/\lambda^2 \quad (\lambda > \omega).$$

In view of (3.25),

$$\|R(\lambda)\| \le C/\lambda^2 \quad (\lambda > \omega). \tag{3.66}$$

Now, if $u \in D(A) \cap D(B)$,

$$\lambda^2 R(\lambda)u + \lambda R(\lambda)Bu + R(\lambda)Au = R(\lambda)P(\lambda)u = u\ ,$$

hence it follows from (3.66) that

$$\lambda^2 R(\lambda)u \to u \quad \text{as} \quad \lambda \to \infty\ . \tag{3.67}$$

Using the uniform bound (3.66) and the denseness of $D(A) \cap D(B)$, we can extend (3.67) to arbitrary $u \in E$.

§VIII.4 Construction of phase spaces.

We show in this section that a phase space $\mathfrak{E}_m$ for the equation

$$u''(t) + Bu'(t) + Au(t) = 0 \tag{4.1}$$

can be constructed if the Cauchy problem for (4.1) is well posed in $t \ge 0$ and Assumption 3.1 is satisfied (as noted in Theorem 2.3 the sole assumption that the Cauchy problem for (4.1) is well posed in the sense of §VIII.1 is insufficient). The space $\mathfrak{E}_m$ is the direct generalization of the space $\mathfrak{E}_m$ in §III.1 for the incomplete equation. For its definition, the following result shall be needed.

LEMMA 4.1. Let the Cauchy problem for (4.1) be well posed in $t \ge 0$ and let Assumption 3.1 be satisfied. Assume that $u \in E$ is such that $C(\hat{t})u$ is continuously differentiable in $t \ge 0$. Then there exist C, ω such that

$$\|C'(t)u\| \leq Ce^{\omega t} \qquad (t \geq 0), \tag{4.2}$$

where C (but not ω) may depend on u.

Proof: We use equality (3.55); if $C(\hat{t})u$ is continuously differentiable in $t \geq 0$ (in fact, in the support of φ) then $(\varphi C)(\hat{t}) + (\varphi' \mathcal{S})(\hat{t})u$ is continuously differentiable as well: we can then differentiate both sides of (3.55), obtaining

$$C'(t)u = (\delta \otimes I + \mathfrak{h}(t;\varphi)) * ((\varphi' C)(t)u + (\varphi C')(t)u + (\varphi'' \mathcal{S})(t)u + (\varphi' \mathcal{S}')(t)u), \tag{4.3}$$

hence (4.2) follows using (3.47). This ends the proof of Lemma 4.1.

The construction of the maximal phase space for (4.1) proceeds now much in the same way as for (III.2.1). The space $\mathfrak{E}_m$ is defined as follows:

$$\mathfrak{E}_m = E_0 \times E \tag{4.4}$$

endowed with any of its product norms. The space E_0 consists of all $u \in E$ such that $C(\hat{t})u$ is continuously differentiable in $t \geq 0$. The norm in E_0 is

$$\|u\|_0 = \|u\| + \sup_{s \geq 0} e^{-\omega s}\|C'(s)u\|, \tag{4.5}$$

where $\omega > \omega'$, ω' so large that (3.2), the first inequality (3.57) and (4.2) hold. We obviously have

$$D_0 = D(A) \subseteq E_0. \tag{4.6}$$

The proof that E_0 is a Banach space is much the same as that for the equation (III.2.1) and we omit it.

THEOREM 4.2. Let the Cauchy problem for (4.1) be well posed in $t \geq 0$ and let Assumption 3.1 be satisfied. Then $\mathfrak{E}_m$ is a phase space for (4.1).

Proof: We must show that

$$\mathfrak{S}(t) = \begin{bmatrix} C(t) & \mathcal{S}(t) \\ C'(t) & \mathcal{S}'(t) \end{bmatrix} \tag{4.7}$$

is a strongly continuous semigroup in $\mathfrak{E}_m$. We prove first that each $\mathfrak{S}(t)$ is a bounded operator in $\mathfrak{E}_m$. In order to do this we take $u \in D_0$ and fix $t > 0$. Due to time invariance of (4.1) the function $u(\hat{s}) = C(t + \hat{s})u$ is a solution of (4.1) thus it follows from formula (1.5) that

$$C(s + t)u = C(s)C(t)u + \mathcal{S}(s)C'(t)u \quad (s,t \geq 0). \tag{4.8}$$

This equality is extended to all $u \in E_0$ as follows: integrate in $0 \leq \tau \leq t$,

$$\int_0^t C(s + \tau)u\, d\tau = C(s)\int_0^t C(\tau)u\, d\tau + \mathcal{S}(s)(C(t)u - u), \tag{4.9}$$

and extend (4.9) to arbitrary $u \in E$ by denseness of D_0; for $u \in E_0$ we differentiate and obtain (4.8).

The analogue of (4.8) for $\mathcal{S}(\hat{t})$ is

$$\mathcal{S}(s + t)u = C(s)\mathcal{S}(t)u + \mathcal{S}(s)\mathcal{S}'(t)u \quad (s,t \geq 0), \tag{4.10}$$

and is shown by applying (1.5) to $u(\hat{s}) = \mathcal{S}(\hat{s} + t)u$, $u \in D_1$; since all operators in (4.10) are bounded we can extend the equality to all $u \in E$. We note in passing that (4.8) itself can be extended to all $u \in E$ in a modified form observing that $\mathcal{S}(s)C'(t)$ must have a bounded extension. We shall not make use of this in what follows.

We prove that each $\mathfrak{S}(t)$ is a bounded operator in $\mathfrak{E}_m$. To do this, we must show that the operators

$$\begin{matrix} C(t) : E_0 \to E_0 & \mathcal{S}(t) : E \to E_0 \\ C'(t) : E_0 \to E & \mathcal{S}'(t) : E \to E \end{matrix} \tag{4.11}$$

are bounded in the spaces indicated. This is rather obvious for $C'(t)$ (from the definition of E_0) and for $\mathcal{S}'(t)$ (from Assumption 3.1). Note also that it follows from Corollary 3.7 and Lemma 4.1 that

$$\|C'(t)\|_{(E_0;E)} \leq Ce^{\omega t}, \quad \|\mathcal{S}'(t)\|_{(E;E)} \leq Ce^{\omega t} \quad (t \geq 0) \tag{4.12}$$

(here and in other inequalities C denotes an arbitrary constant, not necessarily the same in different places).

Continuity of $C(t)$ is proved as follows. Write (4.8) in the form $C(s)C(t)u = C(s+t)u - \mathcal{S}(s)C'(t)u$; apply to an element u of E_0 and differentiate with respect to s. We obtain

$$C'(s)C(t)u = C'(s+t)u - \mathcal{S}'(s)C'(t)u. \qquad (4.13)$$

If follows that $C(t)$ is a bounded operator from E_0 into E_0 and

$$\|C(t)\|_{(E_0,E_0)} \le Ce^{\omega t} \qquad (4.14)$$

for some constant C. Finally, boundedness of $\mathcal{S}(t)$ is shown as follows. Write (4.10) in the form $C(s)\mathcal{S}(t)u = \mathcal{S}(s+t)u - \mathcal{S}(s)\mathcal{S}'(t)u$. We then differentiate this equality term by term. The result is

$$C'(s)\mathcal{S}(t)u = \mathcal{S}'(s+t)u - \mathcal{S}'(s)\mathcal{S}'(t)u. \qquad (4.15)$$

It follows that $\mathcal{S}(t)$ is a bounded operator and

$$\|\mathcal{S}(t)\|_{(E;E_0)} \le Ce^{\omega t} \quad (t \ge 0) \qquad (4.16)$$

for some constant C. We have then completed the proof that each $\mathfrak{S}(t)$ is continuous in $\mathfrak{E}_m$: moreover, there exists a constant C such that

$$\|\mathfrak{S}(t)\|_{(\mathfrak{E}_m)} \le Ce^{\omega t} \quad (t \ge 0) \qquad (4.17)$$

for some constant C, the constant ω being the same in Corollary 3.7 and Lemma 4.1.

The semigroup equation

$$\mathfrak{S}(s+t) = \mathfrak{S}(s)\mathfrak{S}(t) \quad (s,t \ge 0) \qquad (4.18)$$

follows from (4.8) and (4.10) and their differentiated versions (4.13) and (4.15).

The next step is to show that $\mathfrak{S}(\hat{t})$ is strongly continuous. It is enough to prove that

$$\|\mathfrak{S}(h)u - u\|_{(\mathfrak{E}_m)} \to 0 \qquad (4.19)$$

as $h \to 0+$. However, we shall skip this step since we show below that $\mathfrak{S}(t)u$ has a derivative at the origin (in the norm of $\mathfrak{E}_m$) for u in a dense subset of $\mathfrak{S}_m$; this, combined with the uniform bound (4.28) obviously implies (4.19), since

$$\|(\mathfrak{S}(h) - I)u\| \le \|(\mathfrak{S}(h) - I)\mathfrak{v}\|$$

$$+ \|(\mathfrak{S}(h) - I)(u - \mathfrak{v})\| \le \|(\mathfrak{S}(h) - I)\mathfrak{v}\| + 2Ce^{\omega h}\|u - \mathfrak{v}\|.$$

THEOREM 4.3. $\mathfrak{S}(\hat{t})$ is a strongly continuous semigroup with infinitesimal generator $\mathfrak{B}$ given by

$$\mathfrak{B} = \overline{\mathfrak{A}} = \text{closure of } \mathfrak{A}, \tag{4.20}$$

where

$$\mathfrak{A} = \begin{bmatrix} 0 & I \\ -A & -B \end{bmatrix} \tag{4.21}$$

with domain

$$D(\mathfrak{A}) = D(A) \times (D(A) \cap D(B)). \tag{4.22}$$

The function $u(\hat{t})$ is a solution of (4.1) only if

$$\mathfrak{u}(t) = [u(t), u'(t)] \tag{4.23}$$

is a solution of

$$\mathfrak{u}'(t) = \mathfrak{A}\mathfrak{u}(t). \tag{4.24}$$

We have

$$D(\mathfrak{B}) \supseteq D(A) \times D_1 . \tag{4.25}$$

Proof: We begin by showing that $D(A)$ is dense in E_0 in the topology of E_0. To do this we select a "δ-sequence" $\{\psi_n\}$ of scalar functions like that used in the proof of Corollary 3.5 (b), and show that

$$u_n = \int \psi_n(t)C(t)u \, dt \to u \tag{4.26}$$

as $n \to \infty$ for each $u \in E_0$. That (4.26) holds in the topology of E (for any $u \in E$) is obvious. Assume now that $u \in E_0$. Then, using (4.13) we see that

$$C'(s)u_n = \int \psi_n(t)C'(s)C(t)u \, dt = \int \psi_n(t)C'(s+t)u \, dt - S'(s)\int \psi_n(t)C'(t)u \, dt$$

and we check easily that $e^{-\omega s}C'(s)u_n$ converges uniformly in $t \geq 0$ to $e^{-\omega s}C'(s)u$, so that $u_n \to u$ in E_0.

We show next that

$$h^{-1}(\mathfrak{S}(h)u - u) \to \mathfrak{A}u \quad \text{as} \quad h \to 0+ \tag{4.27}$$

in $\mathfrak{E}_m$ for each $u \in D(\mathfrak{A})$. This is equivalent to the following four limit relations as $h \to 0+$:

$$h^{-1}(C(h)u - u) \to 0 \quad \text{in} \quad E_0 \tag{4.28}$$

for $u \in D(A)$,

$$h^{-1}\mathcal{S}(h)u \to u \quad \text{in} \quad E_0 \tag{4.29}$$

for $u \in D(A) \cap D(B)$,

$$h^{-1}C'(h)u \to -Au \quad \text{in} \quad E \tag{4.30}$$

for $u \in D(A)$, and

$$h^{-1}(\mathcal{S}'(h)u - u) \to -Bu \quad \text{in} \quad E \tag{4.31}$$

for $u \in D(A) \cap D(B)$. To prove (4.28) we use (4.13) in the form

$$C'(s)h^{-1}(C(h)u - u) = h^{-1}(C'(s+h)u - C'(s)u) - \mathcal{S}'(s)h^{-1}C'(h)u. \tag{4.32}$$

This expression, as a function of s, is bounded in norm by a constant times $e^{\omega' s}$, ω' the constant described after (4.5). The limit of (4.32) as $h \to 0$ is

$$C''(s)u - \mathcal{S}'(s)C''(0)u = C''(s)u + \mathcal{S}'(s)Au = 0$$

after (3.11). To show (4.29) we write (4.10) in the form

$$C'(s)(h^{-1}\mathcal{S}(h)u - u) = h^{-1}(\mathcal{S}'(s+h)u - \mathcal{S}'(s)u) - \mathcal{S}'(s)h^{-1}(\mathcal{S}'(h)u - u) - C'(s)u, \tag{4.33}$$

which is bounded in norm as well by a constant times $e^{\omega' s}$; its limit as $h \to 0+$ is

$$\mathcal{S}''(s)u - \mathcal{S}'(s)\mathcal{S}''(0) - C'(s)u = \mathcal{S}''(s)u + \mathcal{S}'(s)Bu + \mathcal{S}(s)Au = 0 \tag{4.34}$$

in view of Corollary 3.5(d). Finally, the two limit relations (4.30) and (4.31) are obvious, since $h^{-1}C'(h)u \to C''(0)u = -AC(0)u - BC'(0)u = -Au$ and $h^{-1}(\mathcal{S}'(h)u - u) \to \mathcal{S}''(0)u = -A\mathcal{S}(0)u - B\mathcal{S}'(0)u = -Bu$ for $u \in D_1 \supseteq D(A) \cap D(B)$ (see Corollary 3.5 (c)).

Having proved (4.27), we know that $\mathfrak{S}(\hat{t})$ is a strongly continuous semigroup and that, if $\mathfrak{B}$ is its infinitesimal generator, then

$$\mathfrak{A} \subseteq \mathfrak{B} . \tag{4.35}$$

To improve (4.35) to (4.20) it will be sufficient to prove that

$$u^h = \frac{1}{h}\int_0^h \mathfrak{S}(t)u\,dt \in D(\mathfrak{A}) \tag{4.36}$$

for all $u \in D(\mathfrak{A})$. In fact, if (4.36) is true and $u \in \mathfrak{E}_m$ we may select a sequence $\{u_n\} \subseteq D(\mathfrak{A})$ with $u_n \to u$ in $\mathfrak{E}_m$ (that $D(\mathfrak{A})$ is dense in $\mathfrak{E}_m$ follows from (4.26) and from Corollary 3.5 (b)). Then $(u_n)^h \to u$ whereas $\mathfrak{A}(u_n)^h = \mathfrak{B}(u_n)^h = h^{-1}(\mathfrak{S}(h)u_n - u_n) \to h^{-1}(\mathfrak{S}(h)u - u)$; $u^h \in D(\overline{\mathfrak{A}})$ for any $u \in \mathfrak{E}_m$ and any $h > 0$ and

$$\overline{\mathfrak{A}}u^h = h^{-1}(\mathfrak{S}(h)u - u) . \tag{4.37}$$

Assume that $u \in D(\mathfrak{B})$. Taking into account that the right side of (4.37) tends to $\mathfrak{B}u$ as $h \to 0+$ it follows from the fact that $\overline{\mathfrak{A}}$ is closed that $u \in D(\overline{\mathfrak{A}})$ and $\overline{\mathfrak{A}}u = \mathfrak{B}u$, which completes the proof of (4.20).

The inclusion relation (4.36) can be reduced to the four relations

$$\int_0^h C(t)u\,dt \in D(A) \quad (u \in D(A)), \tag{4.38}$$

$$\int_0^h \mathcal{S}(t)u\,dt \in D(A) \quad (u \in D(A) \cap D(B)), \tag{4.39}$$

$$\int_0^h C'(t)u\,dt = C(h)u - u \in D(A) \cap D(B) \quad (u \in D(A)), \tag{4.40}$$

$$\int_0^h \mathcal{S}'(t)u\,dt = \mathcal{S}(h)u - u \in D(A) \cap D(B) \quad (u \in D(A) \cap D(B)). \tag{4.41}$$

If $u \in D(A)$ we have

$$A\int_0^h C(t)u\,dt = \int_0^h AC(t)u\,dt, \tag{4.42}$$

so that (4.38) holds. On the other hand, if $u \in D(A) \cap D(B) \subseteq D_1$ then $\mathcal{S}(\hat{t})u$ is a solution of (4.1) so that (4.39) holds and

$$A\int_0^h \mathbf{S}(t)u\,dt = \int_0^h A\mathbf{S}(t)u\,dt. \tag{4.43}$$

Obviously, (4.40) is verified; to check (4.41) we note that, since $u \in D(A) \cap D(B) \subseteq D_1$, $\mathbf{S}(\hat{t})u$ is a solution of (4.1) so that $\mathbf{S}(h)u \in D(A)$: on the other hand,

$$B\int_0^h \mathbf{S}'(t)u\,dt = \int_0^h B\mathbf{S}'(t)u\,dt. \tag{4.44}$$

This ends the proof of (4.36).

We check the final statements in Theorem 4.1. In the matter of (4.25), we note that $D(\mathfrak{B})$ is the set of all $\mathfrak{u} = [u,v]$ such that $\mathfrak{S}(t)\mathfrak{u}$ is differentiable in $\mathfrak{E}_m$. In view of the definition (4.7) of $\mathfrak{S}(\hat{t})$ this means that

$$\begin{aligned} &(a)\quad C(t)u + \mathbf{S}(t)v \quad \text{is differentiable in } E_0, \\ &(b)\quad C'(t)u + \mathbf{S}'(t)v \quad \text{is differentiable in } E. \end{aligned} \tag{4.45}$$

Statement (b) is obvious if $[u,v] \in D(A) \times D_1$. As for (a), we use again (4.8) and (4.10). The details should be familiar by now and we omit them. We have concluded the proof of Theorem 4.4.

EXAMPLE 4.4. We show below that, in general,

$$\mathfrak{A} \neq \mathfrak{B}\ ; \tag{4.46}$$

moreover, equality in (4.25) cannot in general be guaranteed, so that there is no total equivalence of the equations (4.1) and (4.24). We consider the space $E = \ell^2$ in Example 2.1, but, for technical reasons, we assume the elements of E to be of the form $u = \{u_n\} = \{u_n; n \geq 2\}$. We define the operators A, B in (2.2) thusly:

$$A\{u_n\} = \{n^3 u_n\},\ B\{u_n\} = \{n(n+1)u_n\}, \tag{4.47}$$

so that we have $\lambda_n^- = -n$, $\lambda_n^+ = -n^2$,

$$C(t)\{u_n\} = \left\{\frac{n^2 e^{-nt} - n e^{-n^2 t}}{n^2 - n} u_n\right\}, \tag{4.48}$$

and

$$\mathbf{S}(t)\{u_n\} = \left\{\frac{e^{-nt} - e^{-n^2 t}}{n - n^2} u_n\right\}. \tag{4.49}$$

It follows that

$$\sigma(t) = \|C(t)\| \le \sup_{n\ge 2} \left|\frac{n^2+n}{n^2-n}\right| = 3 ,$$

$$\tau(t) = \|\mathfrak{S}(t)\| \le \sup_{n\ge 2} \frac{1}{n^2-n} = \frac{1}{2} ,$$

hence the Cauchy problem for (4.1) is well posed in $t \ge 0$. We check easily that Assumption 3.1 is satisfied. The space E_0 consists of all sequences $u = \{u_n\}$ such that

$$\|u\|_0^2 = \|\{u_n\}\|_0^2 = \sum |nu_n|^2 < \infty, \tag{4.50}$$

the norm $|\cdot|$ being equivalent to the general norm (4.5). Consider the space $E_1 \subseteq E$ consisting of all sequences $u = \{u_n\}$ such that

$$\|u\|_1^2 = \|\{u_n\}\|^2 = \sum |n^2 u_n|^2 < \infty. \tag{4.51}$$

Obviously, $D(A) \subset E_1 \subset E_0 \subset E$, each inclusion being bounded and strict.

Take a sequence $\{u_m\} = \{u_{nm}\}$ of elements of $D(A)$ such that

$$u_m \to u \in E_1 \tag{4.52}$$

in the topology of E_1, where u does not belong to $D(A)$. In view of (4.52), both $\{v_m\} = \{nu_{nm}\}$ and $\{w_m\} = \{n^2u_{nm}\}$ converge in E, so that

$$\mathfrak{A}\begin{bmatrix} u_m \\ v_m \end{bmatrix} = \begin{bmatrix} 0 & I \\ -A & -B \end{bmatrix}\begin{bmatrix} u_m \\ v_m \end{bmatrix} = \begin{bmatrix} v_m \\ -Au_m - Bv_m \end{bmatrix} = \begin{bmatrix} v_m \\ -w_m \end{bmatrix}$$

converges in $\mathfrak{G}_m$, proving that $u = [u,v]$ belongs to $D(\overline{\mathfrak{A}}) = D(\mathfrak{B})$. But $u \notin D(\mathfrak{A})$, so that

$$D(\mathfrak{B}) \ne D(A) \times D_1 .$$

REMARK 4.5. The role of part (b) of Assumption 3.1 is somewhat more obscure that of (a), which has been shown to be essential in the construction of phase spaces. It can be shown (see Exercises 1 to 11) that (b) can be given up, but some additional assumptions in D_0 and D_1 must be added that are not especially easy to verify. A version of Theorem 4.2 holds (Exercise 13).

§VIII.5. Miscellaneous comments.

The equation (4.1) has been extensively studied by semigroup methods since the paper of LIONS [1957:1]. Other early contributions are due to MITJAGIN [1961:1] and SOBOLEVSKIĬ [1962:1], [1964:1] where A and B can in fact depend on t. For references see KREĬN [1967:1]; some of the more recent literature is in the author [1983:1]. In these papers, the emphasis lies in reducing (4.1) to a first order system in a product space (by means of ad hoc assumptions on the coefficients A, B) and then using semigroup theory.

The definitions and results presented in this chapter are due to the author; Sections VIII.1, VIII.2 and VIII.3 are taken from [1970:2]. Section VIII.4 is contained (in a somewhat different formulation) in [1981:1]. A much earlier treatment of (4.1) in the spirit of §VIII.4 is due to WEISS [1967:1], where the notion of phase space is introduced in a somewhat more general form: in this formulation of the theory, the Cauchy problem for (4.1) is not assumed to be well posed in the sense of §VIII.1.

It is obvious that one could generalize the theory in this chapter to more general equations (say, differential equations of order n). This has been done for the material in the first three sections in the author [1970:2]. A more interesting line of approach is suggested by the following argument. Assume the Cauchy problem for (4.1) is well posed and that Assumption 3.1 is satisfied. Consider the space $X = D(A) \cap D(B)$ endowed with the "joint graph norm" $\|u\|_X = \|u\| + \|Bu\| + \|Au\|$. Since A and B are closed, X is a Banach space and $\mathcal{P} = \delta'' \otimes I + \delta' \otimes B + \delta \otimes A$ (δ the Dirac delta) is a distribution in the space $\mathcal{D}'_+((X;E))$ of all distributions defined in $-\infty < t < \infty$, with support in $t \geq 0$ and values in the space of operators $(X;E)$. On the other hand, we check on the basis of part (b) of Assumption 3.1 that $\mathcal{S}(\hat{t})$ (extended to $t < 0$ by setting $\mathcal{S}(t) = 0$ there) is a distribution defined in $-\infty < t < \infty$, with support in $t \geq 0$ and values in $(E;X)$. By definition of $\mathcal{S}(\hat{t})$ (and due to the fact that we have $D(A) \cap D(B) \subseteq D_1$) we prove that

$$\mathcal{P} * \mathcal{S} = \delta \otimes I . \tag{5.1}$$

On the other hand, (3.14) implies that

$$\mathcal{S} * \mathcal{P} = \delta \otimes J \tag{5.2}$$

where J (resp. I) is the identity operator in X (resp. E). In this setting, the problem of constructing the propagator $\mathcal{S}(t)$ of (4.1) is that of finding a convolution inverse of $\mathcal{P}$ satisfying suitable properties (such as having support in $t \geq 0$, etc.) and we can pose this problem in relation to an arbitrary distribution $\mathcal{P} \in \mathcal{D}_+((X;E))$, thus being able to treat equations more general than differential; for instance, $\mathcal{P} = \delta'(\hat{t}) \otimes I + \delta(\hat{t}) \otimes A + \delta(\hat{t} - h) \otimes B$ corresponds to the difference-differential equation

$$u'(t) + Au(t) + Bu(t-h) = 0 ,$$

and the distribution $\mathcal{P} = \delta'(\hat{t}) \otimes I + \delta(\hat{t}) \otimes A + h(\hat{t}) \otimes B$, where $h(\hat{t})$ is the Heaviside function ($h(t) = 1$ for $t \geq 0$, $h(t) = 0$ for $t < 0$) corresponds to the integrodifferential equation

$$u'(t) + Au(t) + B\int_0^t u(s)ds = 0.$$

For details on this approach see the author [1976:1], [1980:1], [1983:1], [1983:2].

In Exercises 1 to 4 we suppose that the Cauchy problem for

$$u''(t) + Bu'(t) + Au(t) = 0 \tag{5.3}$$

is well posed in $t \geq 0$ as defined in §VIII.1; we do not require Assumption 3.1.

EXERCISE 1. Let $u \in D(A)$ such that $Au \in D_1$. Show that (3.11) holds (Hint: prove the integrated version

$$C(t)u = u - \int_0^t \mathcal{S}(s)Au\, ds) . \tag{5.4}$$

EXERCISE 2. Let $u \in D_0 \cap D(B)$ be such that $Bu \in D_1$. Show that (3.12) holds. (Hint: prove the integrated version

$$\mathcal{S}(t)u = \int_0^t (C(s)u - \mathcal{S}(s)Bu)ds) . \tag{5.5}$$

EXERCISE 3. Combining Exercises 1 and 2 show that if $u \in D_0 \cap D_1$ is such that $Au \in D_1$, $Bu \in D_1$ then (3.14) holds, that is

$$\mathcal{S}''(t)u + \mathcal{S}'(t)Bu + \mathcal{S}(t)Au = 0 . \tag{5.6}$$

EXERCISE 4. Show that part (b) of Lemma 3.6 is true in the present level of generality, that is

$$P(\lambda) = \lambda^2 I + \lambda B + A \tag{5.7}$$

is one-to-one in a region of the form

$$\operatorname{Re}\lambda \geq \alpha + \beta \log (1 + |\lambda|) . \tag{5.8}$$

In Exercises 5 to 11 we require part (a) of Assumption 3.1, that is

ASSUMPTION 5.1. $\mathcal{S}(\hat{t})u$ is continuously differentiable in $t \geq 0$ for all $u \in E$.

EXERCISE 5. Using Exercise 2 show that the operator $\mathcal{S}(t)B$ (with domain $\{u \in D_0 \cap D(B);\ Bu \in D_1\}$ has a bounded extension $\overline{\mathcal{S}(t)B}$ to all of E given by

$$\overline{\mathcal{S}(t)B} = C(t) - \mathcal{S}'(t) . \tag{5.9}$$

EXERCISE 6. Define $R(\lambda;\varphi)$ as in (3.19),

$$R(\lambda;\varphi)u = \int_0^\infty e^{-\lambda t}\varphi(t)\mathcal{S}(t)u\,dt , \tag{5.10}$$

with $\varphi(\hat{t})$ a test function identically equal to 1 near zero. Using (a slightly extended version of) (5.6) show that if $u \in D_0 \cap D(B)$ is such that $Au, Bu \in D_1$ then (3.37) holds, that is,

$$R(\lambda;\varphi)P(\lambda)u = u + \hat{N}(\lambda;\varphi)u , \tag{5.11}$$

where $N(t;\varphi) = 2\varphi'(t)\mathcal{S}'(t) + \varphi''(t)\mathcal{S}(t) + \varphi'(t)\overline{\mathcal{S}(t)B}$.

EXERCISE 7. Define $R(\lambda)$ as in (3.41) for $\operatorname{Re}\lambda \geq \omega$, ω large enough. Show that for u as in Exercise 6 we have

$$R(\lambda)P(\lambda)u = u . \tag{5.12}$$

EXERCISE 8. For $\operatorname{Re}\lambda \geq \omega$, ω large enough, define $\tilde{\mathcal{S}}(\hat{t})$ by (3.48), $\eta(\hat{t};\varphi)$ given by (3.46). Prove that the Laplace transform of $\tilde{\mathcal{S}}$ equals $R(\lambda)$.

EXERCISE 9. Define

$$Y_\alpha = \begin{cases} t^{\alpha-1}/\Gamma(\alpha) & (t \geq 0) \\ 0 & (t < 0) \end{cases}$$

(convolution by Y_α produces the "antiderivative of order α"). Show that, multiplying (5.12) by λ^{-3} and inverting Laplace transforms we obtain, using Exercise 8,

$$\tilde{\mathcal{S}} * (Y_1 \otimes u + Y_2 \otimes Bu + Y_3 \otimes Au) = Y_3 \otimes u \qquad (t \geq 0)^{(4)} \tag{5.13}$$

for u as in Exercise 6.

EXERCISE 10. Assume that the set of all $u \in D_0 \cap D(B)$ such that $Au, Bu \in D_1$ is dense in the space $X = D(A) \cap D(B)$ endowed with the norm

$$\|u\|_X = \|u\| + \|Au\| + \|Bu\| .$$

Show that (5.13) can be extended to all $u \in D(A) \cap D(B)$.

EXERCISE 11. Show that, if $u \in D_1$,

$$(Y_1 \otimes I + Y_2 \otimes B + Y_3 \otimes A) * \mathcal{S}u = Y_3 \otimes u \qquad (t \geq 0) . \tag{5.14}$$

Combining (5.13) and (5.14), prove that, under the conditions of Exercise 10,

$$\mathcal{S}(t) = \tilde{\mathcal{S}}(t) . \tag{5.15}$$

EXERCISE 12. Under the conditions of Exercise 10, show that formula (3.53) (resp. (3.54)) holds for $\mathcal{S}'(\hat{t})$ (resp. for $\overline{\mathcal{S}(t)B}$). Show that there exist constants C, ω such that

$$\|\mathcal{S}(t)\|, \ \|\mathcal{S}'(t)\|, \ \|\overline{\mathcal{S}(t)B}\| \leq Ce^{\omega t} \qquad (t \geq 0) . \tag{5.16}$$

EXERCISE 13. Under the conditions of Exercise 8 show that, if $u \in E$ is such that $C(\hat{t})u$ is continuously differentiable in $t \geq 0$ then formula (4.3) holds, so that

$$\|C'(t)u\| \leq Ce^{\omega t} \qquad (t \geq 0) . \tag{5.17}$$

with C (but not ω) may depend on u. Show that Theorem 4.2 is valid under the present hypotheses.

EXERCISE 14. We suppose here that the Cauchy problem for (5.3) is well posed in $0 \leq t \leq a$ $(a > 0)$, that Assumption 3.1 is satisfied there and that the set of all $u \in D(A) \cap D(B)$ such that $Bu \in D(A) \cap D(B)$ is dense in E. Then the Cauchy problem for (5.3) is well posed in $t \geq 0$ and Assumption 3.1 is satisfied. Note that all the assumptions in this Exercise are satisfied for the incomplete equation

$$u''(t) + Au(t) = 0 \tag{5.18}$$

under the only assumption that the Cauchy problem for (5.18) is well posed; of course, the result for (5.18) can be proved in a more elementary way by ad hoc methods.

FOOTNOTES TO CHAPTER VIII

(1) We note the inconsistency of notation involved in writing the incomplete equation $u'' + Au = 0$, and not $u'' = Au$ as in Chapters II and III.

(2) Although the argument could be completed using (3.25), the "left-handed" representation (3.41) simplifies some of the arguments.

(3) We might set here $\omega_1 = \min(\omega, \omega')$: for if $\omega' < \omega$, $R(\lambda)$ can be analytically continued to $\operatorname{Re}\lambda > \omega'$ by means of $Q(\lambda)$.

(4) Convolution by Y_3 is employed here to avoid using convolution of distributions.

BIBLIOGRAPHY

R. A. ADAMS

[1975:1] Sobolev Spaces. Academic Press, New York, 1975

L. AMERIO and G. PROUSE

[1971:1] Almost-periodic Functions and Functional Equations. Van Nostrand - Reinhold, New York, 1971

A. V. BALAKRISHNAN

[1960:1] Fractional powers of closed operators and the semigroups generated by them. Pacific J. Math. 10 (1960) 419-437

S. BANACH

[1932:1] Théorie des Opérations Linéaires. Monografje Matematyczne, Warsaw, 1932

H. BART

[1977:1] Periodic strongly continuous semigroups. Ann. Mat. Pura Appl. 115 (1977) 311-318

H. BART and S. GOLDBERG

[1978:1] Characterizations of almost periodic strongly continuous groups and semigroups. Math. Ann. 236 (1978) 105-116

L. R. BRAGG and W. DETTMAN

[1968:1] An operator calculus for related partial differential equations. J. Math. Anal. Appl. 22 (1968) 261-271

P. L. BUTZER and H. BERENS

[1967:1] Semi - groups of Operators and Approximation. Springer, New York, 1967

J. CHAZARAIN

[1971:1] Problèmes de Cauchy abstraits et applications a quelques problèmes mixtes. Jour. Functional Analysis 7 (1971) 386-445

I. CIORĂNESCU

[1974:1] Sur les solutions fondamentales d'ordre fini de croissance. Math. Ann. 211 (1974) 37-46

[1982:1] On periodic distribution semigroups. Integral Equations and Operator Theory 7 (1982) 27-35

C. CORDUNEANU

[1968:1] Almost Periodic Functions. Interscience-Wiley, New York, 1968

G. DA PRATO

[1968:1] Semigruppi periodici. Ann. Mat. Pura Appl. 78 (1968) 55-67

G. DA PRATO and E. GIUSTI

[1967:1] Una caratterizzazione dei generatori di funzioni coseno astratte. Boll. Un. Mat. Italiana (3) 22 (1967) 357-362

E. M. DE JAGER

[1975:1] Singular perturbations of hyperbolic type. Nieuw Arch. Wisk. 23 (1975) 145-171

L. DE SIMON

[1964:1] Un'applicazione della teoria degli integrali singolari allo studio delle equazioni differenziali lineari astratte del primo ordine. Rend. Sem. Mat. Univ. Padova 34 (1964) 205-223

J. W. DETTMAN

[1973:1] Perturbation techniques for related differential equations. J. Differential Equations 14 (1973) 547-558

J. DIXMIER

[1950:1] Les moyennes invariantes dans les semi-groupes et leurs applications. Acta Sci. Mat. Szeged 12 (1950) 213-227

N. DUNFORD and J. T. SCHWARTZ

[1958:1] Linear Operators, part I. Interscience, New York, 1958

[1963:1] Linear Operators, part II. Interscience-Wiley, New York, 1963

H. O. FATTORINI

[1969:1] Sur quelques équations différentielles pour les distributions vectorielles. C. R. Acad. Sci. Paris 268 (1969) A707-A709

[1969:2] Ordinary differential equations in linear topological spaces, I. J.Differential Equations 5 (1969) 72-105

[1969:3] Ordinary differential equations in linear topological spaces, II. J. Differential Equations 6 (1969) 50-70

[1970:1] On a class of differential equations for vector valued distributions. Pacific J. Math. 32 (1970) 79-104

[1970:2] Extension and behavior at infinity of solutions of certain linear operational differential equations. Pacific J. Math. 33 (1970) 583-615

[1970:3] Uniformly bounded cosine functions in Hilbert space. Indiana Univ. Math. J. 20 (1970) 411-425

[1971:1] Un teorema de perturbación para generadores de funciones coseno. Rev. Unión Matemática Argentina 25 (1971) 291-320

[1976:1] Some remarks on convolution equations for vector valued distributions. Pacific J. Math. 66 (1976) 347-371

[1980:1] Vector valued distributions having a smooth convolution inverse. Pacific J. Math. 90 (1980) 347-372

[1981:1] Some remarks on second order abstract Cauchy problems. Funkcialaj Ekvacioj 24 (1981) 331-344

[1983:1] The Cauchy Problem. Encyclopedia of Mathematics and its Applications, vol. 18, Addison-Wesley, Reading, Mass. 1983

[1983:2] Convergence and approximation theorems for vector valued distributions. Pacific J. Math. 105 (1983) 77-114

[1983:3] A note on fractional derivatives of semigroups and cosine functions. Pacific J. Math. 109 (1983) 335-347

[1983:4] Singular perturbation and boundary layer for an abstract Cauchy problem. Jour. Math. Anal. Appl. 97 (1983) 529-571

[1985:1] On the Schrödinger singular perturbation problem. To appear.

[1985:2] On the growth of solutions of second order linear differential equations in Banach spaces. To appear.

[1985:3] The hyperbolic singular perturbation problem: an operator-theoretical approach. To appear.

W. FELLER

[1953:1] On the generation of unbounded semi-groups of bounded linear operators. Ann. of Math. 57 (1953) 287-308

S. R. FOGUEL

[1964:1] A conterexample to a problem of Sz.-Nagy. Proc. Amer. Math. Soc. 15 (1964) 788-790

A. FRIEDMAN

[1969:1] Partial Differential Equations. Holt-Rinehart-Winston, New York, 1969

[1969:2] Singular perturbation for the Cauchy problem and for boundary value problems. J. Differential Equations 5 (1969) 226-261

R. GEEL

[1978:1] Singular Perturbations of Hyperbolic Type. Mathematical Centre Tracts n° 98, Mathematisch Centrum, Amsterdam, 1978

J. GENET and M. MADAUNE

[1977:1] Perturbations singulières pour une classe de problèmes hyperboliques non linéaires. Lecture Notes in Mathematics, vol. 594 (1977) 201-231, Springer, New York, 1977

E. GIUSTI

[1967:1] Funzioni coseno periodiche. Boll. Un. Mat. Ital. (3) 22 (1967) 478-485

J. GOLDSTEIN

[1969:1] Semigroups and second order differential equations. J. Functional Analysis 4 (1969) 50-70

[1969:2] An asymptotic property of solutions of wave equations. Proc. Amer. Math. Soc. 23 (1969) 359-363

[1970:1] An symptotic property of solutions of wave equations, II. J. Math. Anal. Appl. 32 (1970) 392-399

[1970:2] On a connection between first and second order differential equations in Banach spaces. J. Math. Anal. Appl. 30 (1970) 246-251

[1972:1] On the growth of solutions of inhomogeneous abstract wave equations. J. Math. Anal. Appl. 37 (1972) 650-654

[1974:1] On the convergence and approximation of cosine functions. Aeq. Math. 10 (1974) 201-205

J. GOLDSTEIN and J. T. SANDEFUR

[1976:1] Asymptotic equipartition of energy for differential equations in Hilbert space. Trans. Amer. Math. Soc. 219 (1976) 397-406

I. S. GRADSTEIN and I. M. RIDZYK

[1963:1] Tables of Integrals, Sums, Series and Derivatives. (Russian). Gostekhizdat, Moscow, 1963

F. P. GREENLEAF

[1969:1] Invariant Means on Topological Groups and Their Applications. Van Nostrand, New York, 1969

R. GRIEGO and R. HERSH

[1971:1] Theory of random evolutions with applications to partial differential equations. Trans Amer. Math. Soc. 156 (1971) 405-418

J. HADAMARD

[1923:1] Lectures on Cauchy's Problem in Linear Partial Differential Equations. Yale University Press, New Haven, 1923. Reprinted by Dover, New York, 1952

P. R. HALMOS

[1967:1] A Hilbert Space Problem Book. Van Nostrand, New York, 1967

H. HELSON

[1953:1] Note on harmonic functions. Proc. Amer. Math. Soc. 4 (1953) 686-691

E. HILLE

[1938:1] On semi-groups of linear transformations in Hilbert space. Proc. Nat. Acad. Sci. USA 24 (1938) 159-161

[1948:1] Functional Analysis and Semi-Groups. Amer. Math. Soc. Colloquium Pubs. vol. 31, New York, 1948

[1952:1] On the generation of semi-groups and the theory of conjugate functions. Comm. Sém. Math. Univ. Lund Tome Supplémentaire (1952) 122-134

[1952:2] A note on Cauchy's problem. Ann. Soc. Pol. Math. 25 (1952) 56-68

[1953:1] Sur le problème abstrait de Cauchy. C. R. Acad. Sci. Paris 236 (1953) 1466-1467

[1953:2] Le problème abstrait de Cauchy. Univ. e Politecnico Torino Rend. Sem. Mat. 12 (1953) 95-105

[1954:1] Une généralisation du problème de Cauchy. Ann. Inst. Fourier Grenoble 4 (1954) 31-48

[1954:2] The abstract Cauchy problem and Cauchy's problem for parabolic partial differential equations. J. Analyse Math. 3 (1954) 81-196

[1957:1] Problème de Cauchy: existence et unicité des solutions. Bull. Mat. Soc. Sci. Math. Phys. R. P. Roumaine (N. S.) 1 (1957) 141-143

E. HILLE and R. S. PHILLIPS

[1957:1] Functional Analysis and Semi-Groups. Amer. Math. Soc. Colloquium Pubs. vol. 31, Providence, 1957

G. C. HSIAO and R. J. WEINACHT

[1979:1] A singularly perturbed Cauchy problem. J. Math. Anal. Appl. 71 (1979) 242-250

N. JACOBSON

[1953:1] Lectures on Abstract Algebra. Van Nostrand, New York, 1953

T. KATO

[1976:1] Perturbation Theory for Linear Operators. 2nd. ed. Springer, New York, 1976

J. KEVORKIAN and J. COLE

[1981:1] Perturbation Methods in Applied Mathematics. Springer, New York, 1981

J. KISYŃSKI

[1963:1] Sur les équations hyperboliques avec petite paramétre. Colloq. Math. 10 (1963) 331-343

[1970:1] On second order Cauchy's problem in a Banach space. Bull. Acad. Polon. Sci. Ser. Sci. Math. Astronom. Phys. 18 (1970) 371-374

[1971:1] On operator-valued solutions of D'Alembert's functional equation, I. Colloq. Math. 23 (1971) 107-114

[1972:1] On operator-valued solutions of D'Alembert's functional equation, II. Studia Math. 42 (1972) 43-66

[1972:2] On cosine operator functions and one-parameter semigroups of operators. Studia Math. 44 (1972) 93-105

[1974:1] On M. Kac's probabilistic formula for the solution of the telegraphist's equation. Ann. Pol. Math. 29 (1974) 259-272

H. KOMATSU

[1966:1] Fractional powers of operators. Pacific J. Math. 19 (1966) 285-346

S. G. KREĬN

[1967:1] Linear Differential Equations in Banach Spaces (Russian). Izdat. "Nauka", Moscow, 1967

S. G. KREĬN, Ju. I. PETUNIN and E. M. SEMENOV

[1978:1] Interpolation of Linear Operators (Russian). Izdat. "Nauka", Moscow, 1978

S. KUREPA

[1962:1] A cosine functional equation in Banach algebras. Acta Sci. Math. Szeged 23 (1962) 255-267

[1972:1] Uniformly bounded cosine functions in a Banach space. Math. Balkanica 2 (1972) 109-115

[1973:1] A weakly measurable selfadjoint cosine function. Glasnik Mat. Ser. III 8 (28) (1973) 73-79

[1976:1] Decomposition of weakly measurable semigroups and cosine operator functions. Glasnik Mat. Ser. III 11 (31) (1976) 91-95

B. LATIL

[1968:1] Singular perturbations of Cauchy's problem. J. Math. Anal. Appl. 23 (1968) 683-698

P. D. LAX and R. RICHTMYER

[1956:1] Survey of the stability of linear finite difference equations. Comm. Pure Appl. Math. 9 (1956) 267-293

J. L. LIONS

[1957:1] Un remarque sur les applications du théorème de Hille Yosida. J. Math. Soc. Japan 9 (1957) 62-70

E. R. LORCH

[1941:1] The integral representation of weakly almost periodic transformations in reflexive vector spaces. Trans. Amer. Math. Soc. 49 (1941) 18-40

D. LUTZ

[1980:1] Uber Operatorwertige Lösungen der Funktionalgleichung des Cosinus. Math. Z. 171 (1980) 233-245

V. P. MIKHAILOV

[1976:1] Partial Differential Equations (Russian). Izdat. "Nauka", Moscow, 1976.

B. C. MITJAGIN

[1961:1[Differential equations with small parameter in a Banach space. Izv. Akad. Nauk. Azerbaidžan, Ser. Fiz.-Matem. i Tehn. 1 (1961) 23-38

I. MIYADERA

[1951:1] On one-parameter semigroups of operators. J. Math. Tokyo 1 (1951) 23-26

[1952:1] Generation of a strongly continuous semi-group of operators. Tôhoku Math. J. (2) (1952) 109-114

B. NAGY

[1976:1] Cosine operator functions and the abstract Cauchy problem. Period. Math. Hungar. 7 (3) (1976) 15-18

H. S. NUR

[1971:1] Singular perturbations of differential equations in abstract spaces. Pacific J. Math. 36 (1971) 775-780

C. F. MUCKENHOUPT

[1928/29:1] Almost periodic functions and vibrating systems. Jour Math. Phys. M.I.T. (1928/29) 163-199

E. W. PACKEL

[1969:1] A semigroup analogue of Foguel's counterexample. Proc. Amer. Math. Soc. 21 (1969) 240-244

R. S. PHILLIPS

[1951:1] On one-parameter semi-groups of linear transformations. Proc. Amer. Math. Soc. 2 (1951) 234-237

[1953:1] Perturbation theory for semi-groups of linear operators. Trans. Amer. Math. Soc. 74 (1953) 199-221

[1954:1] A note on the abstract Cauchy problem. Proc. Nat. Acad. Sci. USA 40 (1954) 244-248

S. I. PISKAREV

[1979:1] Discretization of an abstract hyperbolic equation (Russian). Tartu Riikl. Uel. Toimetised N° 500 Trudy Mat. i Meh. 25 (1979) 3-23

[1982:1] Periodic and almost periodic cosine operator functions. Mat. Sb. 118 N° 3 (1982) 386-398

C. PUCCI and G. TALENTI

[1974:1] Elliptic (second-order) partial differential equations with measurable coefficients and approximating integral equations. Advances in Math. 19 (1976) 48-105

F. RIESZ and B. SZ.-NAGY

[1955:1] Functional Analysis. Ungar, New York, 1955

H. SLICHCHTING

[1951:1] Grenzschicht-Theorie. Braun, Karlsruhe, 1951

A. Y. SCHOENE

[1970:1] Semi-groups and a class of singular perturbation problems. Indiana Univ. Math. J. 20 (1970) 247-263

M. SHIMIZU and I. MIYADERA

[1978:1] Perturbation theory for cosine families in Banach spaces. Tokyo J. Math. 1 (1978) 333-343

J. SMOLLER

[1965:1] Singular perturbations and a theorem of Kisyński. J. Math. Anal. Appl. 12 (1965) 105-114

[1965:2] Singular perturbations of Cauchy's problem. Comm. Pure Appl. Math. 28 (1965) 665-677

M. SOVA

[1966:1] Cosine operator functions. Rozprawy Mat. 49 (1966) 1-47

[1968:1] Problème de Cauchy pour équations hyperboliques opérationnelles à coefficients constants non-bornés. Ann Scuola Norm. Sup. Pisa (3) 22 (1968) 67-100

[1968:2] Unicité des solutions exponentielles des équations différentielles opérationnelles linéaires. Boll. Un. Mat. Ital. (4) 1 (1968) 629-699

[1968:3] Solutions périodiques des équations différentielles opérationnelles: Le méthode des développements de Fourier. Časopis Pešt. Mat. 93 (1968) 386-421

[1968:4] Semigroups and cosine functions of normal operators in Hilbert spaces. Časopis Pešt. Mat. 93 (1968) 437-458

[1970:1] Équations différentielles opérationnelles linéaires du second ordre à coefficients constants. Rozprawy Československe Akad. Věd. Rada Mat. Prirod. Věd. 80 (1970) sešit 7, 1-69

[1970:2] Équations hyperboliques avec petit paramètre dans les espaces de Banach généraux. Colloq. Math. 21 (1970) 303-320

[1970:3] Régularité de l'evolution linéaire isochrone. Czechoslovak Math. J. 20 (1970) 251-302

[1970:4] Perturbations numériques des evolutions paraboliques et hyperboliques. Časopis Pešt. Mat. 96 (1971) 406-407

[1972:1] Encore sur les équations hyperboliques avec petit paramètre dans les espaces de Banach généraux. Colloq. Math. 25 (1972) 135-161

P. E. SOBOLEVSKIĬ

[1962:1] On second-order differential equations in a Banach space (Russian). Doklady Akad. Nauk. SSSR 146 (1962) 774-777

[1964:1] On second order equations with a small parameter in the higher derivative. Uspehi Mat. Nauk. 19 (1964) 217-219

E. M. STEIN

[1970:1] Singular Integrals and Differentiability Properties of Functions. Princeton University Press, Princeton, 1970

E. M. STEIN and G. WEISS

[1971:1] Introduction to Fourier Analysis on Euclidean Spaces. Princeton University Press, Princeton, 1971

M. H. STONE

[1932:1] On one parameter unitary groups in HIlbert space. Ann. Math. 33 (1932) 643-648

B. SZ.-NAGY

[1938:1] On semi-groups of self-adjoint transformations in Hilbert space. Proc. Nat. Acad. Sci. USA 24 (1938) 559-560

[1947:1] On uniformly bounded linear transformations in Hilbert space. Acta Sci. Math. Szeged 11 (1947) 152-157

T. TAKENAKA and N. OKAZAWA

[1978:1] A Phillips - Miyadera type perturbation theorem for cosine functions of operators. Tôhoku Math. J. 30 (1978) 107-115

C. C. TRAVIS and G. F. WEBB

[1981:1] Perturbation of strongly continuous cosine family generators. Colloquium Math. 45 (1981) 277-285

G. N. WATSON

[1948:1] A treatise on the Theory of Bessel Functions. 2nd. ed. Cambridge University Press, Cambridge, 1948

B. WEISS

[1967:1] Abstract vibrating systems. J. Math. Mech. 17 (1967) 241-255

G. WHITHAM

[1974:1] Linear and Nonlinear Waves. Wiley-Interscience, New York, 1974

D. V. WIDDER

[1931:1] Necessary and sufficient conditions for the representation of a function as a Laplace integral. Trans Amer. Math. Soc. 33 (1931) 851-892

[1946:1] The Laplace Transform. Princeton University Press, Princeton, 1946

K. YOSIDA

[1948:1] On the differentiability and the representation of one-parameter semigroups of linear operators. J. Math. Soc. Japan 1 (1948) 15-21

[1978:1] Functional Analysis. 5th. ed. Springer, Berlin 1978

J. ZABCZYK

[1975:1] A note on C_0 semigroups. Bull. Acad. Polon. Sci. Ser. Sci. Math. Astronom. Phys. 23 (1975) 895-898

E. ZAUDERER

[1983:1] Partial Differential Equations of Applied Mathematics. Wiley-Interscience, New York, 1983

A. ZYGMUND

[1959:1] Trigonometric Series, vol I. Cambridge University Press, Cambridge, 1959